HUMPHRY DAVY

SCIENCE & POWER

DAVID KNIGHT

First published 1992

Blackwell Publishers
108 Cowley Road
Oxford OX4 1JF
UK

238 Main Street, Suite 501
Cambridge, Massachusetts 02142
USA

British Library Cataloguing-in-Publication Data

A CIP catalogue record for this book is available from the British Library.

Library of Congress Cataloging-in-Publication Data

A CIP catalogue record for this book is
available from the Library of Congress.
ISBN 0-631-16816-8

Typeset in 10 on 12 pt Palatino
by Graphicraft Typesetters Ltd., Hong Kong
Printed in Great Britain by T.J. Press Ltd., Padstow, Cornwall

This book is for Sarah, Harriet and Frances, who have lived with Davy for some years; and he was never an easy person to live with.

Oh, most magnificent and noble nature!
Have I not worshipped thee with such a love
As never mortal man before displayed?
Adored thee in thy majesty of visible creation,
And searched into thy hidden and mysterious ways
As Poet, as Philosopher, as Sage?

Not content with what is found upon the surface of the earth, [the chemist] has penetrated into her bosom, and has even searched the bottom of the ocean for the purpose of allaying the restlessness of his desires, or of extending and increasing his power.

If matter cannot be destroy'd
The living mind can never die;
If e'en creative when alloy'd
How sure its immortality!

[The chemist exerts] on a scale infinitely small a power seeming a sort of shadow or reflection of a creative energy and which entitles him to the distinction of being made in the image of God and animated by a spark of the divine mind. Whilst chemical pursuits exalt the understanding, they do not depress the imagination or weaken genuine feeling.

Humphry Davy

Contents

Contents

Preface

Our society depends upon science and yet, to many of us, what scientists do is a mystery. The sciences are not just collections of facts, but are ordered by theory; and this is where Einstein's famous phrase about science being a free creation of the human mind comes in. Science is a fully human activity, and the personalities of those who practise it are important in its progress and often interesting to us. Looking at the lives of scientists is a way of bringing science to life.

The scientists whose lives appear in this series have been chosen because of their eminence, but the aim of the biographer is to place them in their context, writing about the times in which they lived as well as their lives. While their commitment to science, their creativity and their scepticism must be always at the back of the biographer's mind, how they earned a living, made a career and got on with family and friends are essential parts of any biography.

Davy is a wonderful subject for a biographer because he is so accessible. Though parts of his work were abstruse and difficult in his own time, he made it his business to popularize his discoveries, so that his great idea that chemical forces are electrical, making him the Newton of chemistry, seems obvious to us now. His was a first-rate scientific mind, and yet he was not troubled by the problem of two cultures:[1] he was the friend of poets, and wrote poetry – not just the comic verse characteristic of many later scientists – and wrote about religion. He reflected about the science he was practising and over which he came to preside. There are many places in which we can meet him, for his ventures into applied science, and his travels, meant that his life was by no means cloistered. His was a fascinating time, as science began to have an important social role and to become specialized. Bacon, Galileo, Kepler and Descartes had a vision of a scientific revolution, but it was realized only in Davy's time (with the French and the Industrial Revolutions), and in Britain he represented this new phenomenon of powerful science. He held out against narrow expertise, but recognized and forwarded specialized societies within science. His biography is a study of science and power.

A biography is a dialogue, in which one participant is usually (and probably better) dead; it is therefore personal and cannot be definitive. To be fair, the biographer must allow his subject to speak, and I have tried to let Davy have plenty of the conversation. He was careful about the words he chose. A fascinating person to meet and to hear, he was driven both by the urge to understand and by straightforward ambition; having risen to a dizzy height, he was socially uneasy; he was no saint, but he was ready to undertake responsibility, and anxious to be useful and to be appreciated for it.

David Knight
University of Durham

Acknowledgements

I would like to acknowledge the help given to me by the Librarian at the Royal Institution and her assistants. I am very grateful for permission to quote from the Davy and Faraday manuscripts at the Royal Institution, and to use the illustrations provided by the Institution. Finally, I would also like to thank Penwith District Council for permission to quote from Alison Pritchard's *Poetry*.

Acknowledgements

1

Beginning: the Meaning of Life

Near the end of his life Humphry Davy wrote:

> Ah! could I recover any thing like that freshness of mind, which I possessed at twenty-five, and which, like the dew of the dawning morning, covered all objects and nourished all things that grew, and in which they were more beautiful even than in mid-day sunshine, – what would I not give! – All that I have gained in an active and not unprofitable life. How well I remember that delightful season, when, full of power, I sought for power in others; and power was sympathy, and sympathy power; – when the dead and the unknown, the great of other ages and distant places, were made, by the force of the imagination, my companions and friends; – when every voice seemed one of praise and love; when every flower had the bloom and odour of the rose; and every spray or plant seemed either the poet's laurel, or the civic oak – which appeared to offer themselves as wreaths to adorn my throbbing brow.[1]

When he wrote this, Davy was approaching fifty: brilliant and enigmatic, he was one of the most respected and most disliked men of science ever. He had risen from obscurity to the presidency of the Royal Society, from which he had just retired because of failing health. Behind him lay a triumphant career in science, practical and theoretical. No wonder he has attracted biographers.

Every historian is engaged in conversation with predecessors to make sense of the materials that have come down from the past; and the biographer is engaged in dialogue with his subject as well. Davy himself gave us reason to see his life as the transformation of an imaginative, enthusiastic and immensely attractive young man into the most brilliant chemist of the age, and then into a glum and lonely old man, with a brilliant future behind him. In science the great ideas often come early, and those who do not die, like a Nelson or a Wolfe, at the height of their powers run the risk that the second half of their lives will be an anticlimax, even though they may turn out to be able administrators and managers of science, advancing it less directly but no less importantly.

Davy died in 1829 when he was fifty; had he died ten years earlier we would have got from contemporaries a very different, and much more favourable, view of him. By then everyone was a Davyan, accepting that science can and must be applied to increase the comforts and reduce the dangers of life, as he did with the miners' lamp. In 1818 Davy had received a baronetcy for this invention; his researches on potassium and chlorine had changed the course of chemistry; his reputation was secure. Then in 1820 Sir Joseph Banks, who had been President of the Royal Society throughout Davy's life, died in office, and Davy was chosen as his successor. It seemed a further triumph at the time, the summit of his career reached at only forty-one, but his reign was not happy.

By 1829 it was clear that Davy's life was a story not only of prodigious success but also of its costs. One of the ways in which the sciences came to rival the churches in the nineteenth century was as vehicles for social mobility, and Davy's biography is a 'log cabin to White House' story. It might have appealed to Samuel Smiles, but he never wrote it, perhaps because it had too much pure science in it. Henry Mayhew, famous for his studies of those on the margins of Victorian society in *London Labour and the London Poor*, did write a biography of Davy, in the form of the life of a wonderful boy for boys to read.[2] This is close to being a work of improving fiction, but it is hard to resist the true story of a boy from a remote corner of England, whose father was an often-unemployed woodcarver and whose widowed mother had to support the family by opening a milliner's shop, becoming President of the Royal Society.

Mayhew was perhaps wise to concentrate on Davy's youth. Other biographers too have emphasized his Cornish connections – notably in recent years Anne Treneer, whom I once had the pleasure of meeting and whose *Mercurial Chemist* places Davy in his provincial context.[3] It is sobering to contemplate the transition from the delightful young man in Penzance and then Clifton, where he befriended Southey, Coleridge and Wordsworth, to the grumpy and isolated exile in his last years, admired but friendless, frustrated when not sozzled with opium,[4] and apparently proving once again that illness need not enoble. 'You talk of honours;' he wrote to his wife on 1 August 1827, 'I ought to have been made a Privy Counsellor and a Lord of Trade as my predecessor was.'[5]

Davy's early friends had feared that success, especially in London, would corrupt him: and at the Royal Institution (which he made into the first research institute in Britain, supported by his lectures) he did indeed become the ally, or perhaps the tool, of the landed interest.[6] He lectured on agriculture to audiences of 'improving' landowners during a war when food was short; but the Royal Institution then supported his chemical researches. Landed and professional men and their wives and daughters flocked to hear his lectures on science. Even his research seems sometimes to have been done in public, with a rapidity often baffling to onlookers, in a tradition that goes back to Robert Boyle and others in the seventeenth century. Soon no intellectual dinner party was complete without him.

There is no doubt that Davy loved a lord; and when in 1812 he was knighted – the first man of science to be so honoured since Newton – and then married a wealthy bluestocking widow, a cousin of Sir Walter Scott, his transition to the upper classes seemed to have been achieved. But things were not so easily done in Regency England, a very class-conscious society. The first biography, by J. A. Paris, exposed some of Davy's gaucheries.[7] Paris was a doctor, a Cambridge graduate with Cornish connections, who remarked that 'no man ever soared, like an eagle, to the pinnacle of fame, without exciting the envy and perhaps the hatred of those who could only crawl up half way', and who aimed to include 'the common frailties of genius' in a true portrait. His life of Davy was based upon materials supplied by Jane Davy; the marriage had not turned out happily (she was very formal, and in letters to his brother John,[8] Davy referred to her as 'Lady Davy') but they remained affectionate, particularly when, as usually in Davy's last years, they were apart – and then they corresponded frequently.[9]

John Davy thereupon wrote another biography, in response to the sneers of those who considered themselves to be Davy's social if not intellectual superiors and to show how he had risen through merit. After Lady Davy's death, John published a supplementary volume of *Fragmentary Remains*, which include many of Davy's letters to her.[10] He had meanwhile published Davy's *Works*[11] in a handsome nine-volume edition as part of a life's work of vindicating his brother's reputation. Anybody writing about Davy is greatly in his debt.

Davy's marriage was childless, like Banks' and Faraday's, and John was one of the three or maybe four young men to whom he was something like a father, training them in science; the others were his cousin Edmund, who went on to a career in chemistry in Ireland, and Michael Faraday. With Faraday, to whom Mayhew's book is dedicated, we see a relationship that went sour; they were two prickly men, who had come a long way socially, but Faraday (whose father was a blacksmith) had never known Davy except as Sir Humphry. Davy failed to realize that Faraday had, by 1820, grown up as a man of science, and from being a generous father became an oppressive one. Faraday saw Davy in a quite different and cooler light by the 1820s, and found little to respect;[12] he had also become a full member of the strict sect of Sandemanians, while Davy's religion was rather vague, though intense and personal. Like a family quarrel, this was something from which neither man emerged with much credit. Davy knew what older men of science ought to do, as shown in this poem written about eagles in 1821:

> The mighty birds still upward rose,
> In slow but constant and most steady flight,
> The young ones following; and they would pause,
> As if to teach them how to bear the light,
> And keep the solar glory still in sight.
> So went they on till, from excess of pain,
> I could no longer bear the scorching rays;
> And when I looked again they were not seen,
> Lost in the brightness of the solar blaze.
> Their memory left a type, and a desire;
> So should I wish towards the light to rise,
> Instructing younger spirits to aspire
> Where I could never reach amidst the skies,
> And joy below to see them lifted higher,
> Seeking the light of purest glory's prize.
> So would I look on splendour's brightest day
> With an undazzled eye, and steadily
> Soar upwards full in the immortal ray,
> Through the blue depths of the unbounded sky,
> Portraying wisdom's boundless purity.
> Before me still a lingering ray appears,
> But broken and prismatic, seen thro' tears,
> The light and joy of immortality.[13]

But as often in human affairs, practice fell short of theory.

In a series of scientific lives Sir Harold Hartley (himself a distinguished applied scientist with a scholar's feeling for history) tackled Davy.[14] His interest was in Davy's 'scientific work', and particularly in 'trying to explain why, with his great genius, he did not accomplish more'; comparing him with his systematic Swedish contemporary J. J. Berzelius. It is hard to see how Hartley's account, particularly of Davy's electrochemical researches, could be bettered. But at the end of his story of this intensely creative period comes a chapter headed 'Distractions and Marriage'; the book is emphatically focused upon Davy's research. There is nothing wrong with this; a smallish book has to be focused, and what is most interesting to us about Davy is that he was a great chemist; nevertheless he was more than that, and the interest in his life does not cease with his research.

Hartley recognized Davy's romantic temperament, giving strengths as well as weaknesses, but the picture he projects of the man of science seems to come from a later period, at any rate for Britain.[15] He wrote to me: 'Don't get too involved in philosophy & try to chart the major channels of advance.'[16] This approach is limiting: the term 'scientist', we should note, was not coined until after Davy's death, and it is wise to be chary of it. He thought of himself as a natural philosopher or a chemical philosopher developing a world view, and not as a specialist. Science in Britain developed and was expounded in a tradition of natural theology, establishing the existence and wisdom of God,[17] and by the end of his life Davy was happy to belong there.

Instead of taking a scientific career as something normal, we have to look back to a time when spending one's life in science was a curious ambition. Davy was one of the first in Britain to earn his living through scientific research, and to find how and why he lived this way is as interesting and important as to see what he did in the laboratory.[18] Davy took up chemistry where Joseph Priestley had left it, but Priestley had seen himself as primarily a Unitarian minister. One of Priestley's, and later Davy's, philosophical heroes was the empiricist David Hartley, who had (typically for eighteenth-century England) rated scientific research rather low, writing of devotees of science that they were 'remarkable for Ignorance and Imprudence in common necessary Affairs' and given to 'Vanity, Self-conceit, Moroseness, Jealousy, and Envy'. Their scientific 'pleasures of the Imagination . . . come to their Height early in Life, and decline in old Age', and should therefore not be overindulged. The proper study of mankind was man, after all, with due focus on God: and its aim was to find the meaning of life.[19] Davy sought to make the full-time pursuit of science compatible with the search for wisdom.

Sir Harold warned me against treating science as an intellectual game. It is indeed a serious matter, but in Davy's time it was only just ceasing to be a kind of hobby, and successful science even now involves something like play – otherwise it becomes dull, the kind of 'normal science' that philosophers write about.[20] He also told me that he could make nothing of Davy's last book, the posthumously published *Consolations in Travel*, a series of dialogues in which Davy, drawing upon Royal Institution lectures and earlier writings in notebooks, tried to come to terms with providence, disease and death, and to present his world view.[21] In his life of Davy the last ten years occupy only one chapter, 'PRS: Final Years', the terminal sixth of the book, and it is downhill all the way.

I do not believe that Hartley's is the best way to see Davy, nor how Davy, a master of narrative whose tall tales had delighted his school fellows, would have wanted his story to be told. Any drama that reaches its highest point in the middle is unsatisfactory, and if we are to make sense of our own life or anybody else's, we have to see it as developing or evolving to its end. It was not mere boundless ambition that propelled Davy towards the presidency of the Royal Society, and his published dialogues are not just the maunderings of a prematurely old man whose shaping spirit of imagination had deserted him and whose Romantic cult of youth gave him no comfort at fifty. They can be better seen as the confessions of an inquiring spirit;[22] and they show us something of the eloquent dark little man (about five feet seven inches, but seeming shorter)[23] from the Celtic fringe, whose eyes were said to be too fine to be always gazing into crucibles, and who held audiences spellbound.[24]

At the conference that marked the bicentenary of Davy's birth in 1978 June Fullmer presented Davy as a reformer, first of chemistry and then of the Royal Society;[25] and in 1983 David Miller gave a picture of Davy caught between hostile camps within the Society.[26] Davy had presented himself as a new broom after the long years of Banks, anxious to make the Society more like an Academy of Sciences, with a utilitarian emphasis, membership confined to those of scientific distinction, and close links to government and to other, more specialized scientific societies. He hoped the British Museum would be made into something more like its Parisian equivalents, the natural history materials being organized as in Cuvier's Musée d'Histoire Naturelle; and with Stamford Raffles he founded the London Zoo.

British science nevertheless, like British government, survived the 1820s largely unreformed. Davy's constituency was the Royal Institution, with its links to the landed gentry and its programme of high-level popularization to support research; but in the sciences the capacity to popularize is not highly regarded. To Cambridge men, escaping from a century of Newton worship and intent on catching up with French mathematics, the Royal Institution's way seemed amateurish; they longed for a society of gentlemanly specialists. Others, however, liked the clubbish and unspecialized atmosphere of the Royal Society, where those engaged full time in science could meet congenial men of rank with a genuine if mild interest in it. Davy seems to have wanted women to be able to come to some meetings. What was expected of the new President was such that nobody could have met it; Davy's popularity fell and his anger became notorious as he tried to be domineering. Red in the face, he glares at us from the huge portrait that dates from this time on the stairs of the Royal Institution.

To have a man, however able, whose schooling had been sketchy and whose upbringing had been plebeian and provincial running such a society is amazing; what is not amazing is that this experiment of Davy's did not work. He could not play the autocrat as Banks, a great landed proprietor, had done, and his social unease seemed haughtiness. His research had been solitary, and while his science was admired and he had achieved recognition, he was not loved. The important thing with social mobility, as Faraday learned partly from watching Davy, is to know where to stop.

Davy's relaxation had always been fishing, notably for salmon and trout. He was an extremely good fisherman and came to know a good deal about the natural history of fish and the insects associated with them.[27] He also loved to go shooting, especially for snipe, and in 1823 recruited Roderick Murchison to science: 'I fell in with Sir Humphry Davy, and experienced much gratification in his lively illustrations of great physical truths. As we shot partridges together in the morning, I perceived that a man might pursue philosophy without abandoning field sports.'[28] Davy's companions in these impeccably gentlemanly activities were often men of science, including the eminent chemist and crystallographer William Hyde Wollaston, whom he introduced to fishing. It was in these outdoor activities that Davy could be relaxed with friends, though, especially in his retirement, he seems to have been perfectly prepared to fish alone.

Wollaston had been a possible rival for the presidency, having taken over temporarily on Banks' death. A Cambridge man and a physician, he was greatly respected and known as the Pope because his chemical analyses were infallible. But he was shy and cautious, and his was the style contemporaries liked to contrast with Davy's: the former anxious to avoid error, the latter keen to discover truth. His lack of bedside manner made medicine an unpromising career, but he discovered a process for making platinum malleable and therefore available to chemists for crucibles and spatulas, and for electrodes sealed through glass. The fortune he made from this work meant that he could retire from doctoring; but Davy at the end of his life contrasted Wollaston's behaviour with his own refusal to patent the miners' lamp and thereby become wealthy: 'It was not worth his while to have died so rich; but I suppose there is pleasure in accumulating. So will W. die! with perhaps two or three hundred thousand; yet these men might have applied money to the noblest purposes.'[29] In 1820 Wollaston had declined a contest, thereby probably proving himself the wiser man.

Davy's presidency was not without its triumphs: he induced the King, through Robert Peel, to found Royal Medals to be awarded each year in recognition of scientific achievement; he established good relations with specialized societies; and the Council, for the first time in the history of the Royal Society, came to have on it a majority of men who had published some science. He tried to ensure that candidates for Fellowship should have scientific research to their credit, but in this as in other reforms he did not succeed.[30] In 1823 he thought of a scheme for protecting the copper bottoms of warships from corrosion cathodically (a principle now much used), by fastening pieces of the more-reactive metal iron to them. This promised well, but failed because weed attached itself so firmly to the protected copper that the ships sailed badly. The episode did not redound to Davy's credit, and he overreacted to jokes and criticisms.

It would be hard then, as others have noted, to see the presidency as the triumph and climax of Davy's career. Berzelius said of him that his work was 'only brilliant fragments',[31] and this applies to his work as President of the Royal Society as much as to his researches. And the last years of ill health and wandering do not on the face of it look very promising either, especially if we believe that scientific discovery was or should have been the focus of his life. Well aware of his unpopularity, having retired after a stroke and seen himself succeeded by his former patron and Vice-President, the Cornish MP Davies Gilbert (formerly Giddy), who had never done research of any distinction, Davy might have been expected to despair. He did indeed write in his journal on 27 September 1827:

> As I have so often alluded to the possibility of my dying suddenly, I think it right to mention that I am too intense a believer in the Supreme Intelligence, and have too strong a faith in the optimism of the universe, ever to accelerate my dissolution ... I have been, and am taking a care

of my health which I fear it is not worth; but which, hoping it may please Providence to preserve me for wise purposes, I think my *duty*, – G.O.O.O.[32]

The G.O.O.O. was a form of pious ejeculation often used by Davy in this journal. He would have liked Peel to be his successor: 'He has wealth and influence, and has no scientific glory to awaken jealousy, and may be useful by his parliamentary talents to men of science.'[33]

Out of this gloomy but perhaps also hopeful state came Davy's two little books of dialogues, *Salmonia* and *Consolations*, which, with his *Agricultural Chemistry*, were his writings that circulated widest. If we did not have them, we would know much less about him; but beyond that they represent the end of his odyssey. Had he died in 1827, his life would have lacked the shape that these works give it. It is by no means new to see *Consolations* as particularly important in understanding Davy and the natural philosophy of his time;[34] indeed, in his *eloge* of Davy for the Paris Academy of Sciences (therefore one of the first biographies) Georges Cuvier wrote:

Le progrès de l'espèce humaine, le sort qui lui est réservé, celui qui attend chacun de nous, la destination de milliers des globes, dont à peine quelques astronomes apercevoivent une petite partie, y sont le sujet de dialogues où le pöete ne brille pas moins que le philosophe, et où, parmi les fictions variées, une grande force de raisonnement s'applique aux questions les plus sérieuses; on aurait dit qu'une fois sorti de son laboratoire il retrouvait ces douces rêveries, ces pensées sublimes qui avait enchanté sa jeunesse; c'était en quelque sort l'ouvrage de Platon mourant.[35]

Poetic imagination applied to really serious questions, the sublime thoughts of his youth, a dying Plato: it seems as though in Cuvier's view Davy had in his last work moved from science to wisdom.

Davy had himself written a rhapsody to nature:

Oh, most magnificent and noble Nature!
Have I not worshipped thee with such a love
As never mortal man before displayed?
Adored thee in thy majesty of visible creation,
And searched into thy hidden and mysterious ways
As Poet, as Philosopher, as Sage?[36]

And this can, I believe, be used as a guide to his own life. He never stopped being a poet and a natural philosopher – experimenting with an electric fish when finally struck down in Rome – but he increasingly became the sage.

He believed in the importance of his message, dictating a letter to his brother after this stroke, on 23 February 1829:

> I am dying from a severe attack of palsy, which has seized the whole of the body, with the exception of the intellectual organ. . . . I bless God that I have been able to finish my intellectual labours. I have composed six dialogues, and yesterday finished the last of them. There is one copy in five small volumes complete; and Mr Tobin [his companion, whom we shall meet] is now making another copy, in case of accident to that. I hope you will have the goodness to see these works published.[37]

To his wife, he also sent the message that:

> I should not take so much interest in these works, did I not believe that they contain certain truths which cannot be recovered if they are lost, and which I am convinced will be extremely useful both to the moral and intellectual world. I may be mistaken in this point; yet it is the conviction of a man perfectly sane in all the intellectual faculties, and looking to futurity with the prophetic aspirations belonging to the last moments of existence.[38]

John Davy duly complied with his brother's request. The only curious feature of the story is that later, when assembling Davy's *Collected Works*, he should have added to the six an unrelated fragment of dialogue on atoms and elements, which he never intruded into any separate editions of the *Consolations*.[39]

Whatever Davy may have earlier thought about the pleasures of being twenty-five rather than fifty, he was able to face death when it came, seeing it, after suffering endured, as an escape from worn-out machinery and a way to higher things. The *Consolations* begin, like Gibbon's *Decline and Fall of the Roman Empire*, with a dream in the Colosseum – the sweet dreams of Cuvier's obituary – in which Davy meets spirits from other planets, who are living a more spiritual and intellectual life, and believes that he would join them. The universe was progressive, and his life intelligible to him as a progress towards readiness for a higher mode of existence. He would have been delighted to be compared to Plato, a sage rather than just a philosopher, and he had even organized his *Salmonia* as an ennead in the Platonic tradition, though perhaps without planning it.

We shall therefore follow Davy from his Cornish youth and his apprenticeship to a Penzance apothecary-surgeon; through his work on laughing gas in Clifton and his association with Coleridge and Wordsworth at the heart of the Romantic Movement; on to London for lecturing, for scientific discoveries and in doing applied science; becoming President of the Royal Society; writing *Salmonia*; and finally emerging as a Plato on his deathbed in *Consolations*. It was a life in science that although short was complete.

So much has been written about Davy that there are probably no great discoveries of materials to be made. There are wonderful manuscripts at the Royal Institution, notebooks and loose laboratory notes and various letters, some of them beautifully bound by Faraday. A moving document is Faraday's copy of Paris's biography of Davy, into which he bound all sorts of relevant manuscripts, including letters of increasing cordiality; but after 1820 there is nothing. Davy's handwriting was a scribble, and he conducted his researches in romantic disorder and in great bursts of speed after an incubation period.

There are further letters of Davy's at the British Library and elsewhere, and June Fullmer's edition of them has been long and eagerly awaited; we may then hope that they will form a basis for her to write a bigger biography than this one. Letters and jottings have an immediacy that printed matter, the mainstay of that 'public knowledge' which is science, lacks; the copy of *Salmonia* that Davy annotated for the second edition coming between the two! The eminent in Davy's day wrote an enormous number of letters: Banks is supposed to have written at least 50,000. On the other hand, the learned world was small and when in London for the winter months men of science saw each other at societies and at clubs like the Athenaeum, of which Davy was a founder, so they did not need to correspond unless someone was at a distance – which happened about this time with the Devonian controversy in geology,[40] for example.

Davy's lectures made him famous, with one-way traffic in Albemarle Street on Royal Institution discourse nights and a black market in tickets when he went to Ireland; it is hard for us to recover what they can have been like, but from Jane Marcet we find:

> On attending for the first time experimental lectures, the Author found it almost impossible to derive any clear or satisfactory information from the rapid demonstrations which are usually, and perhaps necessarily, crowded into popular courses of this kind. But frequent opportunities having afterwards occurred of conversing with a friend on the subject of chemistry, and of repeating a variety of experiments, she became better acquainted with the principles of that science, and began to feel highly interested in its pursuit. It was then that she perceived, in attending the excellent lectures delivered at the Royal Institution, by the present Professor of Chemistry [1806; Davy], the great advantage which her previous knowledge of the subject, slight though it was, gave her over others ... She had the gratification to find that the numerous and elegant illustrations, for which that school is so much distinguished, seldom failed to produce on her mind the effect for which they were intended.[41]

Davy's lectures must have been to many of the audience a spectacle akin to a visit to the theatre; indeed that is perhaps what lectures always ought to be. Jane Marcet hoped that her book would enable readers to understand what was going on; to those without such a help, the rhetoric of science must have communicated excitement without real understanding. But the excitement of hearing and seeing a man of genius talking about his new discoveries and interpretations is genuine; there he could be, in his own words, the idol of today, the man divine; and it must have been one of Davy's secrets that he did not attempt to cover the whole science systematically. A lecturer can only interest an audience in what interests him.

Most of Davy's contemporaries never got a chance to hear him, and, of course, recorded voices are available to us only from the very end of the nineteenth century. What they saw were publications by and about him; and for his scientific work, what got into the public domain this way is what mattered. Here we are lucky, because in Regency Britain there were not yet 'two cultures', and even papers read to, and then published by, the Royal Society had to be generally intelligible. Davy's style was expansive; he uses too many commas for our taste, no doubt because he punctuated with reading aloud in mind, but his papers are readable. They were read not merely by those few active in chemical research, but by anybody wanting to keep up with what was happening in the intellectual world.

Davy's research was almost all carried out at the Royal Institution,[42] and we are very fortunate that the old laboratory there has now been fitted out as it was in Faraday's day. Davy was happy to vandalize or misuse chemical apparatus, turning it to new purposes in the interests of discovery; and chemical apparatus is fragile anyway. But equipment like his can now be seen in its appropriate setting; and Faraday's *Chemical Manipulation* gives us an excellent guide into practice at the Royal Institution.[43]

With abundant materials, then, and in the hope of seeing Davy's life in the way which at the end gave it meaning for him, we embark upon his childhood in Cornwall, at the very south-western tip of England.

2

Growing Up

When Davy was born, on 17 December 1778, the Industrial Revolution was well under way, and one of his roles was to be the apostle of applied science. If asked to name the first industrial centres, we would think of the Midlands, especiallly Coalbrookdale and then Birmingham; of Lancashire and Yorkshire and their textile mills; and of the mining areas of Northumberland and Durham. But Cornwall, still the wild west of England where the Celtic language had only recently died out, had been a centre of tin mining for two thousand years.[1] Our vision of prehistory, where bronze comes between stone and iron, makes it easy for us to forget how important bronze, an alloy of copper and tin, was into the nineteenth century; what was seen as a great feat of French science at the time of the Revolution of 1789 was the conversion of church bells into cannon made of bronze (gunmetal) of slightly different composition.

The refining of tin was a metallurgical process involving old knowledge, but by the late eighteenth century the Cornish mines had become so deep that they ran into water and needed pumping out. James Watt had recently improved the steam engine, using a separate condenser so that the cylinder remained constantly hot and fuel was thereby saved.[2] Primitive engines had long been in use in the collieries, but because coal at the pithead was cheap, the economy offered by Watt's engine was not important to them. Boulton and Watt found their best market in Cornwall, where they leased their engines for a charge based on the saving of fuel, which had to come from a distance and was therefore expensive. By the early nineteenth century Cornish engineers were improving on Watt's models – sometimes being challenged in the courts for infringing his patents – and Cornish engines became the standard of efficiency against which others were judged.[3]

Davy's childhood was therefore both rural and industrial. Cornwall, though remote from London, was familiar to men of science, for in the middle of the eighteenth century William Borlase, Rector of Ludgvan, wrote two magnificent volumes on its antiquities and natural history, giving us a valuable background to Davy's upbringing.[4] Penzance, at the tip of the county, might seem far from the centre of things, but before the Reform Bill of 1832 Cornwall was over-represented in Parliament and had strong links with the capital as well as with the industrial Midlands.[5] It was a part of the Diocese of Exeter, but the bishop and cathedral were a long way off and John Wesley and later Methodists had been, despite Borlase's efforts, very successful in evangelizing in Cornwall. Davy's later friend and patron, Robert Southey, reports in his classic biography of John Wesley that he addressed 32,000 people there at an open-air meeting at Gwennap in his seventieth year (1772).[6] The Davy family's connections seem to have been with Church rather than Chapel, but Humphry was never later a strongly committed member of any particular congregation.

The Davys were keen to be seen as respectable. Their ancestors had been 'yeomen' and even 'gentlemen', but Humphry's father, Robert, did not inherit all that had been hoped for from his uncle, also named Robert (a friend of Borlase), who had brought him up. In addition, he apparently also lost money through experiments in farming and speculative investments in mining, two fields in which his son was to apply science. He was himself a woodcarver, a craft not in much demand, but in which he seems to have been skilful. He married Grace Millett, from another old Cornish family, and Humphry was the first of their five children. At first the family lived in Penzance, but when Humphry was nine they moved to the only property Robert owned, a small farm at Varfell in Ludgvan, which is about three miles from Penzance and overlooks St Michael's Mount. Then during school terms Humphry lived in the house of John Tonkin, a family friend who became his guardian.

Humphry was first sent to a little school, where Mr Bushell taught reading and writing only, but the old man soon recommended that he be sent to the Grammar School, where at the age of six he duly went. The master here was the Reverend Mr Coryton, who seems to have been no model pedagogue and punished boys by twisting their ears; but Davy in 1802 wrote to his mother:

> Learning naturally is a true pleasure: how unfortunate then it is that in most schools it is made a pain . . . After all, the way we are taught Latin and Greek does not much influence the important structure of our minds. I consider it fortunate that I was left much to myself when a child, and put upon no particular plan of study, and that I enjoyed much idleness while at Mr Coryton's school. I perhaps own to these circumstances the little talents that I have, and their peculiar application. What I am I have made myself; I say this without vanity, and in pure simplicity of heart.[7]

When we worry about the education of able children, we should perhaps remember remarks like these, which were echoed by other eminent men of science in the nineteenth century.[8] Davy at any rate never absorbed much classical learning, but he clearly became intellectually self-propelled.

While at school he found that he was good at writing verses, apparently in Latin as well as English, and wrote Valentines for his contemporaries to send. He was also a great story-teller, holding his audience with tales from the *Arabian Nights* and from local traditions coming from his Davy grandmother. She firmly believed in ghosts, even after young Humphry had impersonated one, and in later life he too was noted as superstitious. He was from boyhood a very keen fisherman, and his mother's marinaded pilchards were his favourite dish. He also took up shooting, stuffing some of the rarer birds he shot.[9] His earliest experience of chemistry was apparently making fireworks, helped by his sister, in an unfurnished room in which an elderly friend, the Reverend Dr Tonkin, exercised on a 'chamber horse' – an armchair on springboards (an adult version of a rocking horse) – which served them as a table.[10]

At the beginning of 1793 he went on (at John Tonkin's expense) to the Grammar School at Truro, the county town of Cornwall. The school was kept by Dr Cardew, whom Humphry considered a most excellent master compared to Mr Coryton and who later wrote of him:

> He gave me much satisfaction, being always regular in the performance of his duties as a school-boy, and in his general conduct. He was, too, I believe, much liked by his school-fellows for his good humour; but he did not at that time discover any extraordinary abilities, or, as far as I could observe, any propensity to those scientific pursuits which raised him to such eminence. His best exercises were translations from the classics into English verse.[11]

He was thought clever, but no prodigy, and in December 1793, just before his sixteenth birthday, he left school.

Returning to Penzance, Davy resided with Tonkin, 'by whom he had been, in a manner, adopted', and after a desultory beginning, adopted a course of self-education, but as yet with no definite career in mind. Throughout this year his father's health had been declining and in December 1794 he died, at the age of only forty-eight, leaving his widow an income of about £150 a year and debts of some £1,300. Davy's mother moved to Penzance and, with a French refugee, opened a milliner's shop. By 1799 the debts were paid off, Mrs Davy got an unexpected legacy of £300 a year and gave up the shop. Her courage and energy had meant that the family were educated. Humphry told his mother not to grieve, for he would do all in his power for his brothers and sisters; in due course he did see John through medical school at Edinburgh, but in 1794 he could not do much to help.

On 10 February 1795 Davy was apprenticed to Bingham Borlase, who, like Tonkin, was an apothecary-surgeon of Penzance and, also like him, mayor of the town.[12] At that time, surgeons and apothecaries – the general practitioners in medicine – learned their craft as apprentices; only the physicians had a university education, though it was becoming possible to attend a hospital and medical school in London and obtain better qualifications, or to go to Edinburgh; Davy seems to have hoped ultimately to study in Edinburgh after qualifying. Later, in 1804, he was to register at Jesus College, Cambridge, for a medical course. This would have conferred more social status but less practical and theoretical knowledge, but Humphry never, in fact, took it. However, in 1795 a respectable career in medicine seemed to be opening up before him, and in Cornwall an apprenticeship was acceptable even among the sons of gentlemen. He determined upon a formidable programme of self-education:

1 Theology
  Religion (taught by Nature)
  Ethics (taught by Revelation)
2 Geography
3 My Profession
  Botany
  Pharmacy
  Nosology
  Anatomy
  Surgery
  Chemistry
4 Logic
5 Language
  English

French
Latin
Greek
Italian
Spanish
Hebrew

6 Physics
The doctrines and properties of natural bodies
Of the operations of nature
Of the doctrines of fluids
Of the properties of organised matter
Of the organisation of matter
Simple Astronomy

7 Mechanics

8 Rhetoric and Oratory

9 History and Chronology

10 Mathematics[13]

The classification of sciences seems rather unusual to us, since 'physics' includes a good deal of what we would think of as biology (a term not yet coined in 1795), but not mechanics; and chemistry is no longer a medical science. It is not plausible to see the sciences as altogether a social construction, but the boundaries between them, and their hierarchy, are different at different times and places.

Already we see here, and in the notebooks that Davy now began to keep, the breadth of his interests. The little books are unorganized, usually begun at different times from both ends and containing all sorts of jottings, thoughts, details of experiments, reflections on religion, and poems. We see Davy in his later teens and early twenties thinking with a pen in his hand, scribbling down things often used in later lectures or in writings like *Salmonia* and *Consolations*. They were clearly a very important part of that programme of 'making himself' that we met in the letter he was to write to his mother in 1802; full of things he could not have learned at school, and handled unsystematically rather than formally as in a course.

The self-educated man, we are told, will have had a very ignorant teacher, but Davy seems to have been fortunate in the people and books he met, though his lack of a formal and systematic training in chemistry was to remain with him. Berzelius considered that Davy's work was no more than 'brilliant fragments' because he had lacked a proper grounding and never had to work across the whole field of chemistry.[14] But against that we can set Davy's enthusiasm – schools, universities and examination systems are all too good at inoculating us against academic subjects. He began with metaphysics, building on an interest in philosophy and religion that had begun at school and was to remain with him throughout life. He read Locke, Berkeley, David Hartley the utilitarian and exponent of the association of ideas, some Hume, Helvetius, Condorcet, and the Scottish 'common-sense' philosophy of Reid and Stewart, making notes that were sometimes critical and original.

He was also reading history and poetry, and began writing poetry at this time, as well as a prose romance featuring Prince Arthur and a Druid, and set at Land's End at night. Music meant little to him: he seems to have been tone deaf, given to humming tunelessly while working, and unable to learn to sing 'God save the King' in tune, and some later found his voice affected, though there can be no doubt of his power as an orator. He both loved the scenery of his native place and sought to understand its structure: 'How often when a boy have I wandered about those rocks in search of new minerals, and when fatigued, sat down upon the turf, and exercised my fancy in anticipations of scientific renown' he said on later being shown a picture of Botallack Mine.[15] He was clearly a dreamy young man, with pronounced intellectual interests and a surprising confidence in his abilities and his glorious future. This was not incompatible with a career in medicine; most of his friends and associates later were medical men, and it was the obvious profession to follow for anybody of scientific tastes. He seems to have been a satisfactory apprentice.

In 1799 his first poems were published in the *Annual Anthology* edited by Southey in Bristol. The best known of them is 'The Sons of Genius', from which these stanzas come:

> While superstition rules the vulgar soul,
> Forbids the energies of man to rise,
> Raised far above her low, her mean control,
> Aspiring genius seeks her native skies.
>
> Inspired by her, the sons of genius rise
> Above all earthly thoughts, all vulgar care;
> Wealth, power, and grandeur, they alike despise, –
> Enraptured by the good, the great, the fair.
>
> Yet not alone delight the soft and fair,
> Alike the grander scenes of nature move;
> Yet not alone her beauties claim their care,
> The great, sublime, and terrible they love.

The sons of nature, – they alike delight
  In the rough precipice's broken steep;
In the bleak terrors of the stormy night;
  And in the thunders of the threatening deep.

Ah! then how sweet to pass the night away
  In silent converse with the Grecian page,
Whilst Homer tunes his ever-living lay,
  Or reason listens to the Athenian sage.

To scan the laws of nature, to explore
  The tranquil reign of mild Philosophy;
Or on Newtonian wings to soar
  Through the bright regions of the starry sky.

From these pursuits the sons of genius scan
  The end of their creation, – hence they know
The fair, sublime, immortal hopes of man,
  From whence alone undying pleasures flow.

Theirs is the glory of a lasting name,
  The meed of genius, and her living fire;
Theirs is the laurel of eternal fame,
  And theirs the sweetness of the muses lyre.[16]

Davy was a son both of genius and of nature, and this poem, written in 1795–6 and full of both the sublime and the beautiful,[17] shows how he had absorbed current ideas and was aware of what we call the Romantic Movement.[18]

Dr Paris tells an anecdote of a shipwrecked French surgeon with whom Davy became friends, and whose instruments became his first scientific apparatus:

> ... this long-neglected and unobtrusive machine ... was destined for higher things; and we shall hereafter learn that it actually performed the duties of an air-pump, in an original experiment on the nature and sources of heat. The most humble means may certainly accomplish the highest ends ... a kite, made with two cross sticks and a silk handkerchief, enabled the chemical Prometheus to rob the thundercloud of its lightnings; but that a worn-out instrument ... should have furnished him who was born to revolutionize the science of his age, with the only means of enquiry at that time within his reach, affords, it must be admitted, a very whimsical illustration of our maxim.[19]

The stories Paris tells are often to be taken with a pinch of salt; they may not be exact, but this one emphasizes Davy's capacity to misuse apparatus creatively, and no doubt this was something observed early in his life, when he turned what was at hand to his own scientific purposes.

The story also brings out Davy's French connections. His mother had kept shop with a French milliner, and apparently Humphry learned French from a refugee priest from La Vendée, M. Dugast. We are told that he proved good at learning the grammar, 'but could not succeed in obtaining the correct pronunciation; and, in fact, notwithstanding his extensive intercourse with foreigners, and his residence in France, he never, in later life, could speak French either with correctness or fluency'.[20] This Churchillian defect did not prevent him from reading French books and journals concerned with science, which was very important at a time when Paris was the centre of things. At a later stage Davy, like a polished gentleman, also learned Italian, but, in common with most of his British contemporaries, he found German beyond him.

Late in 1797, when he was nearly nineteen, Davy began the study of chemistry with Lavoisier's classic *Elements of Chemistry*, in French rather than in the English translation of 1790; and also apparently with William Nicholson's *Dictionary of Chemistry*. To learn a science from a dictionary seems implausible, but the eminent French chemist Macquer had produced a classic example in the middle of the eighteenth century,[21] and Nicholson's was to form the basis of Andrew Ure's, which was widely used in the nineteenth century; they were discursive reference works from which the browser could learn much, though not systematically. Nicholson was a well-known popularizer, who wrote a standard introduction to chemistry, presented as an inductive science in which theory was best avoided or treated as a working hypothesis not to be taken seriously, in the Baconian tradition strong in England.[22] Newton's remarks about both the cause of gravitation and the nature of light promoted this belief, though Newton's own method – intuitive and mathematical – was hardly Baconian in a strict sense; he went far beyond cautious generalizing from observation, but insisted upon inferring real causes rather than using hypotheses.

Humphry wrote 'Davy and Newton' in a notebook, and aspired to be the Newton of his day, and was thus very ready to spot and attack the hypotheses he found in Lavoisier's beautifully organized book, as Newton had done with Descartes' science. Lavoisier wrote an introduction cutting off the science from theories of particles, which he believed metaphysical and untestable; the chemist must instead begin with 'simple bodies', substances that could not be further analysed. This meant that the list of such elements was provisional; a new analytical technique would break down some substances previously seen as elementary. Davy was himself to be one of the great discoverers of chemical elements.

One of Lavoisier's elements was 'caloric', the substance of heat, which he believed to be matter, though probably imponderable or weightless. Joseph Black in Edinburgh had pondered on why the snow did not melt away as soon as the temperature rose above freezing, and had inferred that when ice changes state and becomes water, and again when water becomes steam, a definite quantity of heat is absorbed. He called this 'latent heat', because its absorption does not lead to a rise in temperature.[23] A change of state was thus analogous to a chemical reaction: a definite quantity of ice and of heat reacted together to form water. For Lavoisier, water was a compound of caloric and ice, and steam a compound of more caloric and ice. Liquids and gases could not therefore be real elements; hydrogen, for example (which could not even be liquefied for a century after Lavoisier wrote), was a compound of an unknown basis and caloric. Light was also on his list of elements. In some reactions hydrogen was given off, in others caloric, in others light – perhaps all of them together.

James Watt had worked with Black in Glasgow and they corresponded later.[24] In the winter of 1797/8 Watt's son Gregory, who suffered from tuberculosis, was sent to Cornwall in the hope that the mild winter weather there would be good for his health.[25] He came as a lodger to the Davys, and he and Humphry soon became great friends. Apparently things began coolly, Gregory being reserved and his landlady's son trying to show off to him by talking metaphysics, but when Davy offered to demolish the French theory of heat both became much excited and in chemistry found their affinity for one another.

Watt was a prominent member of the Lunar Society of Birmingham, a group of industrialists and men of science, which also included Josiah Wedgwood, Erasmus Darwin, Joseph Priestley, Matthew Boulton and Richard Lovell Edgeworth; they met at the full moon so they could see their way home.[26] With the political reaction of the 1790s, the sacking of Priestley's house by a mob loyal to 'Church and King', and the ageing of the original members, the society became less active and important, and Manchester became a more important scientific centre than Birmingham, but this connection was nevertheless crucial for Davy's subsequent career. Britain at this time was a country governed by patronage: who you knew was more important than what you knew, and posts were filled through personal knowledge and recommendation rather than by open advertisement. Davy reinforced his Lunar connections when Thomas Wedgwood, a son of the potter, also came to Penzance for a winter, being also consumptive. Like Gregory Watt, he was scientific in his interests, and with Davy he later experimented with what became photography. Thomas Wedgwood and Gregory Watt both died young from their tuberculosis and their greatest discovery turned out to be Davy.

Equally important was Davy's contact with a local patron, Davies Giddy (who later changed his name to Gilbert and succeeded his protégé as President of the Royal Society when Davy resigned in 1827), a Member of Parliament with scientific interests. Giddy met the uncouth young Davy swinging on Borlase's gate and on being told of his interests, exclaimed, 'Chemical experiments! – if that be the case, I must have some conversation with him', and soon perceived his ability, offering him the use of his library.[27] He later introduced Davy to Dr Edwards, who subsequently lectured in chemistry at St Bartholomew's Hospital in London and had a well-equipped laboratory. Humphry was transported with delight on seeing apparatus that he knew about only from descriptions and engravings. Davy also got help from a Quaker, Robert Dunkin, a saddler turned instrument-maker; it seems likely that it was his help, rather than any unassisted reuse of apparatus from a French surgeon, that lay behind Humphry's early experiments.[28]

At Oxford Giddy had gone to the chemistry lectures of Thomas Beddoes, who attracted the largest audiences in the university. Beddoes had a fair prospect of being elected to a new chair of chemistry, but his support for radical causes at the time of the French Revolution put paid to such hopes, and even the idea of a professorship at Oxford was shelved. He had Lunar connections, his wife being an Edgeworth; he was one of the first to introduce Kantian ideas into Britain, and persuaded Coleridge and Wordsworth to visit Germany; but chemistry, especially in its medical connections, was his great interest. In medicine he was a devotee of the ideas of John Brown, for whom health involved the balance of active and passive vital forces, 'sthenia' and 'asthenia'.[29] This led Beddoes to prescribe opium but also to hope that the new gases, or 'factitious airs', prepared by Priestley might be useful agents in the treatment of diseases, especially respiratory ones like tuberculosis. He was also a firm believer in preventive medicine, so convinced that a healthy lifestyle and sensible clothes were of enormous importance that he wrote both a improving novel, *Isaac Jenkins*, and a treatise, *Hygeia*, to make contemporaries aware of it.[30]

Beddoes visited Cornwall to look at its minerals (important in current geological controversy) and to see Giddy, and in April 1798 he received, through Gregory Watt and with Giddy's backing, Davy's account of his experiments on heat and light. He was immensely excited about them, because although a convert (unlike Priestley) to Lavoisier's views on burning, he was pleased to see French theory knocked about; he had indeed republished John Mayow's seventeenth-century studies on combustion to show how far they anticipated Lavoisier's work.

Beddoes was so delighted with Davy's work that he published the 'Essay on Heat and Light' in 1799 in a symposium he edited, *Contributions to Physical and Medical Knowledge, principally from the West of England*. Davy soon after repudiated the paper, as 'my infant chemical speculations', apprentice work that should never have been brought before the public; but the essay shows the freshness and originality of his thinking, and some of the ideas in it still feature in the *Consolations in Travel* composed at the very end of his life.[31] There are few natural philosophers for whom we have such early published work, so we, at least, can be grateful for this essay. Its publication does not seem to have done any harm to Davy's reputation in 1799 or later; on the contrary, it brought him to the notice of those with patronage.

Lavoisier had supposed oxygen gas to be a compound of an unknown basis with caloric; for Davy it was a compound with light and should be called 'phosoxygen'. For Davy,

> Light is a body in a peculiar state of existence. Its particles are so amazingly minute, that they are very little affected by gravitation; and pass unaltered through the pores of diaphanous bodies. They move through space with a velocity almost inconceivable, and communicate no perceptible mechanical motion to the smallest perceptible particles of matter. From the peculiar velocity of light we estimate its quantity of repulsive motion. . . . Light is the source of the most numerous and pleasurable of our perceptions . . . particles of light in the state of repulsive projection coming into contact with the retina, communicate to it portions of their repulsive motion.[32]

Particles of light are attracted by those of ordinary matter in such a way as to yield the laws of refraction 'so admirably explained by the immortal Newton'.

Following Newton, like most British contemporaries, Davy believed that he had established that light was material particles,[33] while his experiment of melting ice by friction proved that heat was just motion. He therefore wrote of oxygen:

> This gas (first discovered by the immortal Priestley) the great Lavoisier supposed to be oxygen combined with caloric, and on this supposition his theory of combustion is founded. The non-existence of caloric, or the fluid of heat, has been proved, and the materiality of light demonstrated.

> Light is liberated during the oxygenation of certain bodies, as the following experiments will prove.[34]

It was not entirely new to show that when things burn, light is given off, but it was very bold to take on Lavoisier (given a slightly less exalted adjective than Priestley) at the central point of his theory. Davy's subsequent success in chemistry was to follow from the study of what we would call chemical energy, and from overturning Lavoisier's theory that oxygen was the agent responsible for acidity; and he never seems to have lost his feeling for the importance of light, though he never developed a thoroughly consistent theory about it.

Davy realized that what he called 'phosoxygen' our oxygen gas, was being continually used up by animals, by combustion and by other chemical processes. Priestley and Jan Ingenhousz had found that plants restored oxygen to the atmosphere, and Davy had extended their work by showing that flowerless pondweeds, 'marine cryptogamia', also did this. But the point of his essay was to connect his theory of light with speculations about the nature of life, a current issue of great importance in electrical and physiological debate with the work of Luigi Galvani,[35] later reflected in Mary Shelley's *Frankenstein*.

Oxygen brought its combined light to the brain, where it was secreted 'in the form of an etherial fluid or gas' and conveyed to the nerves.[36] Because electricity stimulated the nerves, 'we have concluded the nervous fluid to be the electric aura'; sensations or ideas are motions of the nervous ether or light; and the 'irritability' of living matter is a consequence of contained light, which is essential to perceptive existence.[37] The essay ends with a peroration:

> We may consider the sun and the fixed stars, the suns of other worlds, as immense resevoirs of light destined by the great *organiser* to diffuse over the universe organization and animation. And thus will the laws of gravitation, as well as the chemical laws, be considered as subservient to one grand end, *perception*. Reasoning thus, it will not appear impossible that one law alone may govern and act upon matter: an energy of mutation, impressed by the will of the Deity, a law which might be called the law of animation, tending to produce the greatest possible sum of perception, the greatest possible sum of happiness. The further we investigate the phaenomena of nature, the more we discover simplicity and unity of design. An extensive field of sublime investigation is open to us. . . . We cannot entertain a doubt but that every change in our sensations and ideas must be accompanied by some correspondent change in the organic matter of the body. These changes experimental investigation may enable us to determine. By discovering them we should

> be informed of the laws of our existence, and probably enabled in great measure to destroy our pains and to increase our pleasures. Thus would chemistry, in its connection with the laws of life, become the most sublime and important of all sciences.[38]

The confident youthful tone, and the belief in the simplicity and the utility of science, reveal the qualities that made Davy one of the greatest of scientific lecturers, who was, indeed, to make chemistry seem sublime, important and fundamental to all the other sciences. We should not be surprised that Beddoes was happy to publish the essay, nor that Davy soon became friends with Romantic writers.

Beddoes needed an assistant at his Pneumatic Institution and negotiated with Giddy for Davy to be released from his indentures. Davy asked for a 'genteel maintenance' by way of salary; Tonkin was furious that Davy should have been tempted away into a madcap scheme just when he seemed to be on the threshold of a useful career as a medical man in Penzance.[39] The prospects for a livelihood from science were remote and pneumatic medicine looked very much like quackery. But Borlase allowed himself to be persuaded, saying that he freely gave up his indenture on account of Davy's singularly promising talents. Humphry was excited at the prospect of continuing and improving his experiments in Clifton, and on 2 October 1798, two months before his twentieth birthday, he set off from Penzance. Two days later Giddy met him for breakfast in Okehampton in the highest spirits, which were further raised by the mail coach from London arriving decorated with ribbons and laurels, bringing news of Nelson's victory at the Battle of the Nile. Patriotism and chemistry came together at the outset of Davy's career in science as the coach took him on to Bristol.

3

Clifton

Davy retained strong memories of and love for his native county, but he rarely revisited it as his life in science became more and more successful. Bristol, to which he went, had long been one of the largest and most prosperous cities in Britain, a base for the Atlantic trade, notably in tobacco and wine. Liverpool, with cotton, was beginning to overtake it, but Bristol was much more respectable despite the disreputable nature of much of its trading, in slaves and their major product, sugar. From a county based upon primary products, tin and china clay (as well as the fishing Davy remembered with his mother's marinaded pilchards), he had moved to a city of wealthy merchants, near the fashionable resort of Bath, a place priding itself upon culture and polish. The country boy was beginning his great social journey upwards, and in Bristol he was to be fortunate enough to meet some of the most able and exciting intellectuals of the time, and upon equal terms, through Beddoes and his circle. He scribbled:

> by the ardent and incessant exertions of Dr Beddoes the Pneumatic Institution is on the point of establishment. The design of this great object has been repeatedly pointed out by him – to investigate an important branch of medicine which has been heretofore but little considered by medical men & indeed almost treated with contempt by the Friends to Science & to improvement cannot be considered as an unimportant object, & it is therefore to be wished that this description of men would turn their attention to the pursuit.[1]

The Pneumatic Institution began in Beddoes' house in Clifton, a suburb 'commanding a view of Bristol and its neighbourhood, conveniently elevated above the dirt and noise of the city. Here are houses, rocks, woods, town and country in one small spot; and beneath us, the sweetly flowing Avon, so celebrated by the poets. Indeed, there can hardly be a more beautiful spot; it almost rivals Penzance, and the beauties of Mount's Bay', as Davy described it to his mother.[2] Premises were eventually found in Dowery (or Dowry) Square, Hotwells. The project was financed by gifts from the Wedgwood family and from the Durham coal owner and Member of Parliament William Henry Lambton, who was a patient of Beddoes' and whose two sons boarded with him and got to know Davy. John, the elder, later famous for saying that he 'jogged along' on his enormous income, also became an MP, was created Earl of Durham and was responsible for the report that led to Canada evolving from colonial to independent dominion status. Beddoes was not only practising an 'alternative medicine', but was also involved with political radicals in the difficult days as the French Revolution turned to terror, war and military despotism, Paris nevertheless remaining the great centre of science.

Beddoes' wife was the sister of the well-known author Maria Edgeworth and while Beddoes' lectures brought him scientific eminence he also moved happily in literary circles. Indeed, Davy believed that Beddoes had made less of a name for himself in science than he should have done because he had so many enthusiasms and never really settled long at anything. Davy soon got to know Joseph Cottle, the Bristol publisher, who was bringing out the volume that included those first essays on heat and light written in Cornwall, and was also publishing Southey, Coleridge and Wordsworth. Long afterwards Cottle recorded his first meeting with Davy:

> I was much struck with the intellectual character of his face. His eye was piercing, and when not engaged in converse, was remarkably introverted, amounting to absence, as though his mind had been pursuing some severe trains of thought, scarcely to be interrupted by external objects; and from the first interview also, his ingenuousness impressed me as much as his mental superiority. Mr. D. having no acquaintance in Bristol, I encouraged and often received his visits, and he conferred an obligation on me, by often passing his afternoons in my company. During these agreeable interviews, he occasionally amused me by relating anecdotes of himself; or detailing his numerous chemical experiments; or otherwise by repeating his poems, several of which he gave me (still retained); and it was impossible to doubt, that if he had not shone as a philosopher, he would have become conspicuous as a poet.[3]

In Beddoes' house, Davy had access to a well-equipped laboratory; and such was his enthusiasm for chemistry that he even gave pieces of apparatus to Cottle in the hope of converting him to science; he also sent equipment to friends in Cornwall.

The first task in Clifton was to find a house for the Institution. There is a draft letter in a notebook: 'We are negotiating for a house in Dowrie Square just below Clifton which I think will very well answer the purposes of a hospital. We shall I suppose be able to provide for 8 or 10 in patients and as many out ones as we can procure'.[4] By early 1799 they were installed there and the experiments on gases, and the treatments that would accompany and follow them, could begin. Priestley had isolated oxygen; its importance in medicine seemed evident, and in our century has become much greater, for without it life clearly comes to an end. Davy soon investigated another gas, that still has medical uses: nitrous oxide. Here, as often, he began his work in response to implausible claims made by somebody of a certain celebrity: in this case, the American physician, and later Congressman, Samuel Mitchell (or Mitchill). For the French, naming the new simple bodies of chemistry, the gas that made up three-quarters of the atmosphere was deadly, and they called it *azote*. Mitchell believed that nitrogen's combinations with oxygen must be more actively poisonous; nitric oxide and nitric acid clearly filled this bill, but he particularly dreaded nitrous oxide, which would be a very *septon*, the principle of contagion, instantly fatal to anybody who breathed it.

Mitchell's little book *Remarks on the Gaseous Oxyd of Azote and of its Effects*, was published in New York in 1795. Beddoes and Watt had reprinted it as an appendix to their book the *Medicinal Uses of Factitious Airs* in 1794, but it should not have survived the examination of a conscientious referee. It met such a reader in Davy even before he went to Bristol:

> A short time after I began the study of Chemistry, in March 1798, my attention was directed to the *dephlogisticated nitrous gas* of Priestley (nitrous oxide) by Dr. Mitchell's theory of Contagion, by which he attempts to prove that *dephlogisticated nitrous gas*! which he calls *oxide of septon*, was the principle of contagion, and capable of producing the most terrible effects, when respired by animals in the minutest quantities, or even when applied to the skin, or muscular fibre. The fallacy of this theory was soon demonstrated by a few coarse experiments, made on small quantities of this gas procured, in the first instance, from zinc and diluted nitrous acid. Wounds were exposed to its action; the bodies of animals were immersed in it without injury; and I breathed it,

> mingled in small quantities with common air, without any remarkable effects. An inability to procure it in sufficient quantities prevented me, at this time, from pursuing the experiments to any great extent. I communicated an account of them to Dr. Beddoes.[5]

A year later, he returned to this line of research, which led to his first book and to his establishing a name for himself not just as a promising young man, but as an accomplished chemist and a scientific star.

In 1799 'my situation in the Medical Pneumatic Institution, made it my duty to investigate the physiological effects of the aëriform fluids, the properties of which presented a chance of useful agency' as he put it.[6] In March Davy prepared impure samples of the gas and tried breathing it, but it was not until April that he could get it in a pure state, determine its chemical properties and composition, and find its extraordinary effects, as laughing gas, when breathed. The experiments he described with great vividness, forming some of the best subjective accounts of what we call anaesthesia that have ever been penned.

He was aware of the dangers of the experiment, which was one of a series trying various gases: with nitric oxide the effects were very painful, and with carbon monoxide, nearly fatal. Cottle suggested that:

> these destructive experiments, during his residence at Bristol, probably, produced those affections of the chest, to which he was subject through life, and which, beyond all question, shortened his days. Nothing at the moment so excites my surprise, as that Mr. D.'s life should have been protracted, with all his unparalleled indifference concerning it, to the vast age, for him, of fifty years.[7]

We are apt to forget the dangers of the scientific life, especially in days when few safety precautions were taken; and, of course, such precautions are usually the result of unpleasant accidents or accumulated experience – stable doors are generally locked only after horses have bolted.

Davy continued, in his later work on fulminates and on the safety lamp for example, to be careless of danger in his bursts of rapid experimental progress, but he wrote to his brother John, then a medical student in Edinburgh and experimenting on himself with digitalis:

> I hope you will not indulge in trials of this kind. I cannot see any useful result that can arise from them . . . you may injure your constitution without gaining any important result. Besides, if I were in your place, I should avoid being talked of for anything extraordinary of this kind, as you have already fame of a better kind, and the power of gaining fame of the noblest kind.[8]

It may be easier to give such advice than to take it: certainly, as Davy was later to put it 'the business of the laboratory is often a service of danger'.[9]

He used a green silk bag for the experiments, breathing in and out of it after satisfying himself that it would be wasteful to breathe fresh gas at every respiration since only some of that inhaled was absorbed. When on 17 April he breathed four quarts in the presence of Beddoes, he felt a:

> sensation analogous to gentle pressure on all the muscles, attended by an highly pleasurable thrilling, particularly in the chest and the extremities. The objects around me became dazzling and my hearing more acute. Towards the last inspirations, the thrilling increased, the sense of muscular power became greater, and at last an irresistible propensity to action was indulged in; I recollect indistinctly what followed; I know that my motions were various and violent.[10]

He had a good night's sleep after the experiment – the gas left no hangover – and found similar effects the next day on repeating the inhalation in a sceptical frame of mind, proving that the effects were real and must be due to the nitrous oxide.

Later, Davy reported another experiment:

> A thrilling extending from the chest to the extremities was almost immedicately produced. I felt a sense of tangible extension highly pleasurable in every limb; my visible impressions were dazzling and apparently magnified, I heard distinctly every sound in the room and was perfectly aware of my situation. By degrees as the pleasurable sensations increased, I lost all connection with external things; trains of vivid visible images repidly passed through my mind and were connected with words in such a manner, as to produce perceptions perfectly novel. I existed in a world of newly connected and newly modified ideas. I theorised; I imagined that I made discoveries. When I was awakened from this semi-delirious trance by Dr. Kinglake, who took the bag from my mouth, Indignation and pride were the first feelings produced by the sight of the persons about me. My emotions were enthusiastic and sublime; and for a minute I walked around the room perfectly regardless of what was said to me. As I recovered my former state of mind, I felt an inclination to communicate the discoveries I had made during the experiment. I endeavoured to recall the ideas, they were feeble and indistinct; one collection of terms, however, presented itself: and with the most intense belief and prophetic manner, I exclaimed to Dr. Kinglake, *'Nothing exists but thoughts! – the universe is composed of impressions, ideas, pleasures and pains!'* About three minutes and a half only, had elapsed during this experiment, though the time as measured by the relative vividness of the recollected ideas, appeared to me much longer.[11]

Kinglake was one of Beddoes' collaborators in the Institution. Davy reported that in these experiments his cheeks turned purple, but however exciting and alarming the effects were to those who had never seen anything like it except drunkenness, breathing the gas left no lasting after-effects. Davy seems to have become mildly addicted, but cured himself. We should perhaps note that in Beddoes' circle (and amongst Romantics elsewhere) the taking of drugs to extend consciousness was practised;[12] Cottle records experiments with Indian hemp supplied by Sir Joseph Banks, President of the Royal Society;[13] and Coleridge's opium was prescribed for him for toothache, and was a medical standby in those days before its addictive properties were understood. Davy's work straddled any boundaries between physical science, physiology and psychology, and literature.

Between May and July, Davy tells us, he habitually breathed the gas 'occasionally three or four times a day for a week together; at other periods, four or five times a week only. The doses were generally from six to nine quarts; their effects appeared undiminished by habit, and were hardly even exactly similar.'[14] He prudently decided that the 'general effects of its operation upon my health and state of mind, are extremely difficult of description; nor can I well discriminate between its agency and that of other physical and moral causes'.[15] The word 'psychosomatic' was coined by Coleridge, who on his return from Germany was introduced by Cottle to Davy. He and Davy became fast friends; and were involved in one of the Institution's cures.[16]

Beddoes inferred that nitrous oxide would be good in cases of paralysis, but before administering the gas, Davy took the patient's temperature with a small thermometer under his tongue. The man, as Coleridge told Paris,

> wholly ignorant of the nature of the process to which he was to submit, but deeply impressed, from the representations of Dr. Beddoes, with the certainty of its success, no sooner felt the thermometer between his teeth than he concluded that the *talisman* was in full operation, and in a burst of enthusiasm declared that he already felt the effects of its benign influence throughout his whole body – the opportunity was too tempting to be lost – Davy cast an intelligent glance at Mr. Coleridge, and desired the patient to renew his visit on the following day.[17]

After two weeks the man was fully restored to health. Physiological actions are hard to determine, and in 1799 there were not the statistics available to make sense of trials of drugs. Doctors prescribed, and sellers of patent medicines advertised, largely on a basis of anecdotal evidence. Davy was surprisingly careful – he was, after all, only twenty – in making sure what effects were real.

Beddoes believed that testimony was crucial. This was not new: philosophy of science had begun in Britain with Francis Bacon, a lawyer, and in the early days of the Royal Society public experiment had been a route to certainty.[18] Davy was to be unusual among scientists in being prepared throughout his career to conduct at least parts of his research in public: there was always an element of the showman about him and he loved communicating enthusiasm. But the general point should not surprise us. One of the things that might distinguish science from other intellectual, practical and social activities[19] is that it is public knowledge: repeatable and open rather than personal.[20] Children learning science nowadays are made to write up their observations in the passive voice ('a test tube was heated') to bring home to them the impersonal and objective character of what they have been doing.

Things were rather different in Davy's day: he and his circle saw science as creative. All creative activities are disciplined and constrained, and in science the constraints are imposed by the way things are; but there is great scope for the imagination, and for a Davy, who wrote 'I never loved to imitate, but always to invent; this has been the case in all the sciences I have studied.'[21] To experiment in the presence of witnesses provided the control such an impulsive person needed to become a great chemist. In the case of nitrous oxide the results were so astonishing that confirmation was essential, and turned out to be highly pleasurable in general both for those who breathed the gas and perhaps even more for those who watched.

Beddoes persuaded a courageous young lady to breathe the gas. Her remarks were not written down, but we are told that 'to the astonishment of everybody, the young lady dashed out of the room and house, when, racing down Hope-square, she leaped over a great dog in her way, but being hotly pursued by the fleetest of her friends, the fair fugitive, or rather the temporary maniac, was at length overtaken and secured, without further damage'.[22] The damage was presumably to her reputation; the witnesses who recorded their impressions in Davy's book, all less vividly than he did himself, were mostly male.[23]

John Tobin, a medical man and playwright, reported

> suddenly starting from the chair, and vociferating with pleasure, I made towards those that were present, as I wished they would participate in my feelings. I struck gently at Mr. Davy and a stranger entering the room at the moment, I made towards him, and gave him several blows, but more in the spirit of good humour than of anger ... The feelings resembled those produced by a representation of an heroic scene on the

> stage, or by reading a sublime passage in poetry when circumstances contribute to awaken the finest sympathies of the soul.[24]

Kinglake noted that 'its agency was exerted so strongly on the brain, as progressively to suspend the senses of seeing, hearing, feeling and ultimately the power of volition itself', and hoped it would prove 'an extremely efficient remedy, as well in the vast tribe of diseases originating from deficient irritability and sensibility, as in those proceeding from morbid associations, and modifications, of those vital principles'.[25] Peter Roget, now chiefly remembered for his thesaurus but in his lifetime an eminent medical man, reported loss of consciousness. Coleridge found that he 'could not avoid, nor indeed felt any wish to avoid, beating the ground with my feet; and after the mouth-piece was removed, I remained for a few seconds motionless, in great extacy'.[26]

Davy wrote that 'many of the individuals breathed the gas from pure curiosity. Others with a disbelief of its powers'. He also made the famous suggestion, 'As nitrous oxide in its extensive operation appears capable of destroying physical pain, it may probably be used with advantage during surgical operations in which no great effusion of blood takes place.'[27] The caveat was no doubt added because Davy believed that the nitrous oxide combined in some way with blood. It was not until after his death that anaesthetics came into use; and many have wondered why. Some of the reason seems to be the sensational and rather unpredictable character of the effects produced, making it seem a bit like Mesmerism and other contemporary 'quackeries'. Otherwise, the context in which the research was done was the search for remedies, within the framework of Brown's theory, not for an analgesic – opium and alcohol were already familiar and much easier to administer, often mixed as in the formidable 'laudanum'. At all events, Davy himself never pushed his suggestion further, not even when he was President of the Royal Society and the friend and associate of eminent London doctors and could hardly have been completely disregarded.

Davy had at first been uncertain whether nitrous oxide was a 'stimulant' or 'depressant' substance, but his experiments persuaded him that it was the former: in Brown's system it must have been one or the other. In his tentative and cautious conclusion, Davy wrote:

> As hydrocarbonate [water gas, i. e., carbon monoxide and hydrogen] acts as a sedative, and diminishes living action as rapidly as nitrous oxide increases it, on the common theory of excitability, it would follow, that by differently modifying the atmosphere by means of this gas and

> nitrous oxide, we should be in possession of a regular series of exciting and depressing* powers applicable to every deviation of the constitution from health: but the common theory of excitability is most probably founded on a false generalisation.
>
> * That of Brown modified by his disciples [Davy's note].[28]

Here, as usual, Davy was prepared to cast doubt on the orthodox view, what is now often called the 'paradigm'; but it had nevertheless been the scheme within which he and Beddoes were working, and were understood.

Following the comments of the healthy, there are a series of brief case studies on the effects of nitrous oxide 'upon persons inclined to hysterical and nervous affections'.[29] These were all female, and the experiments began from the observation that the gas produced such 'affections in delicate and irritable constitutions'. Davy tried giving one woman air in the bag 'to ascertain the influence of imagination'; she felt no effect, but when given nitrous oxide mixed with rather little air, felt unpleasant sensations and suffered from weakness and debility for a great length of time.[30] Others were more fortunate; but any hope that nitrous oxide might be a valuable addition to the doctors armoury seemed faint. Indeed, the same proved true of the other gases, and Cottle tells us (on Davy's authority) that in the last years of the Pneumatic Institution patients had to be paid to submit to its treatments.[31]

The descriptions of the effects of nitrous oxide on individuals is only a small part of the book, beginning on page 453. The rest is much less sensational. It includes experiments to estimate the capacity of the lungs, which were original, and also a great deal of chemistry. Davy devised ways of preparing pure samples of the various oxides of nitrogen and did careful volumetric analyses of them. This was carrying on the tradition of Joseph Priestley, the greatest British chemist of the generation before Davy's, and a member of the Lunar Society. Davy had come into contact with Watt, Boulton and Edgworth; and Priestley, now an exile in America, wrote to him recognizing an intellectual heir.

Lavoisier had believed that all acids must have some material substance in common, which was the source of their acidity – this is a classically chemical way of thinking – and he believed that oxygen was this crucial substance. Davy found to his astonishment that the compounds of oxygen and nitrogen had extremely different properties both from one another and from their component elements. Nitrogen (he preferred the name to the French *azote*) was suffocating; oxygen, too strong for ordinary breathing; common air, just right for life;

nitrous oxide, intoxicating; nitric oxide, impossible to breathe; and nitric acid, an extremely powerful reagent. Chemical properties clearly did not depend in any simple way upon material components. Davy took pleasure in this, because it was important for his generation to establish that science and political revolution were not linked. The ideology behind the French Revolution was supposed to be scientific: Diderot, d'Alembert and Voltaire had all seen science as a weapon to use against the *ancien régime*. Science, materialism and the Reign of Terror were all connected together in the English mind, and to show that chemistry depended upon still-mysterious forces or powers, rather than matter, not only went better with Romantic beliefs about how the world worked, but also with patriotism.[32]

When the book was published, Davy was, at the age of twenty-one, a scientific star, his work attracting both popular attention and the respect of chemists. The wayward Beddoes had provided the opportunity for the dreamy youth with metaphysical and scientific ambitions to concentrate his mind upon particular problems that could be handled experimentally. The most important aspect of the scientific imagination is the capacity to focus, to ask questions of nature that are both interesting and answerable: contrariwise, in many parts of philosophy we do not expect definite and enduring answers, and it is the argument that is interesting. Galileo's predecessors had asked how motion was possible; he sought and found the quantitative law of falling bodies. Davy isolated and purified the various oxides of nitrogen, producing a standard monograph, and also investigated their real effect on the human body, all by careful experimental work. He had come to Clifton as someone full of possibilities. The price of success in what Davy called chemical philosophy meant a different narrowing of focus – it became evident to him and to others that it would be in science that he would excel. While keeping time for poetry and metaphysics, he became a chemist; medicine, which was to have been his career, gradually dropped out of sight after *Researches* was published. He made a 'Resolution. To work two hours with pen before breakfast on the ['Child of' deleted] Loves of Nature, or the feelings of Eldon – from six till eight – From nine till two in exp[ts] – from four till six reading, seven till ten, metaphysical system (i. e. system of the universe –'.[33]

Nitrous oxide was not the only topic he investigated in this year. On 10 April 1799 he had written to Giddy that one of the children of a Bristol friend, William Coates, 'accidentally discovered that two bonnet-canes rubbed together produced a faint light. The novelty of this phenomenon induced me to examine it.'[34] He found that when the outer skin, or epidermis, was removed, the canes no longer emitted light;

and on chemical analysis found that the skin 'had all the properties of silex', or silica, as in sand and flint. This was surprising, because silica is insoluble in water and one would not have expected it to be found in vegetation.

The skin of reeds and grasses looked similar to that of the bamboo and, indeed, turned out to be chemically similar. Davy found that corn and grasses contained 'sufficient potash to form glass with their flint. A very pretty experiment may be made on these plants with the blow-pipe. If you take a straw of wheat, barley or hay, and burn it, beginning at the top, and heating the ashes with the blue flame, you will obtain a perfect globule of hard glass fit for microscopic experiments.'[35] Dr Paris on this occasion was full of praise, 'endeavouring to infuse into the reader a portion of that admiration I feel on relating them. They furnish a beautiful illustration of that combination of observation, experiment and analogy, first recommended by Lord Bacon, and so strictly adopted by Davy in all his future grand researches.'[36] They also show how accident can be used by the prepared mind to lead to interesting discoveries, something Davy liked to emphasize.[37]

Though he was a full-time researcher in Bristol, Davy worked very hard in spasms and found time for other things, and particularly for friendships. He got to know both Southey and Coleridge, and both reported their experiences with nitrous oxide. They had married sisters, and both were writing poetry and prose. They had given up their plan to set up a 'pantisocracy' on the banks of the Susquehannah River in America (near Priestley), where all would be free and equal, and though still suspected by the authorities of being revolutionaries (which had led to their being spied on), they were moving politically to the right.

No doubt under Coleridge's tuition, Davy recorded a mystical experience outdoors:

> Today, for the first time in my life, I have had a distinct sympathy with nature. I was lying on the top of a rock to leeward; the wind was high, and everything in motion; the branches of an oak tree were waving and murmuring in the breeze; yellow clouds, deepened by grey at the base, were rapidly floating over the western hills; the whole sky was in motion; the yellow stream below me was flowing (agitated by the breeze), is this analogy? – Every thing seemed alive, and myself part of the series of visible impressions; I should have felt pain in tearing a leaf from one of the trees.[38]

He then reflected, as he was to do throughout life, on the incompleteness of scientific accounts of the world: 'Deeply and intimately connected are all our ideas of motion and life, and this, probably, from very early association. How different is the idea of life in the physiologist and the poet!'[39] After toying with them enthusiastically, he came to reject determinism and materialism with heartfelt fervour and in typically Romantic fashion; the purple passages in his unpublished writings from this date give a preview of the eloquence that was to entrance his lecture audiences.

In an account of waking from a bad dream, he wrote:

> You know a moonlight scene is peculiarly delightful to me; I always considered it as beautiful; but so much solitary enthusiasm, so much social feeling, so much of the sublime energy of love, of sorrow, of consolation, have occurred to me beneath the moonbeams, on the shore of that sea where Nature first spoke to me in the murmurs of the waves and winds, in the granitic caves of Michael, that it is now become sublime.[40]

Nitrous oxide also could stimulate him to poetry:

> Not in the ideal dreams of wild desire
>   Have I beheld a rapture-wakening form:
> My bosom burns with no unhallowe'd fire,
>   Yet is my cheek with rosy blushes warm;
> Yet are my eyes with sparkling lustre fill'd;
>   Yet is my mouth replete with murmuring sound;
> Yet are my limbs with inward transports fill'd.
>   And clad with new-born mightiness around.[41]

The references to love do not seem to indicate that Davy had a serious girlfriend; indeed, while he enjoyed the company and admiration of women, he was to marry late and unhappily.

His friendship with Coleridge, who told him and Cottle about his military adventures,[42] led to his being asked by Wordsworth, then in the Lake District (where Davy later visited him at Dove Cottage), to oversee the publishing of the second edition of the *Lyrical Ballads*. Davy persuaded Wordsworth that there was a creative and imaginative character to science, meaning that the natural philosopher was not incomparable with the poet as an interpreter of nature, and Wordsworth accordingly modified his preface.[43] He had written to Davy in July 1800:

> I venture to address you though I have not the happiness of being personally known to you. You would greatly oblige me by looking over the enclosed poems and correcting anything you find amiss in the punctuation a business at which I am ashamed to say I am no adept. I was unwilling to print from the MSS which Coleridge left in your hands, because I had not looked them over. I was afraid that some lines might

> be omitted or mistranscribed. I write to request that you would have the goodness to look over the proof sheets . . . before they are finally struck off . . . I need not say how happy I should be to see you here in my little cabin.[44]

Davy duly supervised the printing, as he was later to do for Southey's now-unread epic, *Thalaba*.

Much as he admired Coleridge and Wordsworth, he was moved by his editorial labours to parody Wordsworth's plain style:

> As I was walking up the street
> In pleasant Burny town
> In the high road I chanced to meet
> My cousin Matthew Brown.
>
> My cousin was a simple man
> A simple man was He
> His face was of the hue of tan
> And sparkling was his eye –
>
> His coat was red for in his youth
> A soldier he had been.
> But he was wounded and with ruth
> He left the camp I ween –
>
> His wound was cured by Doctor John
> Who lives upon the hill
> Close by the rock of grey free stone
> And just above the mill.
>
> He then became a farmer true
> And took to him for aid
> A wench who though her eye was blue
> Was yet a virgin maid.
>
> He married her and had a son
> Who died in early times
> As in the churchyard is made known
> By poet Wordsworths Rymes.
>
> As long as this fair wife did prove
> To him a wife most true
> His red coat He away did shove
> And wore a coat sky blue.[45]

A few months later Coleridge wrote to Davy about a proposal that he and Wordsworth should set up a chemical laboratory, asking for advice, but nothing seems to have come of this intriguing suggestion.

By the middle of 1800 Coleridge had gone first to London and then to the Lakes, and Southey had gone to Lisbon to recover his health and to write a history of the country. Davy's world in Bristol was beginning to come apart, but his reputation was made and it was increased by work that he now began on electricity.

Priestley's work on 'airs' had been open-ended, incomplete and accessible, with a quotation (in Latin) from Bacon on its title page: he had been like an explorer venturing into a new territory.[46] Davy, by contrast, performed a series of decisive experiments with a definite goal in view.[47]. Although he concluded his *Researches* with the words 'Pneumatic chemistry in its application to medicine, is an art in infancy, weak, almost useless, but apparently possessed of capabilities of improvement. To be rendered strong and mature, she must be nourished by facts, strengthened by exercise, and cautiously directed in the application of her powers by rational scepticism', the book represents completed research.[48] Contemporaries were not tempted by this tepid conclusion into further work in pneumatic chemistry, though inhaling nitrous oxide became a craze; and Davy's analyses of the oxides of nitrogen seemed definitive. John Dalton was to make new use of them with his atomic theory, but Davy needed a new field. He was fortunate that Volta inaugurated electrochemistry at just this right moment.[49]

Luigi Galvani believed that the twitching of frogs' legs that he observed was a sign that electricity was generated in the nerves; and electric fish, one of which Davy dissected when very near death in Rome in 1829, seemed to show the same. In the 1790s galvanism formed a branch of fringe medicine, like pneumatic treatments but more popular – after all, we still speak of people being galvanized into action. Alessandro Volta opposed Galvani's interpretation of the experiments; he believed that animal tissues detected rather than generated the electricity. In 1799 he sent to the Royal Society of London his paper describing his 'pile' of discs of two different metals (silver and zinc were best) with damp cardboard layers between them (salt water was better than fresh); from this sandwich he got electricity like that from a weakly charged Leyden jar condenser, but of enormous capacity and needing no recharging. He attributed the effects to the mere contact of dissimilar metals.

The paper was published (in French) in 1800 and it acted, as Davy later put it, as an alarm bell among the experimenters of Europe: it created enormous excitement.[50] Priestley had believed that mechanics was a superficial science, and that chemistry and electricity held the key to the deep understanding of nature:

> Hitherto philosophy has been chiefly conversant about the more sensible properties of bodies; electricity, together with chymistry and the doctrine of light and colours, seems to be giving us an inlet into their internal structure, on which all their sensible properties depend. By pursuing this new light, therefore, the bounds of natural science may possibly be extended, beyond what we can now form an idea of. New worlds may open to our view, and the glory of the great Sir Isaac Newton . . . be eclipsed.[51]

This was to be Davy's programme. The chemist working with electricity was pursuing the fundamental science. The clockwork universe so detested by Romantic thinkers was an obsolete conception; beneath the apparent solidity and stability of matter lay polar forces in equilibrium. Newton had understood gravitation but the new Newton would come to grips with these new forces and create a dynamic science to replace the mechanical world view.

Volta's experiment was repeated as part of the refereeing process by William Nicholson and Anthony Carlisle, who extended it even before the paper was actually published by using the electric current (strictly, '*galvanic*', because the identity of galvanism and electricity was not completely proved before Faraday's work in the 1830s) to decompose water. Davy entered the field with papers sent to *Nicholson's Journal* in 1800. This informal publication, not tied to any scientific society, rapidly printed short papers and also reprinted some from more august periodicals, like the Royal Society's, and carried news of science done abroad. Davy could not share Volta's view that electricity was generated by mere contact: that would be something for nothing. Rather, a chemical reaction generated the current, and, conversely, an electric current could cause a chemical reaction, as in Nicholson and Carlisle's experiment.

Beddoes built a battery of 110 double plates, and Davy found that it did not work with pure water between them:

> It appears that the galvanic pile of Volta acts only when the conducting substance between the plates is capable of oxidating the zinc; and that in proportion as a greater quantiy of oxygen enters its combination with the zinc in a given time, so in proportion is the power of the pile to decompose water, and to give the shock greater. It seems therefore reasonable to conclude, though with our present quantity of facts we are unable to explain the exact mode of operation, that the oxidation of the zinc in the pile, and the chemical changes connected with it, are *somehow* the cause of the electrical effects it produces.[52]

This was a very prescient generalization, and in establishing it six years later Davy was to put himself at the pinnacle of the scientific world.

Experimenting further, he found that in accordance with his general view of the case, concentrated nitric acid was exceedingly effective in the pile. He also found that metals were not necessary, making a pile with zinc and charcoal. On 18 June 1801 Davy's paper extending this work was read at the Royal Society and subsequently published in their *Philosophical Transactions*.[53] In 1803 this paper was mentioned on the certificate that led to his election as a Fellow of the Royal Society shortly before his twenty-fifth birthday.[54] In it, he described an ingenious battery with two fluids and one metal, rather than the other way round: on one side of a plate of tin or zinc was a fluid capable, and on the other, incapable, of reacting chemically with it. Also in Bristol, he seems to have been the first person in the world to make a carbon arc-lamp.[55] Clearly, he was in the forefront of the many people excited by the new development.

In his notebooks for this period, and indeed beyond it, we find a fascinating jumble of poems, experiments, an outline of an epic based on the life of Moses and various fragments that were to find their way into lectures. As yet he was unknown as a lecturer and, while he was acquiring a great scientific reputation, the Pneumatic Institution was in a tottering state and John Tonkin's worries about his protégé's career must have begun to look well-founded. Davy wrote optimistically:

> An active mind, a deep ideal feeling of good, a look towards future greatness has preserved me. I am thankful to the Spirit who is every where, that I have passed through the most dangerous season of my life with but few errors; in pursuits useful to mankind, pursuits which promise to me, at some future time, the honourable meed of the applause of enlightened men.[56]

In September 1800 he wrote to his mother about 'prospects of a very brilliant nature',[57] and then, in January 1801, not long after his twenty-second birthday, he was invited formally by Count Rumford to take up a post at the Royal Institution in London. Beddoes generously put no obstacles in his way. In February Davy went to London to negotiate, and in March moved there to begin a new phase of his life.

4

The Bright Day

Bacon had written that knowledge was power, but the science of the seventeenth and eighteenth centuries had not proved particularly useful. The early phases of industrialization did not depend upon the latest science: Benjamin Franklin, asked about the use of electricity, replied by asking what was the use of a baby? In effect, natural philosophers drew cheques upon the future. The Pneumatic Institution in Bristol had not turned out to be useful as its promoters had hoped; but in London Davy became first the preacher of applied science, and ultimatey a successful exponent of it. In his work we see the transition from the old pattern, whereby the man of science investigated some process or machine to see what was essential, what might be improved and what might be replaced by something cheaper or more accessible, to the modern one, in which new discoveries were employed to do something quite new. His enthusiasm was infectious.

In January 1802 he delivered an Introductory Lecture, drafts of which are in his notebooks, which ended with a splendid peroration:

> In this view, we do not look to distant ages, or amuse ourselves with brilliant though delusive dreams concerning the infinite improveability of man, the annihilation of labour, disease, and even death, but we reason by analogy from simple facts, we consider only a state of human progression arising out of its present condition, – we look for a time that we may reasonably expect – *for a bright day, of which we already behold the dawn.*[1]

This was just what his audience at the Royal Institution wanted to hear; it produced an extraordinary sensation and ensured Davy's and the Institution's success. Rumford and the others who managed the Institution had made the right choice; and Davy was launched on a career as a professional scientist, probably at that time the only person in England apart from the Astronomer Royal who could be thus described.[2] What sort of place was the Royal Institution, which still survives as a centre of research and of accessible lectures, with a splendid library.

It used to be supposed that it was the creation of Benjamin Thompson, the American Tory who, after Independence, left the country and worked in Bavaria. There he was enobled, taking the title Count Rumford from his native place in Massachusetts. He was in charge of cannon-making and of poor-relief, and invented stoves that were very economical of fuel. On coming to England, he brought with him a vision of science – more or less organized common sense – applied to make life easier, and a strong commitment to the education, especially in science, of those working with machinery. He believed that economic progress, and hence happiness, must result from workmen abandoning rule of thumb or mere practice and applying scientific principles.

His cannon-boring had convinced him that heat was the motion of particles, rather than a kind of substance.[3] This was a return to the older view of Bacon and Locke, and away from that of modern chemists like Joseph Black of Edinburgh and Lavoisier, and it was not generally accepted for many years. Davy in his early writings in Penzance, and indeed throughout his career, adhered to the same view of heat as Rumford, and this may have brought him into Rumford's notice and favour. Davy had apparently melted ice by rubbing two pieces of it together, proving that heat was generated by friction, but to the twentieth-century eye it looks as though this could have happened only because his apparatus was poorly insulated. Rumford's demonstration that indefinite quantities of heat can be generated by the friction of a blunt borer on metal stands up better to criticism.

Morris Berman has shown that while Rumford was important in organizing and publicizing the Royal Institution in its first days, the real power lay with the landed interest, who sought new ways of raising better crops in a time of prolonged war and blockade, and the food shortages that went with them.[4] Close study of the records of the Institution reveals the power especially of Sir Joseph Banks and his associates.[5] Banks was a great landowner and a friend of King George III, 'Farmer George'. He had sailed with Captain Cook, and it was he who had botanized at Botany Bay in New South Wales. He was to be the great promoter of British settlement in Australia, and the introduction of sheep there; he was, in effect, the founder of Kew Garden in London; and on 30 November 1778, just a few days before Davy was born, he had been elected President of the Royal Society. He was a fount of patronage in science, and he and his wealthy friends were the power behind the Royal Institution, which, unlike the Royal Society, was to have a laboratory and a great theatre for public lectures. They envisaged the encouragement of a taste for science, especially of a practical kind, amongst the landowning classes, and would have agreed with the remark in one of Davy's dialogues that 'science is nothing more than the refinement of common sense making use of facts already known to acquire new facts.'[6]

Much later in the century this remark was rephrased by T. H. Huxley ('Darwin's bulldog') as 'Science is, I believe, nothing but *trained* and *organized common sense*'[7]; but neither Davy nor he behaved always as though Science were common sense, both being anxious to go beyond facts and use science to develop a world view, and both being bold users of analogy. Nevertheless, this Baconian vision of accessible science was a powerful one, and in the nineteenth century perhaps separated chemists and naturalists from thsoe engaged in more mathematical disciplines. Galileo had admired Copernicus for defying common sense, and physics has ever since been a science of paradoxes.

The lecture theatre at the Royal Institution had a separate entrance to the gallery, as in London theatres, and it was possible for two very different social groups to attend the lectures. But Rumford's vision of technical education for craftsmen never became a reality there. Those involved in industry did not want the details of their processes to become public knowledge and do not seem to have been persuaded of the advantages of a better-educated labour force. Moreover, the Institution had to pay its way. The first lecturer in chemistry had been Thomas Garnett, who had come with a reputation based upon mineral water analyses and public lectures, but he did not care for London life, fell ill and ultimately died. Davy began as his assistant. Thomas Young was also lecturing there, and his lectures on natural philosophy and the mechanical arts were subsequently published; they are a very valuable guide to the historian about the state of the physical sciences in the opening years of the century, but they are not an easy read.[8] Young could not hold the kind of audience attracted to the Royal Institution. High-level popularization of science was not, and is not, a straightforward task, even for a polymath like Young, who worked on the Rosetta Stone as well as in optics, physical and physiological.

Paris tells us that Davy at first struck Rumford as gauche, but soon proved his flair for communicating scientific enthusiasm; and he attributes Davy's early success in part to support from the Tepidarian Society, to which Davy was elected on 7 April 1801.[9] They were apparently a group of 'twenty-five of the most violent republicans of the day', who met to drink tea at Old Slaughter's Coffee House in St Martin's Lane. Davy's associations with Beddoes and Coleridge might well have brought him into such a group, and Thomas Underwood, the radical who gave Paris this information, was a proprietor of the Institution and had recommended Davy to Rumford, but violent republicanism was not very likely to make friends or influence people at the Royal Institution, and Davy's vision of society involved technical progress bringing increased wealth and leisure, with a continuing unequal division of property, which does not sound very like the Tepidarians' programme. In contrast to Rumford, Coleridge wrote to William Godwin in 1801 of Davy's enchanting manners', hoping 'they would not bring too many idlers to harrass and vex his mornings'.[10]

At first, Davy pursued his own research at the Royal Institution, but one of the reasons for his invitation there seems to have been that he had in Bristol given a lecture on 'chemistry and vegetation', and on 29 June 1801 the Managers resolved 'that a course of Lectures on the Chemical Principles of the Art of Tanning be given at the Royal Institution, by Mr. Davy, to commence the second of November next, and that respectable Persons of the Trade, who shall be recommended by Proprietors of the Institution be admitted to those lectures gratis.'[11] They agreed that Davy could have three months leave of absence 'for the purpose of making himself more intimately acquainted with the practical part of the business';[12] his work in applied science had begun.

At Bristol Davy had become a friend of Thomas Poole, of Nether Stowey, a tanner widely known and respected, a great friend also of Coleridge and of the Wedgwoods;[13] to Poole, Davy was to dedicate his last book, 'in remembrance of thirty years of continued and faithful friendship'.[14] In 1800 tanning was an extremely important industry, because leather had such a very wide variety of uses wherever a strong and flexible material was required. Its quality was variable, and the oak bark and galls needed in the tanning process were getting into short supply in this time of war and expanding population. It was a disagreeable and smelly trade, at least to later eyes, in which hides were steeped in pits, scraped, and rubbed with brains, the faeces of dogs and other unpleasant substances. It clearly cried out for the attention of somebody with trained common sense, augmented with chemical knowledge.

The course of lectures led to a major scientific paper. Davy's work, which was published in the Royal Society's *Philosophical Transactions* for 1803 and was largely responsible for his being awarded the Society's Copley Medal in 1805, was essentially a vindication of the best existing practice, coupled with a suggested alternative source of the active astringent substance.[15] His 'chief design was, to attempt to elucidate the practical part of the art; but in pursuing it, I was necessarily led by general chemical inquiries concerning the analysis of the different vegetable substances containing tannin, and their peculiar properties'.[16] He concluded of workmen in tanneries that 'in general, they seem to have arrived, in consequence of repeated practical experiments, at a degree of perfection which cannot be very far extended by means of any elucidations of theory that have as yet been made known.'[17] His chemical investigations conferred authority upon the most advanced practitioners and confirmed the idea that scientific method was not something arcane or recondite.

He began with the work of the French chemist Seguin, who had identified tannin, and the first part of the paper is concerned with analyses of natural products, which did not always give consistent results. Tannin precipitates gelatine, but Davy found it essential to avoid an excess of gelatine, explaining this in terms of the novel theory of C. L. Berthollet, who believed that the relative masses of the substances present affected the course of chemical reactions.[18]

Davy experimented on galls and on the bark of different kinds of trees; he found tannin in tea, and Poole found it in Port wine.[19] The richest source was an Indian plant called catechu, used there as an astringent to chew with betel-nut. Davy used samples supplied by Banks, from Bombay and from Bengal, and found that they were effective in tanning; he even wore a pair of shoes, one tanned with oak in the usual way and the other with catechu.[20] Because much of India was under British rule, this discovery seemed very promising commercially, but, in the event, little seems to have come of it.

Davy determined the proportions of tannin in his various samples, and, after it was removed with gelatine, the quantity of mucilage and of soluble 'extractive matter'; the latter seemed to be very important for good tanning. When, as with galls, there was little of it, the process went quickly but the leather was hard and liable to crack; bark gave a softer leather. Dilute solutions, as the best tanners had found, gave the best leather (though the business took longer and therefore tied up more capital); Davy thought that this was because weaker solutions contained more of the extractive matter. Davy's friend Charles Hatchett tried to take the investigation further by preparing a substitute for tannin,[21] but after Davy's death the *Encyclopedia Metropolitana* published in 1831 an account of tanning in which it is clear that things were still as Davy had left them in 1803, and Davy's work was seen as the predominant scientific contribution to practice.[22] Hatchett was, like Davy, someone who easily crossed between science and humanities, being Treasurer of the Club founded by Dr Johnson and Sir Joshua Reynolds, and being admired by Coleridge.

Even by the time of the Great Exhibition of 1851 chemistry had done no more for tanners, and after his single contribution to the subject, Davy did no more work on it. Knowing when an investigation cannot be profitably pursued further is an important gift in science. Davy was, however, not only set to this work, but also to give courses of lectures under the auspices of the Board of Agriculture on applications of chemistry there. The Board represented the landed interest, and its objective was to increase agricultural output: in 1798 T. R. Malthus pointed out that population would tend to increase geometrically, and food supply at best arithmetically, so that there would always be hunger. More hopeful people believed that if farms could be brought up to the existing highest standard, then there would be enough food. Men of science took over the direction of cheese and butter making from women practising traditional skills,[23] and Arthur Young, the energetic Secretary to the Board, and Sir Thomas Bernard, the eminent philanthropist whose finger was said to be in every pie, were prominent with Banks in promoting county surveys to show what could be done.[24]

Chemistry promised fertilizers and insecticides, and Davy's lectures were very successful. Bernard provided land for experiments, and Davy was invited to such great agricultural celebrations as the sheep-shearing at Woburn, where on one occasion his health was drunk. He also helped to direct experiments at Woburn on different kinds of grasses;[25] and in 1813 his lectures were publised as *Elements of Agricultural Chemistry* in a handsome quarto edition clearly aimed at the upper end of the book-buying public.[26] It was soon followed by a second edition in more modest format, and subsequent editions continued to sell right down to the middle of the century. Indeed, when Davy's *Collected Works* came out in 1839–40, *Agricultural Chemistry* was divided between volumes seven and eight so that it should not affect the sales of the latest separate edition. The book remained the standard work until superseded in the hungry forties by the writings of J. B. Boussingault[27] and Justus von Liebig.[28]

Agricultural chemistry began with soil analysis, and Davy designed apparatus for doing this (a set can be seen in the History of Science museum in Oxford), and also undertook to do analyses himself. In

1805 the Board of Agriculture provided a laboratory near the Royal Institution for this purpose; and in 1806 after Davy's lectures the audience moved to the laboratory to see the experiments demonstrated more informally. Davy's laboratory notebooks were lovingly preserved by Faraday, but are incomplete because parts were lost before Faraday, came upon the scene.[29] These notebooks detail analyses, for example, in May 1806 on bottles of soil sent by Arthur Young from various places, and his little miscellaneous notebooks also describe soil analyses, some for Lord Dundas and for the Duke of Malborough (sic).[30] His agricultural work brought him into contact with the great improving landlords of the day, and thus into the highest social sphere. Banks also introduced him to Thomas Andrew Knight, an eminent plant physiologist with whom Davy did many experiments, and with whom he fished. Knight became one of Davy's closest friends, to whom he dedicated the fourth edition of *Agricultural Chemistry*. He was also mentioned in Davy's will along with Poole and Hatchett, and various medical men whom we shall meet later; Davy bequeathed him a seal, 'not a ring', engraved with a fish.[31]

Agricultural Chemistry consists of eight lectures and an appendix, which reports experiments on grasses done at Woburn. After an introduction, Davy examines the powers and nature of matter; the organization and chemistry of plants; soils; the atmosphere; vegetable and animal manures; mineral manures; and burning, irrigation and 'convertible husbandry' founded on crop rotation. The lectures are accessible, being in plain rather than technical language, calculated to make gentlemen of an intellectual turn of mind keen to apply science on their own estates. The book contains a geological diagram and some sections through plants seen through the microscope; these last were copied from the work of Nehemiah Grew in the late seventeenth century and illustrate the slow progress of this branch of science.

We see Davy's way of proceeding in his discussion of whether animal manure should be allowed to rot until it has become 'short muck' of about half its original weight or should be applied fresh. He firmly believed, with Arthur Young, that it should be put on fresh, and had converted the famous 'Mr Coke' of Norfolk. His reasoning was chemical: the ammonia given off, and the heat of fermentation, were valuable for improving the soil. He reported that in October 1808:

> I filled a large retort capable of containing three pints of water, with some ho[t] fermenting manure, consisting principally of the litter and dung of cattle; I adapted a small receiver to the retort, and connected the whole to a mercurial pneumatic apparatus, so as to collect the condensible and elastic fluids which might rise from the dung. The

> receiver soon became lined with dew, and drops began in a few hours to trickle down the sides of it. Elastic fluid likewise was generated; in three days 35 cubical inches had been formed, which when analysed, were found to contain 21 cubical inches of carbonic acid, the remainder was hydrocarbonate mixed with some azote, probably no more than existed in the common air in the receiver. The fluid matter collected in the receiver at the same time amounted to nearly half an ounce. It had a saline taste, and a disagreeable smell, and contained some acetate and carbonate of ammonia.[32]

We might notice the importance of the taste and smell of things: chemistry is the science of 'secondary qualities', of colours, tastes and smells; it is that which we are likely to remember of it, and in this case it was associated with what must have been a rather unpleasant experiment.

Davy went on to do another experiment closely following from this, and highly convincing to practical men:

> Finding such products given off from fermenting litter, I introduced the beak of another retort filled with similar dung very hot at the time, into the soil amongst the roots of some grass in the border of a garden; in less than a week a very distinct effect was produced upon the grass; upon the spot exposed to the influence of the matter disengaged in fermentation, it grew with much more luxuriance than the grass in any other part of the garden.[33]

It was therefore wasteful and foolish to let these valuable substances escape into the atmosphere rather than apply the dung fresh and get them into the soil. Davy's researches demonstrated that the best practice had a chemical rationale, and could even lead to general reflections on nature and her cycles:

> The fermentation and putrefaction of organised substances in the free atmosphere are noxious processes; beneath the surface of the ground they are salutary operations . . . and that which would offend the senses and injure the health, if exposed, is converted by gradual processes into forms of beauty and of usefulness.[34]

This relationship between death and decay on the one hand, and life on the other, is prominent in Davy's *Consolations*, though in a different context from that of manure.

In revising the book in 1827 for its fourth edition, Davy wrote that his object 'has been principally to dwell upon practical principles and practical applications of science; and it is in the farm and not in the laboratory that these can be put to the test of experiment'.[35] Davy's work on organic manures convinced him of the value of vegetable matter in the soil, both for nourishing plants and for improving the texture, as farmers well knew. His approach to agricultural chemistry can be contrasted to the much more confident laboratory-based proceeding of Liebig, for whom manures should simply be mineral substances needed by the plants but absent in the soil, and who had no time for the 'humus' theory. We who live with the problems of the excessive use of synthetic fertilizers can sympathize with the cautious 'common sense' of Davy's science.

Although there were some setbacks, as when Lord Egremont found that a fertilizer recommended by Davy for steeping turnip seeds seemed to have killed them all off, Davy's work on tanning and on agriculture was generally seen as a great success. By 1806 he had established his position in the Royal Institution and well beyond it. In a rapid promotion that would delight the academic of the present day, he was made full Lecturer in Chemistry in June 1801, and Professor the following May. In November 1803 he was elected a Fellow of the Royal Society, and in 1807 one of its two Secretaries. His introductory lecture brought him immense social success: thereafter he was constantly invited out to dinner, 'and after dinner he was much in the habit of attending evening parties, devoting the evening to amusement; so that, to the mere frequenters of such parties, he must have appeared a votary of fashion rather than of science'.[36] Poole remembered that 'when he first lectured at the Royal Institution, the ladies said, "Those eyes were made for something besides poring over crucibles."'[37] Paris recorded that Davy was wont to rush from the laboratory to dinner, putting a clean shirt on top of his dirty one in his haste, and that he had been known to be wearing no less than five shirts and pairs of stockings at once.

Such anecdotes attach themselves to absent-minded professors; his brother indignantly repudiated it, admitting only that Humphry wore two shirts 'for warmth' in cold weather, and that 'fashion in dress and appearance was of trifling consideration to him; he consulted rather ease and convenience.'[38] He added from his own knowledge of life at the Royal Institution that Humphry was 'in the disposal of his time, far from systematic, directed rather by circumstances than guided by precise rules. When in town, he generally entered the laboratory after breakfast, about ten or eleven o'clock, and, if uninterrupted, remained there till three or four.'[39] Four o'clock was about when one dressed for dinner, eaten then around five o'clock, and 'it was very unusual for my brother to revisit [the laboratory] after he had dressed for dinner, and before breakfast I do not believe he ever entered its precincts. He was never, to the best of my knowledge, in the habit of abridging

greatly his hours of rest, which were commonly seven or eight.[40] The night before a lecture, he dined at home upon fish and rehearsed with his assistants so that the spectacle would be a success. His surviving lecture notes are full, but would be difficult to read from, and he must have delivered them from memory, maintaining the eye contact with the audience so essential in communicating enthusiasm and interest.

In January 1805 Davy was interrupted in his laboratory by a friend of Tobin's, who told him that Tobin's new play (*The Honey Moon*), to be produced at Drury Lane the next night, still lacked the customary prologue, although various poets had been pressed to write something. Davy quitted the laboratory and in two hours wrote a poem including the lines:

> Hence Genius draws his novel copious store;
> And hence the new creations we adore:
> And hence the scenic art's undying skill
> Submits our feeling to its potent will;
> From common accidents and common themes
> Awakens rapture and poetic dreams;
> And, in the trodden path of life, pursues
> Some object cloth'd in Fancy's loveliest hues –
> To strengthen nature, or to chasten art,
> To mend the manners or exalt the heart.[41]

This prologue was duly spoken. Davy was thus at the centre of culture in London, with admirers everywhere, and his old friends were worried that he might be corrupted by the adulation he was receiving. Davy thought that he was mature enough to be safe, but some old friendships were duly broken by death or by distance, geographical or social.

In 1804 Gregory Watt died, and although Davy had met him as an invalid he was shattered by his death, writing a long letter about it to his friend Thomas Clayfield in Bristol:

> He ought not to have died. I could not persuade myself that he would die; and until the very moment when I was assured of his fate, I would not believe he was in any danger. His letters to me, only three or four months ago, were full of spirit, and spoke not of any infirmity of body, but of an increased strength of mind . . . in man, the faculties and intellect are perfected, – he rises, exists for a little while in disease and misery, and then would seem to disappear, without an end, and without producing any effect . . . there is some arrangement of things which we can never comprehend, but in which his faculties will be applied. The caterpillar, in being converted into an inert scaly mass, does not appear to be fitting itself for an inhabitant of air, and can have

> no consciousness of the brilliancy of its future being. We are masters of the earth, but perhaps we are the slaves of some great and unknown beings . . . We suppose that we are acquainted with matter, and with all its elements, and yet we cannot even guess at the cause of electricity, or explain the laws of the formation of the stones which fall from meteors. There may be beings, – thinking beings, near us, surrounding us, which we do not perceive, which we can never imagine. We know very little; but, in my opinion, we know enough to hope for the immortality, the *individual immortality of the better part of man*.[42]

Davy added that Clayfield, when reading the letter to Tobin, should protest if Tobin began to argue against the immortality of man. Davy had clearly come some way from the radical and materialistic circles in which he had moved in Bristol, and which in London were perhaps represented by the Tepidarians. Imperceptible thinking beings may remind us of the 'Ancient Mariner'; Coleridge, in despair about his marriage and in the throes of opium addiction, had gone to Malta in April 1804, cheered on his way by a splendid letter from Davy:

> In whatever part of the World you are, you will often live with me, not as a fleeting idea but as a *recollection* possessed of creative energy, as an *Imagination* winged with fire inspiriting and rejoicing. You must not live much longer without giving to *all men* the *proof of power* which those who know you feel in admiration. Perhaps at a distance from the applauding and censuring murmurs of the world, you will be able to execute those great works which are justly expected from you; you are to be the historian of the Philosophy of feeling. – Do not in any way dissipate your noble nature. Do not give up your birth-right.[43]

To the delight of Coleridge's friends, the expedition was not disastrous, even if it did not fulfil Davy's hopes, and in February 1805 Davy wrote to Poole:

> There has been no news lately from Coleridge; the last accounts state that he was well in the autumn, and in Sicily. On that poetic ground we may hope and trust that his genius will call forth some new creations, and that he may bring back to us some garlands of never-dying verse. I have written to urge him strongly to give a course of lectures on Poetry at the Royal Institution, where his feeling would strongly impress, and his eloquence greatly delight.[44]

These are not the writings of one entirely forgetting his old friends; but there can be no doubt that the polish and the confidence that Davy acquired, his pleasure in the company of the gentry and aristocracy, and his reputation as the coming man of British science must have affected his relationships. He was now, for example, the patron of

Coleridge, who had come to his lectures to increase his stock of metaphors. He was even able to repay some Wedgwood patronage when in his Introductory Lecture for 1805 he showed one of their Etruscan vases to demonstrate how aiming for excellence in things made for the wealthy could improve practice right through the arts and manufactures, bringing in this case a general improvement in pottery as well as 'honourably acquired wealth' to the manufacturers.[45]

Davy kept up less well with Beddoes than with Poole. He came to feel that Beddoes should not have published his early speculations, and by the time Beddoes died at forty-eight in 1808, soon after writing Davy a letter deploring his own lack of concrete achievement, Davy had moved far from him in scientific interests (and eminence) and political beliefs. At the end of his life Davy wrote an assessment of Beddoes that has some of the sourness of a man exorcising the ghost of a father figure:

> Reserved in manner and almost shy, but his countenance was agreeable. He was cold in conversation, and apparently much occupied with his own peculiar views or theories. Nothing was a stronger contrast to his apparent coldness in description than his wild and active imagination which was as Darwin's. He was little enlightened by experiment, and I may say, attentive to it. He had great talents and much reading; but had lived too little among superior men. On his death bed he wrote me a most affecting letter regretting his scientific aberrations. I remember one expression 'like one who scattered abroad the Avena fatua of knowledge from which neither brand, nor blosssom nor fruit has resulted. I require the consolation of a friend'. Beddoes had talents which would have exalted him to the pinnacle of philosophical eminence if they had been applied with discretion.[46]

Generous recognition of the talents of the young Davy deserved better than this from him, but he had at the time of Beddoes' death been really shattered.[47]

Davy found new friends among the superior men he met in London, that magnet for the ambitious. The Royal Society was at that time very formal in its meetings: the Secretary read out the papers, no discussion followed and those actively pursuing scientific research were a small minority. But it had dining clubs associated with it, where important informal conversations went on. One of these was the Royal Society Club, where Davy said that you could get a poor dinner in good company. Another was the Chemical Club, to which Davy and William Hyde Wollaston belonged but which was in the doldrums by 1807, and was, in effect, succeeded by the Animal Chemistry Society,[48] which met quarterly[49] from 1808. In these groups Davy got to know medical men especially, and with some of them he went fishing, his source of real relaxation.

His time in London seems to have been exciting and satisfying enough, but on Christmas Day 1803, in a coach between Bath and Clifton, he looked back with nostalgia and world-weariness:

When in life's first golden morn,
  I left my stormy native shore,
My pathway was without a thorn,
  With roses it seemed covered o'er.

Ambition thrill'd within my breast,
  My heart with feverish hope beat high:
Hope alone disturb'd my rest,
  Hope only bade me heave a sigh.

In pride of untried power my mind
  A visionary empire saw, –
A world, in which it hoped to find
  Its own high strength a master law.

Its love was wild, its friendship free,
  Its passions changeful as the light
That on an April day we see, –
  Changeful, and yet ever bright.

Years of pain have pass'd away,
  Its former lineaments are gone;
Hope gives it now a gentler ray,
  Ambition rules it not alone.

The forms of holy truth severe
  And the fair thoughts with which it glows;
And if it ever feels feels a tear,
  That tear in purest passion flows.

Fled is its anguish, and its joys
  Are such as reason may approve;
No storms its quietness destroys,
  Yet is it ever warm with love.

Its pleasures Fate and Nature give,
  And Fate and Nature will not fly;
It hopes in usefulness to live,
  In dreams of endless bliss to die.[50]

We need not take this vision of a sadder and a wiser man too seriously, but Davy must have had to grow up very fast in these years in London in which he became so polished so quickly.

Davy had always taken a great interest in geology, very much a rising science at the beginning of the nineteenth century. The speculative inquiries of the later eighteenth century, with Neptunists like A. G. Werner emphasizing the role of water and Plutonists like James Hutton attributing most phenomena to volcanoes, gave way to a more empirical and descriptive science. On his holidays during the early years at the Royal Institution Davy took the opportunity to geologize, and he gave what he claimed were the first public series of lectures in London on the science, at the Royal Institution.[51] His tours included visits to Cornwall, to Wales, to Scotland, and to Ireland to see the Giants' Causeway, the nature of basalt being particularly significant in the Neptunist/Plutonist dispute because of its apparently igneous and recent character. In Cornwall his companion was Underwood; in Wales it was another friend, Samuel Purkis, who recorded that:

> we visited every place possessing any remains of antiquity, any curious productions of nature or art, and every spot distinguished by romantic and picturesque scenery. Our friend's diversified talents, with his knowledge of geology and natural history in general, rendered him a most delightful companion in a tour of this description. Every mountain we beheld, and every river we crossed, afforded a fruitful theme for his scientific remarks. The form and position of the mountain, with the several strata of which it was composed, always procured for me information as to its character and classification; and every bridge we crossed, invariably occasioned a temporary halt, with some appropriate observations on the productions of the river and the diversion of angling.[52]

Such conscientious tourism, in the company, as it were, of a guest lecturer sounds rather exhausting, but Davy was genuinely good company. He kept diaries, and impressed his hosts by his youth and by his failure to show off when confronted by provincial know-alls. He sometimes thanked his hosts in verse, as in the Hebrides:

> Whoever, glowing with the holy love
> Of wild magnificence of ample form,
> Has visited these islands, where, upraised'
> Above the vast Atlantic, boldly stand
> The giant monuments of elder time,
> The pillar'd caves of Staffa, and the rocks
> Of fair Iona, let him kindly bless
> That peaceful and that hospitable shore,
> Where stands the house of Ulva; for its halls
> Are graced by virtue, elegance, and taste,
> By social joy and welcome from the heart.

> In them the weary traveller finds repose,
> And quits them with regret and gratitude.[53]

In Ireland in 1806 his companion was George Bellas Greenough, first President of the Geological Society of London and an exact contemporary. He had studied natural history under J. F. Blumenbach at Göttingen rather than pursuing law, and was left sufficient wealth to devote himself to geology, also being a Member of Parliament for a time. His emphasis was upon the mineral character of rocks rather than upon the fossils they contained, and Davy with his chemical background seems to have shared this interest, which also accorded with the practical concerns of the Royal Institution. Davy apparently returned from his summer holidays invigorated, and the autumn was the time (before the winter season began at the Royal Institution and Royal Society) when he did some of his most exciting researches.

By the autumn of 1806 Davy had done the tasks set him by the Managers of the Royal Institution, and his lectures had brought in large and fashionable audiences, including many women. He was now in a position to take up again the research on galvanism that he had begun at Bristol. He had been happy to be doing useful science – and neither he nor his contemporaries in Britain made the distinction between 'pure' and 'applied' science that later became commonplace – but to be self-propelled rather than directed or commissioned is agreeable, anyway some of the time. And this research was to have a rather different character from the highly empirical studies he had done so far: he would now be forcing nature to give him the answer he wanted. It would also propel him into a position of international prominence in the world of science.

5

Electric Affinity

By the middle of 1806 Davy, at the age of twenty-seven, had established himself as a leading member of the scientific community in Britain. So far his lectures had been sparkling and his researches highly competent, but he had been concerned, in tanning and in agriculture, with problems that had no great bearing upon fundamental theory. Like most of those engaged in science most of the time, he was wrestling with questions difficult to answer: he was doing 'normal science'.[1] Chemists then, as now, spent their time upon analyses and sometimes syntheses that may be illuminating but are not profoundly surprising. Only sometimes do they, or others engaged in other sciences, concern themselves with questions difficult to ask, requiring a new perspective and leading to a new picture of things in a scientific rervolution. Davy's work on electrochemistry produced this kind of excitement among his contemporaries: he had long had a vision of himself as a second Newton, and he persuaded others to share it.

Newton had explained gravity by postulating an attraction between every particle of matter in the universe, and had tried to account for the facts of chemistry in terms of forces also. Because even gold, the densest metal, contracts on cooling, its particles cannot be in contact at room temperature: there must be void between them and, therefore, short-range repulsive forces that keep them apart. Newton's successors, and notably Joseph Priestley, came to believe that the particles might be extremely minute, mere physical points, so that all the solid matter in the solar system might perhaps be contained within a nutshell, and the forces the real concern of the natural philosopher. Clearly the force responsible for chemical unions must be different from gravity, for that is universal whereas only some substances will react together. The nature of chemical affinity, a term borrowed from human relationships as illuminated in a novel by Goethe, was the great problem.

Friedrich Schelling, the philosopher, believed that force rather than matter was the underlying reality.[2] Solid objects endure, like columns of smoke or waterfalls do, through the flux of their constituent particles. For plants and animals this had recently been appreciated by physiologists: we know we are the same person we were ten years ago, but actually almost all our atoms will be different. We are not just the sum of our particles. This idealism made the materialism of the later eighteenth century, embraced in an idiosyncratic Christian form by Priestley, seem irrelevant. For Priestley, matter was active: it possessed powers. Lavoisier, similarly, had seen acidity as the result of the presence of the element oxygen. For Newton, on the other hand, powers such as gravity were not inherent to matter, which was inanimate, brute and inert; and Davy followed Newton, his work apparently confirming the separation of powers.

Schelling had believed that the fundamental force or power in nature was polar, like magnetism with its north and south poles, and electricity with its positive and negative charges. He also believed that it was unchanging in amount. It could appear in different forms or aspects, but was never created or destroyed in the operations of nature or of man. Coleridge was aware of German idealistic thinking generally and probably passed on some idea of Schelling's *Naturphilosophie* to Davy in Bristol, where it may have formed part of Davy's mental background. He was nevertheless critical of Johann Ritter and Hans Christian Oersted, disciples of Schelling who did important scientific work,[3] as metaphysicians rather than genuine inductive followers of Bacon and Newton.

In August 1806 Davy recalled his time in Clifton in a poem to Mrs Beddoes:

> Think not that I forget the days,
> When first, through rough unhaunted ways,
> We moved along the mountain side.
> Where Avon meets the Severn tide;
> When in the spring of youthful thought
> The hours of confidence we caught,
> And Nature's children, free and wild,
> Rejoiced, or grieved, or frown'd, or smiled,

As wayward fancy chanced to move
Our hearts to hope, or fear, or love.

Since that time of tranquil pleasure
Eight long years have filled their measurte,
And scenes and objects grand and new
Have crowded on my dazzled view: –
Visions of beauty, types of heaven,
Unask'd for kindness freely given;
Art, Nature, in their noblest dress –
The city and the wilderness;
The world in all its varying forms,
Contentments, clouds, ambition's storms.[4]

That autumn he returned to research begun in Beddoes' Institute.

Like his eminent older contemporaries Henry Cavendish and William Hyde Wollaston, Davy seems to have believed from the start that 'Franklinic' electricity (from the clouds, or from rubbing glass or amber in a 'machine', as described by Priestley),[5] animal electricity in the 'torpedo' or electric eel,[6] and galvanism[7] were different effects of the same power or agent. Cavendish managed to make a contraption that gave a shock rather like an electric fish does, and Wollaston decomposed water using very fine-pointed terminals dipping into it from an electrical machine, but both fell short of definite proof of the identity of the various electricities, and so did Davy. Davy also believed that an electric current was generated by a chemical reaction in Volta's cell or pile, and that in Nicholson and Carlisle's experiment an electric current made a chemical reaction go (the decomposition of water into its elements), though he later came to include an aspect of Volta's 'contact' theory.

At first at the Royal Institution his time was his own and he carried on with his electrical experiments, but then the tanning and agricultural researches meant that he had little time for such work. Meanwhile, Volta's alarm-bell, as Davy called it, had aroused many others to work in this new field, where understanding what was going on was very difficult and where consistent results were hard to obtain: batteries were soon polarized, their power seemed to vary from day to day and standards of chemical purity were doubtful. Some believed that electricity was a kind of fluid; our 'juice'. Franklin had thought that. Others argued that there were two fluids, one positive and the other negative. This view was popular in Continental Europe, and was supported by experiments in which a spark was passed through paper and found to leave a burr on both sides, unlike a bullet. Lavoisier's list of simple substances included caloric (heat) and light as weightless elements, and it seemed possible that electricity might be another. It might therefore enter into chemical combination.

The experiment of Nicholson and Carlisle did not prove the decomposition of water, for the oxygen and hydrogen might be compounds of water with positive and negative electricity. Anyway, when repeated, the experiment turned out to have curious features: around the positive pole, where the oxygen was emitted, the water turned acidic, while around the negative pole it became alkaline. Moreover, the gases were not given off in the exact proportion in which they combined to form water. One suggestion was that 'galvanates' were being formed and dissolving in the water.[8] Eminent and able chemists in many countries tried to distinguish what we (following Davy) can see as the main effect from side reactions, but no clear picture emerged.

Davy came to believe that a purely chemical theory of Volta's pile would not do: he saw 'contact' as disturbing equilibrium, which was then restored by chemical action.[9] The continuing or permanent effect was thus due to chemical reaction, initiated by contact. Decomposition resulted from the attractions of the poles, but for Davy (unlike most of his contemporaries) the series of decompositions and recombinations began in the middle of the vessel. He then seems to have envisaged a kind of chain of reactions culminating in substances being discharged at the poles, but his theory was never very clearly expressed. Faraday later wrote of Davy's great paper of 1806:

> The facts are of the utmost value, and, with the general points established, are universally known. The *mode of action* by which the effects take place is stated very generally, so generally, indeed, that probably a dozen precise schemes of electro-chemical action might be drawn up, differing essentially from each other, yet all agreeing with the statement there given.[10]

Davy was like that. He was averse to detailed, cut-and-dried theory, and given to imaginative analogical thinking. Nevertheless there was enough close reasoning in this paper and its successors to inaugurate the electrical interpretation of chemistry, a process that is still going on and could not, therefore, have been expected to emerge fully armed from the head of its originator.

Davy was no dabbler, writing and then crossing out a *cri de coeur*: 'Were a description indeed to be given of all the experiments I have made, of all the difficulties I have encountered, of the doubts that have occurred, the hypotheses formed –'.[11] The Laboratory Notebook at the Royal Institution seems to indicate for September and October 1806 a programme of chemical analyses using galvanism; on 23 October, for example, he attempted 'to decompose phosphorus by the galvanic fluid', and on the 28th, water.[12] On 31 October/1 November there is a memorandum 'to send to Mrs O. Griselda' and a doodle of a head; but we find numbered experiments in a steady programme, with each day something done.

Many scientific experiments are open-ended: what is going to happen is unclear, and they are done to learn in Nature's school. Others are done to test a theory. Here there is a desirable outcome, and if at first it is not attained, then we may go on trying, refusing to take Nature's no for an answer. That is what Davy was doing in these classic experiments. He could not believe that electricity was forming compounds or anything like that; he was convinced that it was simply analysing water (or other substances), and that when the experiment was properly done oxygen and hydrogen only, and in the proper proportions, would result. This involved him in Bacon's 'putting nature to the question': interrogating, maybe torturing, her until she gave the reply he wanted.

In terms of Aristotelian chemistry we might say that Davy needed to look at the earth, water and air concerned in the experiment. Part of the alkali produced turned out to be soda, which is a component of glass. The glass of this day was known to be far from inert, even when clear rather than green glass was used; indeed, Lavoisier had shown that when water was boiled for a long time in a glass vessel the 'earth' produced was not generated from the water but came from the glass. The Royal Institution had wealthy backers, so Davy was never stinted for apparatus and he switched to doing his electrolysis in vessels not of glass but of agate. Many of his experiments were done using two cups connected by what we would call a bridge, which was made of woven asbestos fibre, as in the best lamp wicks of the day, called 'amianthus', which Davy carefully washed. One pole was in each cup.

With agate, much less soda was generated, and the quantity fell with each experiment. On 30 October Davy recorded:

> The two agate tubes every precaution being taken after 24 hours acid & alkaline but the acid did not affect Nitrat of Silver nor acetate of Barytes but acted on silver & that here is every reason to believe Nitrous Acid – by much likewise. The alkali did not appear more vivid during evaporation but rather less vivid & left scarcely a sensible residuum which had no decided effect upon turmeric or even litmus!!![13]

Davy had cones of gold made in place of the agate tubes in order to be sure that his apparatus really was inert.

Since there were still some traces of acid and alkali, he redistilled his water from a silver still to ensure that it was completely pure, but this did not eliminate the acid and alkali altogether. The acid was nitrous acid, and the only alkali was now volatile – ammonia. It occurred to Davy that these might arise from nitrogen dissolved in the water reacting with the oxygen and hydrogen produced in the electrolysis. Priestley had found that hydrogen seemed to displace nitrogen from water, so, after evacuating his apparatus, Davy admitted hydrogen. With inert apparatus, really pure water (making it a slight mystery to us how the electrolysis worked) and with atmospheric nitrogen excluded, he demonstrated that the passage of an electric current through water yields oxygen and hydrogen only, and in the right proportions. Nature had confirmed his prediction.

The Royal Society had in 1775 received a bequest of £100 from Henry Baker 'for an oration or discourse to be spoken or read yearly by some one of the Fellows of that Society on such part of Natural History or Experimental Philosophy at such time and in such manner as the President and Council for the time being shall please to order and appoint;[14] and to be asked to deliver the Bakerian Lecture is a very great honour. Davy's experiments had aroused sufficient interest and excitement for him to be invited to do it, and on 20 November 1806 he described his researches in this form.

It is a classic of chemical literature, and contemporaries appreciated it at once. J. J. Berzelius, himself working in the same field and to become the great pundit of early nineteenth-century chemistry, wrote 'that it must be placed among the finest memoirs with which chemical theory has been enriched',[15] and the eminent Scottish chemist Thomas Thomson, 'I consider this not merely the best of all his own productions, but as the finest and completest specimen of inductive reasoning which appeared during the age in which he lived.[16] Even more to the point, it was awarded the prize for the best work on electricity that the Institut in Paris was offering on the instructions of Napoleon, open to citizens of any nation. Chemistry had been a French science since Lavoisier had revolutionized it in the 1780s; now Davy had established that it need not be so, and in his field he was the world leader.

It may seem that to tidy up Nicholson and Carlisle's experiment done six years before, and show that the acid and alkali were accidental rather than essential features of the electrolysis of water, is not an amazing achievement. What has to be remembered is that men of science all over Europe had been following up Volta's discovery throughout those years and none of them had reasoned and experimented with the clarity of Davy. And in his Bakerian Lecture he went on to electrify chemists with his ideas on affinity. He began in a tone of restrained triumph:

> The chemical effects produced by electricity have been for some time objects of philosophical attention; but the novelty of the phenomena, their want of analogy to known facts, and the apparent discordance of some of the results, have involved the inquiry in much obscurity. An attempt to elucidate the subject will not, I hope, be considered by the Society as unfitted to the design of the Bakerian Lecture. I shall have to detail some minute (and I fear tedious) experiments; but they were absolutely essential to the investigation. I shall likewise, however, be able to offer some illustrations of appearances which hitherto have not been fully explained, and to point out some new properties of one of the most powerful and general of material agents.[17]

He described his experiments, done with the battery of 100 double plates of copper and zinc six inches square, rather more fully than in the surviving laboratory notes; reporting that, after twice exhausting the apparatus, filling it with hydrogen, and connecting it to the battery:

> The process was continued for 24 hours, and at the end of this time neither of the portions of water altered in the slightest degree the tint of litmus. It seems evident that water, chemically pure, is decomposed by electricity into gaseous matter alone, into oxygene and hydrogene. The cause of its decomposition, and of the other decompositions which have been mentioned, will be hereafter discussed.[18]

He proposed the use of electricity in chemical analysis, using cups made of different minerals for his experiments, including some from Ireland and lava from Etna. Then he tried to account for the passage of chemical elements through solutions and bridges, where he saw a succession of decompositions and recompositions throughout the fluid, like a kind of folk dance resulting in the arrival of different elements at the two poles.

At the end came his summing up and attempt to derive general conclusions. He found that when electrically charged, metals behaved differently: positively charged silver was reactive, and negatively charged zinc inert. Chemical properties were not thus simply inherent in material elements, but depended upon electrical states. Ordinarily, bodies had definite 'electrical energies' (a term not them used in science in an exact way); and when different substances were brought into contact they would be found to be (weakly) charged.[19]

> Amongst the substances that combine chemically, all those, the electrical energies of which are well known, exhibit opposite states . . .

> supposing perfect freedom of motion in their particles or elementary matter, they ought, according to the principles laid down, to attract each other in consequence of their electrical powers. In the present state of our knowledge, it would be useless to attempt to speculate on the remote cause of the electrical energy, or the reason why different bodies, after being brought into contact, should be found differently electrified; its relation to chemical affinity is, however, sufficiently evident. May it not be identical with it, and an essential property of matter?[20]

The tension in Davy between bold, 'romantic' speculation and the cautious Baconian scientific method appropriate to good Englishmen is well illustrated in that paragraph, but it was his excited generalizing, controlled by experiment, that got through to his admiring contemporaries. Opposite electric charges held the particles of matter together, and in principle this affinity might be quantified, realizing a long-standing Newtonian dream.

Davy hoped that 'many applications of the general facts and principles to the processes of chemistry, both in art and in nature' would 'readily suggest themselves to the philosophical inquirer', expecting that electrochemical analyses would be a very valuable innovation. He thought that atmospheric electricity might have chemical effects and be related to chemical changes, earthquakes and slower geological processes, ending with the remark that

> natural electricity has hitherto been little investigated, except in the case of its evident and powerful concentration in the atmosphere. Its slow and silent operations in every part of ther surface will probably be found more immediately and importantly connected with the order and oeconomy of nature; and investigations on this subject can hardly fail to enlighten our philosophical systems of the earth; and may possibly place new powers within our reach.[21]

Investigation of the slow and silent powers was to come later, with A. C. Becquerel in France; Davy was himself to use electricity as a tool for analysis. But his suggestive remarks about the importance and ubiquity of electricity, and its connection with chemical affinity, resonated like the queries Newton had put at the end of his *Opticks*.

Early in 1807 Davy was elected one of the two Secretaries of the Royal Society. This honour carried with it the duty of editing *Philosophical Transactions*, which kept him busy in London for some of the summer. Nevertheless he found time to go to Cornwall to visit his family, but not to see Poole, to whom the wrote on 28 August, asking him to help in getting Coleridge to agree to give a course of lectures at the Royal Institution:

> The Managers . . . are very anxious to engage him; and I think he might be of material service to the public, and of benefit to his own mind, to say nothing of the benefit his purse might also receive. In the present condition of society, his opinions in matters of taste, literature, and metaphysics, must have a healthy influence; and unless he soon become an actual member of the living world, he must expect to be hereafter brought to judgement 'for hiding his light'.[22]

He went on with political reflections, considering that Bonaparte (the award of whose prize he had not yet heard about) had given up the idea of invasion. Britain's colonial empire he thought 'must fall in due time, when it has answered its ends', and the wealth of Britain must diminish, but 'when we had fewer colonies than Genoa, we had Bacons and Shakespeares' and that spirit (muscle, bone and nerve, upon which wealth was the mere fat) would preserve liberty and independent thought.

In September he seems to have got down again to a burst of research, culminating on 19 October in the decomposition of potash, which on 19 November formed the subject of his second Bakerian Lecture. The October experiment was a dramatic one, and Humphry's cousin, Edmund Davy, who had come to work with him (and later was Professor of Chemistry in Dublin), reported that Humphry danced about the room in ecstatic delight at the end of it. On that day Davy reports in an excited scribble the 'Capt. Exp^{t}. proving the decomp^{n} of *Potash*'.[23] It was one thing to use an electric current to analyse rocks by making cups for electrolysis from them, but another to prove that alkalis were composed of a very light and malleable metal, with oxygen – the element Lavoisier had, after all, named as the generator of acidity, the very opposite property.

The problem was that caustic potash, the alkali prepared from burned plants, was a non-conductor of electricity, and if it were dissolved in water, then the water and not the alkali was decomposed. Solid dry potash was an insulator, and Davy tried to melt it and electrolyse the fused solid. This did not work, so he then tried using the heating effect of the electric current to do the melting, hoping that the current would then go on and achieve the decomposition of the alkali. He slightly damped the potash and put it into a tube with a platinum wire sealed through its closed end, the open end being immersed in mercury. When the

> platina was made neg – no gas was formed & the mercury became oxydated – & a small quantity of alkaligen was produced round the platina wire as became evident from its [word blotted] inflamation by the action of water – then the mercury was made the neg: gas was

> developed in great quantities from the pos: wire, ['which gas' deleted] & none from the neg mercury and this gas proved to be pure oxygene.[24]

In the notebook he remarked of the alkaligen, 'the substance is analogous to those ['described in the dreams' deleted] imagined to exist by the ['early' deleted] alchemical visionaries'.[25] He called it 'basis of potash', asking himself 'Are the bases of the fixed Alkalies simple bodies? I perhaps shall be asked',[26] but at that point he gave no answer, wondering later whether they were the limits of analysis.

On 19 November 1807, exactly a month after the capital experiment had been so breathlessly written up, he delivered his second Bakerian Lecture; his tone was confident, as became a great man at the height of his powers, with astonishing experiments to describe and the scientific world at his feet. In the previous year he had

> ventured to conclude . . . that the new methods of investigation promised to lead to a more intimate knowledge than had hitherto been obtained, concerning the true elements of bodies. This conjecture, then sanctioned only by strong analogies, I am now happy to be able to support with some conclusive facts. . . . In speaking of novel methods of investigation, I shall not fear to be minute. When the common methods of chemical research have been employed, I shall mention only results. . . . when general facts are mentioned, they are such only as have been deduced from processes carefully performed and often repeated.[27]

He described the experiments vividly enough, reporting how with potash in igneous fusion 'a most intense light was exhibited at the negative wire, and a column of flame, which seemed to be owing to the developement of combustible matter, arose from the point of contact'.[28] He reported after the successful electrolysis that his substance from potash 'remained fluid at the temperature of the atmosphere at the time of its production', indistinguishable from mercury in appearance, whereas that from soda was solid;[29] which indicates that the former cannot have been pure, but must have contained soda. He also added that 'the globules often burnt at the moment of their formation, and sometimes violently exploded and separated into smaller globules, which flew with great velocity through the air in a state of vivid combustion, producing a beautiful effect of continued jets of fire':[30] laboratory life in the early nineteenth century could be a very exciting business, and there were few safety precautions.

The new bodies combined with oxygen to regenerate potash and soda. They seemed to be naturally possessed of positive electrical energy, conductors of electricity and heat, and with silvery metallic lustre. But, unlike all other metals, they were very light: Davy found that his basis of potash 'did not sink in doubly distilled naptha, the specific gravity of which was about .770, that of water being considered as 1.'[31] Careful weighings of globules and comparisons with mercury indicated a specific gravity of about 0.6, 'so that it is the lightest fluid body known'.[32]

When the basis was dropped on to water,

> the phenomena seem to depend on the strong attractions of the basis for oxygene and of the potash formed for water. The heat, which arises from two causes, decomposition and combination, is sufficiently intense to produce the inflammation. Water is a bad conductor of heat: the globule swims exposed to air; a part of it, there is the greatest reason to believe, is dissolved by the heated nascent hydrogene; and this substance being capable of spontaneous inflammation, explodes, and communicates the effect of combustion to any of the basis that may be yet uncombined.[33]

He found it extraordinarily reactive, combining with glass and readily reducing metallic oxides when heated in contact with them. His analysis of potash gave 86.1 of basis to 13.9 of oxygen; our figure would be about 78.2 to 16, which, considering the minute quantities and relatively short time he had had to work, indicates that he was a very competent analyst. But it was his attempts to sketch what was going on in chemical reactions, as between his basis and water, that really caught the imagination.

Then came the tricky problem of classifying these bases, or alkaligens. Classifying is something we tend to think of as a mere prelude to real science, but we should not; it involves theory, and the attempt to see how the world really is ordered.[34] Davy asked:

> Should the bases of potash and soda be called metals? The greater number of philosophical persons to whom this question has been put, have answered in the affirmative. They agree with metals in opacity, lustre, malleability, conducting powers as to heat and electricity, and in their qualities of chemical combination. Their low specific gravity does not appear a sufficient reason for making them a new class, for amongst the metals themselves there are remarkable differences in this respect, platina being nearly four times as heavy as tellurium; and in the philosophical division of the classes of bodies, the analogy between the greater number of properties must always be the foundation of arrangement.[35]

This is an Aristotelian position, classifying according to family resemblances rather than one or two particulars; but we should remember that Davy's contemporaries had recently had to wrestle with the classification of the duck-billed platypus, *Ornithorhynchus paradoxus*,[36] and were involved in acrimonious debate over the new

French natural method and its advantages over the artificial system of Linnaeus.

If the alkaligens were metals, then they could be named:

> . . . it will be proper to adopt the termination which, by common consent, has been applied to other newly discovered metals, and which, though originally Latin, is now naturalized in our language, Potasium and Sodium are the names by which I have ventured to call the two new substances: and whatever changes of theory, with regard to the compositon of bodies, may hereafter take place, these terms can scarcely express an error; for they may be considered as implying simply the metals produced from potash and soda. I have consulted with many of the most eminent scientific persons in the country, upon the methods of derivation, and the one I have adopted has been the one most generally approved. It is perhaps more significant than elegant.[37]

Whereas Lavoisier and his associates had evolved a new language for chemistry that was loaded with theory, especially over oxygen, Davy and his 'Baconian' friends believed that facts and theory could be separated, and that this should be reflected in the terms used in science.

Davy believed that 'the mature time for a complete generalization of chemical facts is yet far distant'; grudgingly admitting that in using Lavoisier's antiphlogistic theory his 'motive for employing it has been rather a sense of its beauty and precision, than a conviction of its permanency and truth'.[38] Davy saw himself as bringing about a new revolution in chemistry, going deeper into the nature of the chemical process than Lavoisier had done: 'the knowledge of the powers and effects of the etherial substances may at a future time [destroy] the more refined and ingenious hypothesis of Lavoisier', as the discovery of gases had upset Stahl's phlogiston theory.[39] The ethereal powers were heat, light, electricity and magnetism.

In a footnote to the published lecture, Davy actually toyed with reviving the phlogiston theory of his friend Priestley:

> A phlogistic chemical theory might certainly be defended, on the idea that the metals are compounds of certain unknown bases with the same matter as that existing in hydrogene; and the metallic oxides, alkalies and acids compounds of the same bases with water; – but in this theory more unknown principles would be assumed than in the generally received theory. It would be less elegant and less distinct.[40]

Because we have been brought up with the idea that phlogiston was proved to be illusory by Lavoisier, it is a surprise to find the old theory appearing about twenty years later in such a context, but triumphing over the French was something never very far from Davy's mind, and so was emphasis upon the doubtful status of hypotheses. Indeed, perhaps because of his caution about large scale theorizing, he was often at his best demolishing the theories of others.

At the end of the lecture, Davy turned to two questions that were to absorb him further. Ammonia, the volatile alkali, behaved chemically very like potash and soda, and yet its composition was apparently settled, for it was a compound of hydrogen and nitrogen. Davy thought that, like the two 'fixed' alkalis, it might contain oxygen, and began experiments to see whether it did. He also began work on the alkaline earths, which by 30 June 1808 had led him to the isolation of calcium, barium, strontium and magnesium, using a mercury cathode and obtaining first an 'amalgam' or alloy with mercury. He also procured a curious substance which he believed to be ammonium amalgam, confirming the analogies between ammonia and what he now spelled potassium.[41]

He was not the first to isolate the alkaline earths, having been anticipated by Berzelius. In the hour of his triumph he had been consulted about the ventilation of Newgate gaol,[42] which he thereupon visited, and went down with 'gaol fever', typhus. He was very seriously ill; medical bulletins were posted at the Royal Institution. Eventually he rallied and recovered, but for most of the winter he was unable to do any research or lecturing. In convalescence he completed a poem apparently begun earlier:

> Lo! o'er the earth the kindling spirits pour
> The flames of life that bounteous Nature gives;
> The limpid dew becomes a rosy flower.
> The insensate dust awakes, and moves, and lives.
>
> All speaks of change: the renovated forms
> Of long-forgotten things arise again;
> The light of suns, the breath of angry storms,
> The everlasting motions of the main.
>
> These are but engines of the Eternal will,
> The One intelligence, whose potent sway
> Has ever acted, and is acting still,
> Whilst stars, and worlds, and systems all obey;
>
> Without whose power, the whole of mortal things
> Were dull, inert, an unharmonious band,
> Silent as are the harp's untuned strings
> Without the touches of the poet's hand.
>
> A sacred spark created by His breath,
> The immortal mind of man His image bears;

Electric Affinity

A spirit living 'midst the forms of death,
  Oppress'd but not subdued by mortal cares!

A germ, preparing in the winter's frost
  To rise, and bud, and blossom in the spring;
An unfledged eagle by the tempest toss'd,
  Unconscious of his future strength of wing.

The child of trial, to mortality
  And all its changeful influences given;
On the green earth decreed to move or die,
  And yet by such a fate prepared for heaven.

Soon as it breathes, to feel the mother's form
  Of orbed beauty through its organs thrill,
To press the limbs of life with rapture warm,
  And drink instinctive of a living rill.

To view the skies with morning radiance bright,
  Majestic mingling with the ocean blue,
Or bounded by green hills, or mountains white
  Or peopled plains of rich and varied hue.

The nobler charms astonish'd to behold,
  Of living loveliness, – to see it move,
Cast in expression's rich and varied mould,
  Awakening sympathy, compelling love.

The heavenly balm of mutual hope to taste,
  Soother of life, affection's bliss to share;
Sweet as the stream amidst the desert waste,
  As the first blush of Arctic daylight fair.

To mingle with its kindred, to descry
  The path of power; in public life to shine;
To gain the voice of popularity,
  The idol of today, the man divine.

To govern others by an influence strong,
  As that high law which moves the murmuring main,
Raising and carrying all its waves along,
  Beneath the full-orbed moon's meridian reign.

To scan how transient is the breath of praise,
  A winter's zephyr trembling on the snow,
Chill'd as it moves; or, as the northern rays,
  First fading in the centre, whence they flow.

To live in forests mingled with the whole
  Of natural forms, whose generations rise,

In lovely change, in happy order roll,
  On land, in ocean, in the glittering skies.

Their harmony to trace; the Eternal cause
  To know in love, in reverence to adore;
To bend beneath the inevitable laws,
  Sinking in death, its human strength no more!

Then, as awakening from a dream of pain,
  With joy its mortal feelings to resign;
Yet all its living essence to retain,
  The undying energy of strength divine!

To quit the burdens of its earthly days,
  To give to Nature all her borrow'd powers, –
Etherial fire to feed the solar rays,
  Etherial dew to glad the earth with showers.[43]

As the poem indicates, his mother seems to have been the most important woman in Davy's life, not only at this time, when he was the 'man divine', but indeed right on to her death in 1826. This phrase recalls the Methodist hymn, 'All hail the power of Jesus' name' (which Davy may have known) where it is applied to Jesus, but the poem's cadences are from the eighteenth century. The poem also reflects Davy's continuing religious feeling that we are immortal souls confined within material bodies; a dualism quite different from Priestley's materialism, according to which the dead would be restored to life in a resurrection of the body by a miracle at the Day of Judgement. Davy's belief was fortified by his science, when he seemed to demonstrate that chemical properties, like human character, do not depend simply upon material constituents.

Berzelius had already been working on electrochemistry and had Davy not done his great work of 1806–8 it is reasonable to say that Berzelius would have made the discoveries in due course. But his personality[44] was very different from Davy's; he was a great systematizer.[45] Davy wrote of him '*Berzelius* was the worthy countryman of Scheele, and certainly one of the great ornaments of the age. Indefatigable in labour, accurate in manipulation, no one has worked with more profit. His manner was not distinguished, his appearance rather coarse, and his conversation was limited to his own subjects.'[46] This was very different from the polished and poetic Davy; had Berzelius made Davy's discoveries they would have come out a bit differently because the style and world view of the two was so different. The rhetoric by which they convinced their fellows was correspondingly dissimilar; science is both personal and objective knowledge.

Compared to Berzelius, and indeed to his French contemporaries, Davy may seem amateurish: 'Jack would be a gentleman', and science was as good a route as any to this destination in a Britain that was still unreformed and lacking any professional structure for science. But there was nothing amateurish about Davy's electrical researches. Despite grumbles from various quarters, the class of metals was duly widened to include potassium and sodium. Following his work, which was compared in the *Edinburgh Review* to that of Newton, not only was chemical affinity seen as somebow due to electricity, but we could also say that his papers were a great milestone on the way to the postulation of conservation of energy and that understanding of electricity and matter which forms the basis of twentieth-century physics. Electricity had seemed something to do with thunderstorms and parlour tricks; Davy made it central to our understanding of the processes of nature, a force as important as gravity. We are all Davyans now. And although he could not quantify his ideas, he had achieved his ambition to be beside Newton on his pedestal.

6

Forces, Powers and Chemistry

Davy's work on the oxides of nitrogen had indicated how very different in chemical and physiological properties the different compounds of two elements were. Chemical composition seemed a poor guide, or no guide at all, to the behaviour of things, and Davy believed instead that the powers or forces associated with matter were crucial. His electrochemical studies had confirmed this belief, which for Davy also went with rejection of the materialism that Priestley had taught[1] in favour of the Newtonian doctrine that spirit was a necessary category. This made discussion of 'animal chemistry' particularly interesting.

Lavoisier had demonstrated, apparently, that the vital flame did indeed depend upon a kind of combustion within the living organism, which absorbs oxygen from the air, and emits carbon dioxide just as a candle does.[2] This was the first clear chemical account of a physiological process, although, of course, it left many loose ends to be followed up. The second account was botanical, when Jan Ingenhousz and Priestley investigated what we call photosynthesis, where green plants in sunlight absorb carbon dioxide and give off oxygen. One of Davy's first experiments had been to extend this to water plants. The two processes together made up a cycle.

Chemistry was taught as a part of medical courses, as hospital training began to replace the simple apprenticeship that apothecaries and surgeons had previously undertaken, like Davy in Penzance. At Guy's Hospital in London, which was innovatory in this respect, a course was taught by William Babington, William Allen and Alexander Marcet.[3] Babington was an apothecary who gained an MD degree at Edinburgh and was appointed Assistant Physician at the hospital. To him Davy dedicated *Salmonia* 'in remembrance of some delightful days passed in his society, and in gratitude for an uninterrupted friendship of a quarter of a century', and, like Knight and other fishing friends, bequeathed a seal. Allen was a Quaker, a pharmacist whose business (Allen and Hanbury's) was pre-eminent and supplied, for example, the Royal Institution with chemicals. He was a saintly man, dedicated to peace and social justice, and strongly empirical in his science; and also a friend of Davy. Marcet, who came from Geneva, was an expert on bladder stones and their chemical analysis. His memory is overshadowed by that of his wife, Jane, whose *Conversations on Chemistry*[4] first appeared in 1807 to provide background for those attending lecture-courses such as Davy's and went into many editions. The young Faraday, not deterred by its being aimed at girls, read it and was duly attracted to the science.

Physiology did not develop in England as fast as elsewhere in Europe,[5] but in 1808 the Animal Chemistry Club, or Society, was set up.[6] The Geological Society, in which Babington and Greenough were the moving spirits, was independent of the Royal Society. This outraged Banks, who feared for the collapse of his empire or (just as plausibly) that science would be fragmented into a mass of specialisms if this sort of thing were to continue. Davy, as an important figure in the Royal Society and with his way to make in science, thereupon (despite his great interest in the science and his lecturing on it) had to resign from the Geological Society, which flourished in its independence as a gentlemanly body given to genuine debate at its meetings.

The Animal Chemists formed a different kind of group, a subset of the Royal Society and thus perfectly agreeable to Banks. They met quarterly, at the house of the eminent anatomist Everard Home in London in the winter, and at Hatchett's home in Roehampton in the summer. Those who were not Fellows of the Royal Society could attend meetings as Probationers, but could play no part in the running of the group. Davy was a prominent member. Sir Benjamin Brodie, an eminent physician and, later, President of the Royal Society, recalled that whereas Davy generally set out 'to display himself to advantage', he was at these meetings able to relax and be at ease. It was a close-knit circle of about nine people, meeting very informally. We can, perhaps, get some idea of the kind of conversations they had from the dialogues in *Salmonia* and *Consolations*, and from Brodie's little book *Psychological Inquiries*,[7] which is very similar to Davy's in format, style and tone, though naturally these published conversations between characters given Greek names must be more stilted then what was actually said.

Davy proposed that the Society should sponsor a translation of Berzelius's *Animal Chemistry*, but nothing came of it.[8] Berzelius was to become the greatest chemist of the 1820s and was to feel that his work had been neglected in Britain, so it was a pity that this suggestion from Davy, who was at other times pleased to put Berzelius in his place, did not bear fruit. Marcet, Wollaston and William Prout do not seem to have belonged to the club despite their interests in physiological chemistry, but Davy's friend J. G. Children, with whom he entered into a partnership to make gunpowder, was a member. The club seems to have been in general agreement that the phenomena of life cannot be 'reduced' to, or fully explained by, the laws of chemistry. Such a view was normal in England, and when after the Battle of Waterloo full scientific relations were restored with France, there was widespread shock at the materialistic doctrines that had been developed there. The young, notably in Edinburgh, took up such ideas with enthusiasm, but in London William Lawrence was charged with blasphemy when a course of medical lectures he had delivered was published in 1820. Davy remained sure, as we shall see in *Consolations*, that life is more than chemistry; indeed, that in organisms matter comes under different powers and obeys new laws.

In 1803 John Dalton had been invited to come up from Manchester and lecture at the Royal Institution; probably Davy had had something to do with the invitation, as did Hatchett and possibly Wollaston.[9] Dalton seems to have mentioned his ideas about atoms, which were published only from 1807, first in the standard textbook by the Scottish chemist Thomas Thomson, who had become an enthusiastic convert, and only after that by Dalton himself. Dalton was a Quaker, little impressed with the glitter of London life and, to Davy's astonishment, happier in relative provincial isolation in Manchester, which was still at that time a rough-edged industrial boom town and not yet a great intellectual centre. Dalton's scientific contemporaries all believed that matter was made up of extremely minute particles, but whereas they, with Newton, supposed that they were all composed of the same stuff, namely matter, Dalton suggested that each element had atoms irreducibly different from those of other elements. He thus took up Lavoisier's idea of elements but abandoned Lavoisier's sceptical notion that atoms were merely metaphysical entities.

Dalton discussed with Davy the oxides of nitrogen, and found that with his idea that elements combined atom to atom in definite simple and multiple proportions, he could make sense of Davy's analyses. Davy had stopped at the facts: Dalton was to go further and make real science out of them with his nascent theory. On the other hand, Dalton had supposed that the compositon of the atmosphere was kept constant because like particles repelled each other; this explained why gases diffuse themselves through a space. Davy showed that this could not be true because hydrogen can be kept indefinitely in an open but inverted gas jar; if its particles repelled each other, they would come out at the bottom and disperse themselves.

Dalton's theory involved a large number of irreducibly different particles[10] – about thirty in 1808, but the number rose steadily as new elements were isolated – and this offended Davy's sense of the simplicity of nature. For determining composition, Dalton did invoke simplicity: when only one compound of two elements was known, it should be assumed to be 'binary'. Water was thus HO. Davy, because two volumes of hydrogen combined with one of oxygen and were generated from water in his famous experiment, preferred, in effect, H_2O as the formula. However much Dalton might differ from Lavoisier, Davy saw his writings on atoms as guesswork. His Bakerian Lecture of 1807 ended with the sentence: 'It would be easy to pursue the speculative part of this enquiry to a great extent, but I shall refrain from so occupying the time of the Society, as the tenour of my object in this lecture has not been to state hypotheses, but to bring forward a new series of facts.'[11]

The same issue of the Royal Society's *Philosophical Transactions* published the papers by Thomson[12] and Wollaston[13] that confirmed Dalton's laws of definite and multiple proportions and, in Thomson's eyes at least, his atomic theory. Davy, as Secretary of the Royal Society, had a large responsibility for what it published, and it seemed to Thomson that Davy had been full converted to Dalton's way of thinking. In fact, Davy resisted the chemical atomic theory, retaining a belief in physical atoms like those of Newton,[14] and toying with the view that atoms were no more than point centres of force, as Boscovich and then Priestley had proposed. He was one among many sceptics.[15]

The great advantage of Dalton's theory was that 'atomic weights', the relative weights of atoms usually compared to that of hydrogen, could be computed. But this involved hypothesis. Water was composed of about eight ounces, grams or tons of oxygen to each one of hydrogen. If its formula were HO, then the atom of oxygen weighed eight times that of hydrogen: their atomic weights were 8 and 1. If on, the other hand, one adopted H_2O, then the figures came out as 16 and 1. Debate over what came to be called 'one volume' and 'two volume' systems persisted until into the 1860s, long after the deaths of Davy and Dalton. Disagreement over the formulae of the simplest chemical substances could not be resolved without bringing in beliefs then untestable.[16]

In his *Elements of Chemical Philosophy*, which was published in 1812, Davy summarized his work, especially in electrochemistry.[17] He referred throughout to the chemical elements as 'undecompounded bodies' to indicate that they might well be further analysed if some even more powerful method were to the invented. The analogy of ammonium and potassium seemed to indicate that the latter might be a compound too, and it offended Davy's taste in world-making that it should be done with such similar but different building blocks as sodium and potassium, or strontium and barium. Instead of atomic weights, he referred simply to 'numbers'. These were akin to Wollaston's 'equivalents', proposed in 1814 in a famous paper in which he also described a slide-rule to enable the chemist to make calculations.[18] Equivalent weights were simply based upon analyses, but whereas atoms could be imagined variously arranged in different substances, mere numbers cannot be. Creative science requires imagination, as Davy well knew in his electrochemistry.

Wollaston's opposition to chemical atomism may have been based simply upon his empiricism, but Davy's clearly was not:

> There is, however, no impossibility in the supposition that the same ponderable matter in different electrical states, or in different arrangements, may constitute substances chemically different: there are parallel cases in the different states in which bodies are found, connected with their differnt relations to temperature. Thus steam, ice, and water, are the same ponderable matter; and certain quantities of ice and steam mixed together produce ice-cold water. Even if it should be ultimately found that oxygen and hydrogen are the same matter in different states of electricity, or that two or three elements in different proportions constitute all bodies, the great doctrines of chemistry, the theory of definite proportions, and the specific attractions of bodies must remain immutable . . . That the forms of natural bodies may depend upon different arrangements of the same particles of matter has been a favourite hypothesis advanced in the earliest era of physical research, and often supported by the reasonings of the ablest philosophers. This sublime chemical speculation sanctioned by the authority of Hooke, Newton, and Boscovich, must not be confounded with the ideas advanced by the alchemists concerning the convertibility of the elements into one another. The possible transmutation of metals has generally been reasoned upon, not as a philosophical research, but as an empirical process.[19]

Davy knew Peter Woulfe, an FRS and alchemist, who invented a wash-bottle for gases and was wont to fix little pieces of paper with prayers on them to his apparatus; and in 1783, in Davy's boyhood, the young James Price MD, FRS, had killed himself after failing to substantiate a transmutation experiment when challenged to do so by Banks and others.

Davy urged his readers not to confuse 'hypothetical views concerning the elements founded upon distinct analogies' with 'an art without principles, the beginning of which was deceit, the progress delusion, and the end poverty'.[20] This is hardly fair to alchemy, a source of powerful images to George Herbert, to Goethe in Davy's own time, and to Jung; and which Davy and other Romantics, and later the more prosaic Faraday, found a subject of some fascination because beneath the strange practices they saw a perception of real unity. Analogy was to Davy, as to many contemporaries in the wake of Bishop Butler of Durham's great book, *The Analogy of Religion*, the key to safe hypothesizing. He concluded his book more positively:

> A few undecompounded bodies, which may, perhaps, ultimately be resolved into still fewer elements, or which may be different forms of the same material, constitute the whole of our tangible universe of things. By experiment they are discovered, even in the most complicated arrangements; and experiment is as it were, the chain that binds down the Proteus of nature, and obliges it to confess its real form and divine nature. The laws which govern the phenomena of chemistry, produce invariable results; which may be made the guide of operations in the arts; and which insure the uniformity of the system of nature, the arrangements of which are marked by creative intelligence, and made constantly subservient to the production of life, and the increase of happiness.[21]

God for Davy was the guarantor of a simple, harmonious and ultimately intelligible world, making science a reasonable activity, as it would not be if the world were governed by pure chance or were a black box producing one damned thing after another: we should not suppose that for him and his British contemporaries, horrified as they were by the way the French Revolution developed, religious sentiments in scientific publications were a meaningless rhetoric.

Davy's speculations about the elements were taken up by William Prout[22] (who also later wrote a book proving the existence of God from chemistry) in his famous hypothesis that all the elements were in some sense polymers of hydrogen,[23] which J. J. Thomson thought he had confirmed in discovering the electron in 1897. In supporting the physical atoms or corpuscles of Boyle and Newton, rather than the chemical atoms of Dalton, Davy was prominent in a line of those active in what later came to be called physical chemistry, and Faraday similarly same to support a very minimal atomism akin to that of Boscovich and Priestley.[24]

Thomas Thomson (no relation of J. J.) became a keen supporter of both Daltonian and Proutian ideas, clashing in the 1820s with Berzelius over whether atomic weights were, in fact, all exact multiples of that of hydrogen. At about the same time the Royal Society gave a medal to John Dalton in 1826 for his atomic theory. Davy was president. In his speech he dwelt on the long history of atomism, unnecessarily defending Dalton against charges of plagiarism, but emphasizing how 'he first laid down, clearly and numerically, the doctrine of multiples; and endeavoured to express, by simple numbers, the weights of the bodies believed elementary'.[25] Praise for Thomson, Wollaston, Gay-Lussac, Berzelius and Prout then follows, before a rather half-hearted return to Dalton:

> I hope you will understand that I am speaking of the fundamental principle; and not of the details as they are found in Mr Dalton's system of chemical philosophy . . . It is in the nature of physical science, that its methods offer only approximations to truth, and the first and most glorious inventors are often left behind by very inferior minds, in the minutiae of manipulation; and their errors enable others to discover truth.[26]

Dalton had discovered the simple principles according to which bodies combine, in definite and multiple proportions by weight:

> thus laying the foundation for future labours, respecting the sublime and transcendental parts of the science of corpuscular motion. His merits, in this respect, resemble those of Kepler in astronomy. The causes of chemical change are as yet unknown, and the laws by which they are governed; but in their connexion with electrical and magnetic phenomena, there is a gleam of light pointing to a new dawn in science.[27]

Davy hoped for a Newton of chemistry in about another century, which was rather modest of him. Clearly he believed that his own studies of forces were of much more fundamental importance than Dalton's views, which were associated with 'boldness and peculiarity' until their hypothetical parts had been purged away by Wollaston, whose table of equivalents 'separates the practical part of the doctrine from the atomic or hypothetical part, and is worthy of the profound views and philosophical acumen and accuracy of the celebrated author'.[28] The 'transcendental' applied to a science meant those parts of it not open to direct empirical test: its deeply entrenched theories, like those of Newtonian mechanics.[29]

John Davy, editing the text for the *Collected Works*, expressed pained surprise that anybody [Charles Babbage] could have found the award of the medal insulting to Dalton, but to modern readers Davy's address is condescending rather than respectful, a polished Londoner facing a naïve Mancunian, the Newton of chemistry patting the Kepler on the back.[30] Davy's Council had actually broken the rules for the medal to give it for work done so long before, and that is no doubt why he stressed Dalton's successors; he did not see Dalton as a rival or even an equal, and was more aware of his great metropolitan society recognizing provincial merit. But, equally, the only genuine science he saw in Dalton's work was the laws; the ideas about atoms might have been a scaffolding that helped Dalton in their construction, but it should now be removed.

In 1826 this was not an unusual position, for, despite its usefulness, chemical atomic theory seemed to generate numerous anomalies, only gradually cleared up. Davy's emphasis on the powers that modify matter (rather than on the characteristics of supposed atoms) was more unusual and more his own. It went with his strong opposition to materialism. Just as man's body and mind were completely distinct, so the properties of things did not depend in any simple way on their material components, but, more importantly, on the powers associated with them. For Davy, as for Newton writing about gravity, these were not innate and essential; indeed, he had shown how modifying electrical charges modified chemical properties among the metals.

In 1808, 1809 and 1810 Davy was again asked to deliver the Bakerian Lecture, but while he had curious and interesting things to describe and discuss in the first two of these, they were not performances at the level of those of 1806 and 1807. By 1810 his researches had once again shaken the whole theoretical structure of chemistry. This time he had demolished Lavoisier's idea that acidity depended upon the presence of a material component, oxygen; and had established the elementary nature of chlorine, making oxygen share its throne with another element that seemed about equally reactive and electronegative. Hitherto chemistry had been combustion-dominated. Phlogiston was the principle of flammability, present in metals and anything that burnt, and absent from the calx or ashes. Oxidations were the main class of reactions known in which heat and light (Davy's first preoccupations) were emitted in great quantities; and water, potash and soda were the first bodies to be decomposed electrically, when in each case oxygen proved to be the negative component. Lavoisier's chemistry placed oxygen in a class, the supporters of combustion, on its own.

Lavoisier's theory that oxygen was the cause of acidity was very plausible on two counts. The first was the belief, characteristic of most chemistry, that properties result from components. This idea lay behind the organic chemistry of the nineteenth century, where researchers analysed natural products in search of their active constituent. These were mixtures, but the feeling was general that, even though in compounds the picture was not so simple, the elements present determined the properties at least in a general way. Second, there was inductive, experimental evidence to back Lavoisier's judgement. Nitrous and nitric acids, investigated after all by Davy, did contain oxygen, and so did sulphurous and sulphuric acids; and the more oxygen they contained, the stronger they were as acids. Black's carbonic acid, the acetic acid of vinegar and the acids that give a sharp taste to fruits all contain oxygen. It was perhaps curious that water was neutral, but the answer was that hydrogen was evidently not an acidifiable base. The only clear exception seemed to be hydrogen sulphide, the bad-egg gas, and that was a weak acid, and therefore the kind of anomaly that could be lived with.

In 1774 C. W. Scheele had isolated the greenish choking gas that results from the action of marine, or 'muriatic', acid (from sea salt) upon what he called 'manganese' and we call manganese dioxide. The work was described in his *Chemical Essays*, F. X. Schwediauer's translation of which Beddoes had edited and published in 1786.[31] Scheele believed that the phlogiston in the acid was absorbed by the manganese, and that the gas was therefore dephlogisticated marine acid. He found that 'the marine acid separated from phlogiston, one of its constituent parts, unites with water in a very small quantity only, and gives it a slight acid taste; but, whenever it is enabled to combine with phlogiston, it assumes its former nature, and again becomes a true muriatic acid.'[32] He determined the properties of the gas, which he found to be extremely reactive; in particular it has a very strong attraction for phlogiston, attacking metals and bleaching flowers.

C. L. Berthollet was an associate of Lavoisier and, after the latter's death, the leading chemist of France.[33] He introduced chlorine bleaching for cloth as a commercial process (rapidly taken up in Lancashire, where bright sunlight cannot be depended upon), but given his part in the overthrow of the phlogiston theory, he could not accept Scheele's explanation of what was going on. In a paper dated 1785 but actually published in 1788, he proposed an alternative view.[34] Scheele's 'manganese' contained a great quantity of vital air, or oxygen; this it communicated to the marine acid, and the green gas was therefore oxygenated muriatic acid. Under this name it was generally known until 1810.[35]

Muriatic acid was a very strong acid. It was, therefore, supposed to contain oxygen combined with an unknown basis, which might be an element. Curiously, Scheele's gas was, as he had noticed, less acidic than muriatic acid itself, whereas the general experience of chemists was that more oxygen led to more acidity. But this was another anomaly that could be lived with.

Berthollet taught at the École Polytechnique in Paris, where for the first time research and teaching were combined in higher education in science. His star pupil was Joseph Louis Gay-Lussac,[36] born eleven days before Davy and one of his great rivals.[37] Coming from a professional family, he had a quite different route into science: he won a place at the École Polytechnique by competitive examination, got a thorough scientific education and became the protégé of Berthollet.[38] He was rapidly propelled into the First Class of the Institut, as the Academy of Sciences was then called, and into important academic posts. He became well known for a balloon trip of 1804 in which he got to a height of over 20,000 feet and established that the composition of the air was the same as at ground level, although it was much thinner. He worked extensively on the properties of gases and in 1808 published the law that gases react in simple ratios by volume.[39]

Perhaps because his balloon-like rise beneath the aegis of Berthollet had caused envy, Gay-Lussac was very cautious in his science, adopting a Baconian programme of looking for laws rather than for theories, and being deferential towards his patron. In 1809 he published, with L. J. Thenard, a paper describing experiments made with potassium and oxymuriatic acid, and other reactions of this gas and muriatic acid, which they tried to decompose. It seemed as if the acid 'was the only one of all the gases which contained water', and they concluded that:

> several other acids, such as sulphuric acid and nitric acid, cannot exist in their state of greatest concentration without water, which appears to be the bond which unites their elements; but water plays a much more important part in muriatic acid. In fact, oxygenated muriatic acid is not decomposed by charcoal, and it might be supposed, from this fact and those which are communicated in this Memoir, that this gas is a simple body. The phenomena which it presents can be explained well enough on this hypothesis; we shall not seek to defend it, however, as it appears

> to us that they are still better explained by regarding oxygenated muriatic acid as a compound body.[40]

The word 'acid' in the chemistry of the time meant what we would call 'acid anhydride'. Sulphuric acid was therefore our sulphur trioxide, and carbonic acid our carbon dioxide; and Gay-Lussac was quite right to note that they require the presence of water to behave as acids. He maintained his conclusion, probably giving too much weight to the opinions of Berthollet, as late as 1811; by then against Davy, who had come to the opposite conclusion. Again, we cannot doubt that had Davy died from his attack of typhus, the elementary nature of chlorine would eventually have been accepted following Gay-Lussac's work; but Davy accelerated the process and put it in his own context of matter and forces.[41]

There was no serious problem for Davy, despite war and blockade, in finding out what his French rivals were up to. Davy read French easily and French scientific journals and the *Moniteur* newspaper, seem to have reached London rapidly; hold-ups seem to have been worse from Sweden or Germany. His electrochemistry had made him one of the leading members of an international chemical community, bound with ties of rivalry and respect.

Davy was a master of scientific rhetoric. Faraday's maxim was 'Work, Finish, Publish', and this is certainly an essential part of being successful in science. But the fear of being a Mendel, who died long before his work was recognized, haunts scientists. Davy was good both at seeing what was new in what he had done and then, especially in his lectures at the Royal Institution but also in his papers at the Royal Society, getting his message across. By 1808 he was so eminent that none of his science could be disregarded. The Bakerian Lectures of 1808 and 1809 essentially report work in progress; they must have been rather fascinating to listen to, as Davy wrestled with the nature of ammonia and its amalgam, the problem of acidity and the general question of composition and properties. Involved in all this was the possibility that he might revive phlogiston. The chemical revolution of Lavoisier looked all in doubt in these years, as the initiative seemed to have passed from France to Britain.

Among the compounds with which Davy worked was hydrogen telluride, which smells even viler than hydrogen sulphide. On 23 September 1809 he wrote to J. G. Children:

> I wish you great sport in Pheasant shooting; but I trust you have had still nobler game in your laboratory. – I doubt not you have found before this as I have done that the substance which we took for

> Sulphuretted Hydrogene is telluretted Hydrogene, very soluble in water, combinable with alkalies & earths, a substance affording another proof that Hydrogene is an oxide.[42]

The reasoning here is that because hydrogen sulphide and hydrogen telluride are acids, their only common component (hydrogen) must contain the principle of acidity, oxygen. Similar reasonings about ammonia and its analogy to potash led to the idea that nitrogen might be an oxide, and Davy tried to analyse that too.

Then came a description of an experiment: 'I have kept charcoal white Hot by the Voltaic apparatus in dry oxymuriatic acid gas for an hour, without effecting its decomposition, – this agrees with what I had before observed with a *red heat*. – It is as difficult to decompose as Nitrogene except when all its elements can be made to enter into new combinations.'[43] Heated charcoal was the standard method of removing oxygen from a gas, and it was perplexing that it had not done the trick. Here Davy was still adhering to the view of acidity that he had met at the outset of his career when he first read Lavoisier's textbook.

Thanks to the generosity of the supporters of the Royal Institution, Davy had been provided with a bigger and better Voltaic battery, with which these experiments were performed. On 11 July 1808, in the wake of Berzelius's isolation of barium and the news that the French were constructing a huge battery, Davy had written to the Managers of the Royal Institution:

> A new path of discovery having been opened in the agencies of the electrical battery of Volta, which promises to lead to the greatest improvements in Chemistry and Natural Philosophy, and the useful arts connected with them; and since the increase of the size of the apparatus is absolutely necessary for pursuing it to its full extent, it is proposed to raise a fund by subscription, for constructing a powerful battery, worthy of a national establishment, and capable of promoting the great objects of science. Already, in other countries, public and ample means have been provided for pursuing these investigations. They have had their origin in this country; and it would be dishonourable to a nation so great, so powerful, and so rich, if, from the want of pecuniary resources, they should be completed abroad. An appeal to enlightened individuals on this subject can scarcely by made in vain. It is proposed that the instrument and apparatus be erected in the Laboratory of the Royal Institution, where it shall be employed in the advancement of this new department of science.[44]

The Managers did not hesitate in responding to this patriotic appeal. They 'felt it their indispensible duty to communicate the same to every member of the Royal Institution, lest the slightest delay might furnish an opportunity to other countries for accomplishing this great work, which originated in the brilliant discoveries recently made at the Royal Institution',[45] and those present themselves subscribed. Chemistry in Davy's hands was becoming a 'big science' needing expensive toys, and, in the British tradition, these were provided privately and not by the government.

On 12 July 1810 Davy's paper on oxymuriatic acid was read at the Royal Society.[46] He began with a reference to Scheele: 'The illustrious discoverer of the oxymuriatic acid considered it as muriatic acid freed from hydrogene, and the common muriatic acid as a compound of hydrogene and oxymuriatic acid.' Then he turned to experiments of his own:

> One of the most singular facts that I have observed on this subject, and which I have before referred to, is, that charcoal, even when ignited to whiteness in oxymuriatic or muriatic acid gases, by the Voltaic battery, effects no change in them; if it has been previously freed from hydrogene and moisture by intense uignition in vacuo. This experiment . . . led me to doubt of the existence of oxygene in that substance, which has been supposed to contain it above all others in a loose and active state; and to make a more rigorous investigation than had been hitherto attempted for its detection.[47]

He found that tin and oxymuriatic acid combined to give the same product as that formed with muriatic acid and tin. In the current theory salts were seen as formed of two parts, a metallic oxide and an acidic oxide: calcium carbonate something like $CaO.CO_2$, and sulphate $CaO.SO_3$,[48] though there was no agreement about exact formulae. It ought, therefore, to have been possible to displace oxide of tin from the muriate with ammonia, but it simply formed a new compound.

The key was in the paper of Gay-Lussac and Thenard, who:

> have proved by a copious collection of instances, that in the usual cases where oxygene is procured from oxymuriatic acid, water is always present, and muriatic acid gas is formed: now, as it is shewn that oxymuriatic acid gas is converted into muriatic acid gas, by combining with hydrogene, it is scarcely possible to avoid the conclusion, that the oxygene is derived from the decomposition of water, and, consequently, that the idea of the existence of water in muriatic acid gas is hypothetical, depending upon an assumption which has not yet been proved – the existence of oxygene in oxymuriatic acid gas.[49]

Few substances, in Davy's opinion, had less claim to be considered as acid than oxymuriatic acid:

> As yet we have no right to say that it has been decompounded; and as its tendency of combination is with pure inflammable matters, it may possibly belong to the same class of bodies as oxygene. May it not be a peculiar acidifying and dissolving principle, forming compounds with combustible bodies, analogous to acids containing oxygene, or oxides, in their properties and powers of combination; but differing from most of them, in being for the most part, decomposable by water?[50]

A new world was opening up before him, where the interpretation of experiments, some of them going back to before he was born, was the important thing. Nature never explains itself, and even the most formal or dramatic experiments never entail a theory. While Davy never developed a unified and coherent theory of acidity, contemporaries who thought he was returning to Scheele's phlogiston theory had got it wrong. Davy believed that Scheele had been essentially right, and Berthollet wrong, about the status of the green gas, but phlogiston now disappeared from Davy's chemistry.

In November 1810 Davy again delivered the Bakerian Lecture, and by now he had got his ideas straighter. This meant that, like Lavoisier, he could in the end rename the substances he was working with:

> To call a body which is not known to contain oxygene, and which cannot contain muriatic acid, oxymuriatic acid, is contrary to the principles of that nomenclature in which it is adopted; and an alteration of it seems necessary to assist the progress of discussion, and to diffuse just ideas on the subject. If the great discoverer of this substance had signified it by any simple name, it would have been proper to have recurred to it; but, dephlogisticated marine acid is a term which can hardly be adopted in the present advanced æra of the science. After consulting some of the most eminent chemical philosophers in this country, it has been judged most proper to suggest a name founded upon one of its obvious and characteristic properties – its colour, and to call it *Chlorine*, or *Chloric* gas [from χλωρος]. Should it hereafter be discovered to be compound, and even to contain oxygene, this name can imply no error, and cannot necessarily require a change.[51]

This was the first naming of another new family of elements recognized by Davy. He suggested a further convention, that compounds of this element might be labelled with the suffix 'ane', in various modifications; but this never caught on. His view was that muriatic acid might retain its name.

He concluded in lofty terms:

> What I have advanced, I advance merely as a suggestion, and principally, for the purpose of calling the attention of philosophers to it. As chemistry improves, many other alterations will be necessary; and it is

> to be hoped that whenever they take place, they will be independent of all speculative views, and that new names will be derived from some simple and invariable property, and that mere arbitrary designations will be employed, to signify the class to which compounds or simple bodies belong.[52]

We duly owe it to Davy that we speak of iodine, bromine and fluorine. He recognized chemistry as a progressive science, in which Lavoisier's revolution had been one among others rather than a definitive and permanent change of course.

In his lectures at the Royal Institution he was less inhibited in his triumphant assault upon French dogmatism:

> The opinions of Berthollet have been received for nearly thirty years; and no part of modern chemistry has been considered as so firmly established, or so happily elucidated; but we shall see that it is entirely false – the baseless fabric of a vision . . . Oxymuriatic acid is not an acid, any more than oxygen; but it becomes acid, like that substance, by combining with inflammable matter . . . The confidence of the French enquirers closed for nearly a third of a century this noble path of investigation, which I am convinced will lead to many results of much more importance than those which I have endeavoured to exhibit to you. Nothing is so fatal to the progress of the human mind as to suppose that our views of science are ultimate; that there are no mysteries in nature; that our triumphs are complete, and that there are no new worlds to conquer.[53]

Davy was still groping for a theory of acidity, and indeed he never really found a satisfactory one – his was not a very systematic mind, and he had overturned some systems himself.[54] In 1812 he wrote that 'the characteristic acid belonging to the combinations of chlorine is formed by the union of that body with hydrogen; and sulphur likewise forms an acid with hydrogen',[55] but he never went over to the view that hydrogen is the cause of acidity. He seems rather to have seen two categories of acids, oxygen and hydrogen acids; muriatic acid, $HC1$, had no element in common with sulphuric acid, SO_3, and acidity was the outcome not of matter in a compound but of a particular concatenation of forces.

In 1814 he remarked that:

> the chemists of the middle of the last century had an idea, that all inflammable bodies contained phlogiston or hydrogen. It was the glory of Lavoisier to lay the foundations for a sound logic in chemistry, by shewing that the existence of this principle, or of other principles, should not be assumed where they could not be detected.[56]

Although he could not at that point yet believe, against 'the analogies of nature', that diamond and charcoal could be chemically identical, he had come out against the idea of property-bearing elements.

Coleridge greeted Davy's dynamical chemistry, concerned with powers rather than matter, enthusiastically:

> Water and flame, the diamond, the charcoal, and the mantling champagne, with its ebullient sparkles, are convoked and fraternized by the theory of the chemist. This is in truth the first charm of chemistry, and the secret of the almost universal interest excited by its discoveries. The serious complacency which is afforded by the sense of truth, utility, permanence, and progression, blends with and enobles the exhilarating surprise and the pleasurable sting of curiosity, which accompany the propounding and the solving of an enigma. It is the sense of a principle of connection given by the mind, and sanctioned by the correspondency of nature. Hence the strong hold which in all, ages chemistry has had on the imagination. If in *Shakespeare* we find nature idealized into poetry, through the creative power of a profound yet observant meditation, so through the meditative observation of a *Davy*, a *Woollaston*, or a *Hatchett*;
>
> By some connatural force,
> Powerful at greatest distance to unite
> With secret amity things of like kind, [Milton]
>
> we find poetry, as it were, substantiated and realized in nature: yea, nature itself disclosed to us, ... as at once the poet and the poem![57]

The time had now come for Davy to take stock. Trinity College, Dublin, awarded him a doctorate; he wrote up his agricultural lectures and his chemical labours into books; he prepared to travel to France to collect his prize; and he began to look seriously for a wife. In about five years he had transformed chemistry, confirming the ideas of those who thought it the basic science, describing a dynamical rather than a mechanical world: now his life was to change its pattern. Long hunted by ladies making up intellectual parties, he was now himself the hunter of an eminent bluestocking.

7

A Chemical Honeymoon, in France

On 2 March 1812 Davy wrote to Jane Apreece, 'I have passed a night sleepless from excess of happiness. It seems to me as if I began to live only a few hours ago'.[1] To his mother he wrote with a curious mixture of joy and stiffness:

> My dear Mother
> You may possibly have heard reports of my intended marriage. Till within the last few days it was mere report. It is now, I trust, a settled arrangement. I am the happiest of men in the the hope of union with a woman equally distinguished for talents, virtues, and accomplishments, and who is the object of admiration in the first society in this country. You, I am sure, will sympathise in my happiness. I believe I should never have married but for this charming woman, whose views and whose tastes coincide with my own, and who is eminently qualified to promote my best efforts and objects in life. Remember me to my aunts with affection, and to my sisters with kind love.
> I am, your affectionate son.[2]

To his brother he actually named his fiancée:

> I have been very miserable; the lady whom I love best of any human being, has been very ill. She is now well, and I am happy. Mrs. Apreece has consented to marry me, and when this event takes place, I shall not

envy kings, princes, or potentates. Do not fall in love. It is very dangerous! My case is a fortunate one. I do not believe there exists another being possessed of such high intellectual powers, just views, and refined taste, as the object of my admiration.
I am, my dear brother, Ever most affectionately yours.[3]

Davy was not a great letter writer, but he was not usually as stilted as this. These inarticulate and affected letters were the product of his urgent wooing of, and then marriage to, an heiress and bluestocking a year younger than he was, and suddenly entering upon a very different kind of life at thirty-two. This was in great contrast to Gay-Lussac, who fell in love at first sight with a girl working in a shop: they married and had five children and more or less lived happily ever after, which cannot be said of the Davys.

On 8 April 1812 Davy was knighted by the Prince Regent, and on the 11th married to Jane Apreece at her mother's house at Portland Place, London, by the Bishop of Carlisle. Jane Davy was a 'brunette of the brunettes', a friend of Sydney Smith and a remote cousin of Walter Scott. She was the only daughter of Charles Kerr of Kelso in Scotland, in the West India trade, and in 1799 had married Shuckburgh Apreece, whose father was a baronet from Huntingdonshire. She had been widowed in 1807 and moved to Edinburgh and opened the doors of her house to the cleverest and brightest of its residents'.[4] The vivacity of her conversation made her much admired; and Sir Henry Holland reported that her parties

> gained for a time the mastery over all others. Coming suddenly to the Scotch capital as a young and wealthy widow, with the reputation and fashions of a continental traveller at a time when few had travelled at all, acquainted with Madame de Staël, and vaguely reported to be the original of Corinne,[5] then fresh in fame, this lady made herself a circle of her own, and vivified it with certain usages new to the habits of Edinburgh life . . . The story was current of a venerable professor stooping in the street to adjust the lacing of her boot.

She was described as 'small, with black eyes and hair, a very pleasing face, an uncommonly sweet smile, and when she speaks, has much spirit and expression in her countenance'.[6] She seemed rather formal, but was clearly very intelligent and apparently widely read in modern languages and Latin.

Perhaps Jane hoped that Sir Humphry would stoop from time to time to unloose the latchet of her shoe. At all events Davy's friends and relations did not take to her. Dr Paris wrote of Davy

> assuming a new station in society, which induced him to retire from those public situations which he had long held with so much advantage to the world, and with so much honour to himself. How far such a measure was calculated to increase his happiness I shall not enquire; but I am bound to observe, that it was not connected with any desire to abandon the pursuit of science, nor even to relax in his accustomed exertions to promote its interests. It was evident, however, to his friends, that other views of ambition than those presented by achievements in science, had opened upon his mind; the wealth he was about to command might extend the sphere of his usefulness, and exalt him in the scale of society; his feelings became more aristocratic – he discovered charms in rank which had before escaped him, and he no longer viewed Patrician distinction with philosophic indifference.[7]

Ever since Dick Whittington the classic route for the country boy making good in London involved marrying an heiress, and for Davy the arduous business of being a professional man of science could now come to an end. He gave up lecturing at the Royal Institution, to the great regret of the Managers, but was duly given the title of Honorary Professor of Chemistry and continued to research in the laboratory there. He wrote up his work as *Elements of Chemical Philosophy, part 1, volume 1*, dedicated to his wife as an assurance that he would not abandon science, and he published his agricultural lectures,[8] in what looks like a farewell to arms.

Other parts of the *Elements* never appeared. Davy described there his own work, notably on electrolysis and on acids, but was not a man to write a textbook of the whole science as the plan of the book implied he would. He made some notes for a second edition of the *Elements*, but, perhaps to his chagrin, it was never required. His successor at the Royal Institution, W. T. Brande, was the kind of 'normal scientist' who wrote what became a standard textbook but did no research of fundamental importance. Berzelius came to England at this time, bringing Davy his Diploma of election to the Swedish Academy of Sciences. He had to call more then once, and was one of the first to find Davy haughty.[9] Davy gave him a copy of the *Elements*, with the invitation to comment in case there were to be a second edition, but when Davy heard of Berzelius' great list of critical notes (no author ever really welcomes that), relations between the two became very cool.

Jane Davy outlived Humphry by a quarter of a century. After her death in 1855 John Davy reflected upon the marriage: a casual acquaintance passed into friendship, and that into love based on each side on sincere admiration, hers for his genius, and his for her charm and sympathy. 'Never, I believe, was admiration more genuine of its kind, or more lasting; indeed it continued, it may be inferred, judging from their closing correspondence, to the very last.'[10] And yet such mutual admiration did not lead to an easy, supportive and loving relationship, and John Davy wrote that there

> was an oversight, if not a delusion, as to the fitness of their union; and that it might have been better for both if they had never met; and mainly for this reason, that the lady, in spite of all her attractions in mixed society, was not qualified for domestic life, for becoming the *placens uxor*, being without those inestimable endowments which are requisite for it – the agreeable temper, the gentle loving affections which are rarely possessed, which are hardly compatible with an irritable frame and ailing body, such as her's were (for her misfortune) in a remarkable degree.[11]

They had no children, which must have been a great disppointment for both of them: Davy was 'of a loving disposition, and fond of children, and required the return of love – required (who does not?) to be beloved, to be happy'.[12] When we think of the large families of the past we are apt to forget all the childless marriages that there were, and the cheerless homes to which this could lead. Heiresses, as Francis Galton was later to point out, have no brothers and sisters to divide their fortune with, and therefore may well be infertile; Whittingtons must beware.

Having an ample fortune and no children, it is not surprising that Jane Davy should have struck John as 'perhaps too independent and self-willed'; and these qualities are, after all, necessary if one is to preside over a successful salon, taming the lions of Edinburgh. We need not go all the way with John Davy's view of gender roles. Davy's life at the Royal Institution in the years before his marriage had not been very domesticated either. He wrote to his mother in 1808: 'Now I live very little in the Royal Institution; I never dine there, and when I do not dine with some of my friends, I dine at a coffee-house.'[13] It is one thing to tire of a rootless bachelor existence, where one is highly sought after at dinner parties, and of giving courses of lectures to admiring but exhausting throngs, but another to set up house and have all the give and take of married life, especially for two people already in their thirties accustomed to adulation. What is quite clear is that the Davys did not manage it.

As the years went by, the unhappiness of their marriage became notorious and the bad temper of both parties remarked upon. There is no suggestion that they were unfaithful to each other: Regency

London delighted in gossip about cuckoldry and 'crim. con.', but the Davys do not seem to have been victims of that. The mortifications that Davy had to endure were to do with practical inventions that did not work out, in agriculture and, in 1811, when attempting to devise a system of ventilation for the House of Lords that involved discussions with Lord Liverpool, soon to become Prime Minister on the assassination of Spencer Percival. In a philistine country like England the man of science who tries to apply it can expect teasing, even pillorying, if it goes wrong; but if it also brings professorships and knighthoods and conversation with the great, that should not permanently sour the temper.

Davy's professional life in the years just before his marriage had been steadily building up. As Secretary of the Royal Society he received an income of 100 guineas a year in addition to his salary of £400 from the Royal Institution; he did two triumphant tours of Ireland, where there was a black market in tickets for his lectures and he received fees of over £1000; and he received 1000 guineas for the copyright of his *Agricultural Chemistry*, and 50 guineas more for each subsequent edition. He had also become involved with J. G. Children in the manufacture of gunpowder. Lavoisier had been instrumental in improving the French gunpowder, and in 1811, with a war on, it must have seemed an obvious way in which a chemist could contribute to his country's strength. Davy seems to have seen himself as a consultant, a gentlemanly role for a natural philosopher, whereas Children and his partner, Burton, wanted him as a full partner, whose name would be very helpful in selling the product. This led to some awkwardnesses and Davy rather stiffly extricated himself from what was a not very successful business enterprise, but it is good to be able to report that it did not put an end to his friendship with Children.

In July 1812 Davy reported to Children researches on gunpowder performed while visiting Harewood House:

> I have my little apparatus, which will enable me to pursue my experiments on gunpowder. There is one conclusion very obvious resulting from the new facts, – a *perfect* gunpowder ought to contain no more charcoal than is necessary to convert the oxygen of the nitre into carbonic acid. Sulphur forms from nitre just as much elastic fluid as charcoal, *i.e.* if similar quantities of nitre be entirely decomposed, one by charcoal, and one by sulphur, and if the sulphurous gas and the carbonic acid gas be compared, their volumes will be equal. The advantage of forming carbonic acid gas is, that it is more readily disengaged from the alkali. Now it is a question, whether sulphur will decompose *sulphate* of potash, – it will decompose the carbonate; of this we are sure. There

> ought then to be just as much sulphur as will form sulphuret of potash with the potash; 191 of nitre, 28.5 of charcoal, and 30 of sulphur, are the true proportions for forming nothing by sulphuret of potash and elastic matter.[14]

In a subsequent letter Davy hoped Children 'was making progress in our manufactory, and wished he had some of the powder because 'the black-cock and grouse would feel its efficacy'.[15]

On 14 October 1812 the Davys' extended honeymoon tour of Scotland was drawing to a close, and he wrote to Children from Edinburgh that 'lady D. is very much indisposed, and anxiety for her hastens my journey to town', adding that he had had a letter from Ampère reporting the discovery in Paris of a 'combination of chlorine and azote', which was a fluid, exploding at the temperature of the hand, and which had cost its discoverer [Dulong] an eye and a finger. Davy reports 'I have tried in my little apparatus with ammonia cooled very low, and chlorine, but had no success'.[16] We should note that Davy had taken on this honeymoon his chest of apparatus, so that he would not be deprived of the opportunity to do chemical research. The scale of operations of the chemist was getting smaller: the large quantities and furnaces of the eighteenth century were giving way to test-tubes and spirit lamps.

On 24 October he wrote again to Children:

> My wife is much better, except that she has a swollen foot. I have never seen her in such good health and spirits. She is resolved to lead a home life of perfect quiet for six weeks, and I fear you will not be able to tempt her to quit her fire-side, though there is no visit she would make with greater pleasure; but lameness does not suit the country; and for one so enthusiastically fond of nature, it would be vexatious to be in the country, and not to be able to enjoy hills, and meads, and woods. But I am ready to come to my business whenever you think I can be useful, I shall set to work to make gunpowder with as much ardour as Miles Peter – I hope with similar results. I shall not be able to endure a very long separation from my wife, but for three or four days I am at your command. I have been working yesterday and today on some new objects; and we are to have a meeting on Wednesday, at one o'clock, at the Institution, to try to make this compound of azote and chlorine, and to try some other experiments. Afterwards we (Angling Chemists) propose a dinner at Brunet's. If you can come to town on that day, I will promise to return with you.[17]

The nitrogen trichloride Davy was so keen to prepare was one of the first high explosives to be investigated, and its preparation duly turned out to be alarming, though we may hope that on this occasion it did not spoil his bachelorish evening. Later, an explosion with it was to provide the occasion for Faraday to experiment on Davy's behalf, transforming him from amanuensis to assistant; as Davy wrote to John on 16 November 1812, of an experiment done at Children's home:

> I have discovered the mode of making the combination of azote and chlorine. It is by exposing chlorine to a very weak solution of ammonia, or to a solution of nitrate of ammonia, or of oxalate of ammonia. It must be used with very great caution. It is not safe to experiment upon a globule larger than a pin's head. I have been severely wounded by a piece scarcely bigger. My sight, however, I am informed, will not be injured. It is now very weak. I cannot see to say more than that I am, Your very affectionate brother.[18]

It was not until April 1813 that he could write again to John: 'I am quite recovered, and Jane is very well, and we have enjoyed the last month in London. I have been hard at work.'[19]

On 12 August 1812 Davy had written to Thomas Clayfield: 'Having given up lecturing, I shall be able to devote my whole time to the pursuit of discovery',[20] but that was not how things worked out. Research and teaching then and since are, for those who are really good at them, very closely linked, and not to have to explain what one is up to before a student audience is bad for anybody in academe. Lectures like Davy's, presenting his current research, are exciting, quite unlike those that, ignoring the invention of printing some time ago, summarize textbooks; and his had a strong element of theatre about them, again unlike talks that lecturers read. Dramatic traditions, which persist into the present, grew up in Davy's time at the Royal Institution; and preparing his lectures must have helped him in his scientific thinking.

However, his audiences were not students. Hopes Davy may have had that gentlemen in his audiences might take up scientific research were not fulfilled; for them, science was a spectator sport, and not something in which they saw themselves participating. Unlike Berthollet, and unlike chemical lecturers at medical schools, Davy was confined to popular lecturing. The social climate in England was still unsuited to serious chemistry as a liberal education, and outside medicine there was little demand for it as a part of training. All lecturing ought to be entertaining, but to be, like Davy, engaged entirely in entertaining and in providing topics for conversation must have been very wearying, and the hope that escaping from it might lead to more research was not unreasonable. Nevertheless, he also realized that 'it is only by making truths popular that their existence can be secured . . . prejudices conquered cannot be *re-established*', and 'If the mathematical sciences are efficacious in their power of strengthening the reason and refining the judgement, the chemical ones are still more eficacious, because they approach nearer the facts of common life.'[21] There can be no doubt that he was an enthusiastic popularizer and advocate of utility, even though he also wrote 'No man ever yet made great discoveries in science, who was not impelled by an abstracted love.'[22]

In March 1813 the young Michael Faraday was engaged as amanuensis and then assistant. Davy's laboratory technique did not need improving unless he happened to have been recently blown up, but his handwriting and keeping of records certainly did. Davy's relationship with Faraday is important enough to deserve a separate chapter, but we should note here that Faraday was offered the job fortuitously because his predecessor had become involved in a fight and been dismissed, and that the Davys were already planning to go to France to collect his prize and to meet his French colleagues and rivals. This was an extraordinary opportunity. Since the Reign of Terror and the outbreak of war between Britain and France in 1792, there had only been a brief interlude, the Peace of Amiens of 1802, when Britons could visit France and most of the Continent. Davy set out 'provided with a portable chemical apparatus'.[23]

Davy had visited Wales, Scotland and Ireland, geologizing, fishing and shooting, on his holidays from the Royal Institution. He wrote copious letters and kept journals on these travels, like a conscientious tourist, and this prepared him to do likewise in France. On the whole, he preferred the works of God to those of man; in good Romantic vein, it was the beauties of nature that moved him, though the ruins in Rome also proved a great stimulus. His contemporaries, no doubt because travel abroad was so unusual, avidly reported on diligences, postilions and the discomforts of inns in journals from which their biographers printed substantial extracts, and in this Davy was no exception.

At that time a passport was not something granted by one's own government, but by that of the country that one wished to visit; and although 'the Emperor had sternly refused his passport to several of the most illustrious noblemen of England', it was forthcoming for Davy.[24] Jane Davy had a maid to accompany her, and Davy, as befitted a knight and married man, had a manservant – for whom the thought of going to France was so appalling that at the last moment his courage failed him and he resigned. The role of Jeeves thus fell upon Faraday, who did not relish it, mostly because Jane Davy thereupon treated him as a mere servant. Davy himself had little use for a valet, and needed Faraday as an assistant and to make travel arrangements. On 13 October 1813 the party left London for Plymouth. On the way, Davy wrote to his mother:

> We are just going to the Continent upon a journey of scientific inquiry, which I hope will be pleasant to us, and useful to the world. We go rapidly through France to Italy, and from that to Sicily; and we shall return through Germany. We have every assurance from the governments of the countries through which we pass, that we shall not be molested, but assisted. We shall stay probably a year or two. . . . As I am permitted to pass through an enemy's country, there must be no politics in any letters to me; and you had better not write except through the channel I shall hereafter point out. . . . Tell Grace not to be afraid, though I am going to France.[25]

If this was the first his mother heard about the scheme, she may not have been wholly reassured. To visit an enemy country was no light undertaking, for this was not merely a war of kings and there was a general hearty dislike of everything French in the patriotic circles in which Davy moved in England. We should note that Davy preferred to see his expedition as one passing through France on to places more agreeable, making a Grand Tour. Faraday wrote in his journal: 'Tis indeed a strange venture at this time, to trust ourselves in a foreign and hostile country, where also so little regard is had to protestations of honour, that the slightest suspicion would be sufficient to separate us for ever from England, and perhaps from life.'[26] Davy had told Poole that he hoped 'through the instrumentality of men of science, to soften the asperity of national war',[27] but it did not really work out that way.

At Plymouth they boarded a cartel and sailed to Morlaix in Brittany; Faraday was delighted when seasickness forced Jane Davy to go below and quieted that voice which had delighted the salons of Edinburgh and London. A cartel was 'a ship commissioned in time of war to exchange prisoners of any two hostile powers; also to carry any particular request or proposal from one to another; for this reason the officer who commands her is particularly ordered to carry no cargo, ammunition, or implements of war, except a single gun for the purpose of firing signals.'[28] On their arrival Davy wrote again to his mother on 22 October:

> We are safe at Morlaix. A sea voyage is always disagreeable, and I have not yet recovered the effects of mine. We were a day and two nights at sea. We found here our passport from Paris, and we shall set

> out tomorrow. The weather is rather unfavourable; but it has rained so much, I hope the clouds are exhausted. Pray write to my brother, and say to him that we are quite safe. As soon as I arrive in Paris, I shall endeavour to find a mode of hearing as often as I can of your health and safety.[29]

On arrival, Faraday reported, they were searched, an unusual experience for true-born Englishmen: 'he then felt my pockets, my breast, my sides, my clothes, and lastly, desired to look into my shoes; after which I was permitted to pass', and could hardly help 'laughing at the ridiculous nature of their precautions'.[30] The French who unloaded the carriage the Davys had prudently brought with them from England and all the rest of their belongings seemed to Faraday 'totally destitute of all method and regularity', and the kitchen appalled him:

> I think it is impossible for an English person to eat the things that come out of this place except through ignorance or actual and oppressive hunger; and yet perhaps appearances may be worse than the reality, for in some cases their dishes are to the taste excellent and inviting, but then they require, whilst on the table, a dismissal of all thoughts respecting the cookery or the kitchen.[31]

In Paris, longing for a fellow countryman, they met Underwood, who had been arrested and detained in France after the Peace of Amiens; he sympathized with the government, was on good terms with the Empress Josephine, and was, in effect, at liberty. On 30 October they went with him to the Louvre, where in the Galerie Napoleon were displayed the greatest works of art from all over Europe.[32] Underwood informed Paris after Davy's death that Davy had walked with rapid step along the gallery and had not directed his attention to a single painting.

> On arriving opposite to Raphael's picture of the Transfiguration, Mr. Underwood could no longer suppress his surprise, and in a tone of enthusiasm he directed the attention of the philosopher to that most sublime production of art, and the chef-d'oeuvre of the collection. Davy's reply was as laconic as it was chilling – 'indeed, I am glad I have seen it; –' and then hurried forward, as if he were desirous of escaping from any critical remark upon its excellencies.[33]

His attitude to the sculpture had been equally frigid: 'while the marble glowed with with more than human passion, the living man was colder than stone!'[34]

Paris sententiously remarked that we 'have here presented to us a philosopher, who, with the glowing fancy of a poet, is insensible to the divine beauties of the sister arts!,[35] but the story does not prove that Davy was a philistine. He was confronted by rooms full of loot collected by his country's mighty enemy, and invited by Underwood to forget all that in the pleasures of aesthetic contemplation. No patriot fresh from England could do so, as we see from Faraday's report of the occasion:

> I saw the Galerie Napoleon to-day, but I scarcely know what to say of it. It is both the glory and the disgrace of France. As being itself, and as containing specimens of those things which proclaim the power of man, and which point out the high degree of refinement to which he has risen, it is unsurpassed, unequalled, and must call forth the highest and most unqualified admiration; but when memory brings to mind the manner in which the works come here, and views them only as the gains of violence and rapine, she blushes for the people that even now glory in an act that made them a nation of thieves.[36]

Faraday was an admirer of painting and a friend of Turner; Davy was probably much less sensitive to the visual arts, but his behaviour in the Louvre does not prove it.

The traffic of cartels and smugglers between Britain and France was enough to get scientific information back and forth, but not to keep British fashions up to date. When Jane Davy went for a walk in the Tuileries Gardens wearing 'a very small hat, of a simple cockle-shell form' when Parisiennes were wearing voluminous bonnets, she attracted an excited crowd, was ordered to leave the park and escorted back to her carriage upon the arm of a gallant officer of the Imperial Guard.[37] Meanwhile Davy had been the guest of honour at a meeting of the First Class of the Institut, and French savants called upon him at the Hotel des Princes in the Rue Richelieu where the Davys were staying.

Davy had especially hoped to meet Ampère, from whom he had heard about nitrogen trichloride, and on 23 November Ampère came with two other chemists, Nicholas Clémont and his father-in-law C. B. Desormes, eminent for work on specific heats. They brought with them a sample of a new substance 'in small scales with a shining lustre, colour deep violet, almost black', with the remarkable property 'that when heated it rises in vapour of a deep violet colour'.[38] Davy at once got to work on it, considering it at first a compound of chlorine and an unknown body. Soon, however, he came to believe that it contained no chlorine, but was an analogue of that element. On 3 December Davy and Faraday went to M. E. Chevreuil's laboratory at the Jardin des Plantes to work on the new substance. Faraday 'observed nothing particular in this laboratory, either as different from the London laboratories or as peculiarly adapted to the performances of processes or experiments. It was but a small place, and perhaps only part of the establishment appropriated to chemistry.'[39] They were aware that not only were they in a hostile country where they must keep their end up, but also that they were in a centre of scientific excellence in a different league from London. On December they went to the École Polytechnique to hear Gay-Lussac give a lecture, illustrated by rough diagrams and experiments, to some two hundred pupils.

There were dinner parties with the various groups of men of science, and on 11 December Faraday borrowed a voltaic pile from a new acquaintance for Davy to experiment on the new substance. Gay-Lussac was known to be at work upon the new substance, and on 6 December had discussed it at the Institut, saying that it might be an element or an oxide. Davy sent off a paper to the Royal Society, dated 10 December 1813, describing his researches, recognizing the affinities of the new substance to chlorine, and calling it iodine. He concluded:

> In my last paper,[40] presented to the Society two months ago, I ventured to suggest that it was probable, that new species of matter, which act with respect to inflammable bodies, like oxygen, chlorine, and fluorine, would be discovered. I had not hoped, at that time, to be able so soon to describe the properties of a body of this kind, which forms an acid with hydrogen, like chlorine and fluorine, and which in some of its combinations resembles oxygen. This new fact will, I hope, do something towards settling the opinion of chemists respecting the nature of acidity, which seems to depend upon peculiar combinations of matter, and not on any peculiar elementary principle.[41]

Davy saw this investigation as part of his research programme, which had begun with the work on chlorine; but to Gay-Lussac it seemed as though Davy was muscling in on his territory and getting a major paper published when his hosts had been laboriously working up a fully detailed account of the new great discovery of French chemistry. Davy, who in his notes about the men of science he met in Paris placed Gay-Lussac 'at the head of the living chemists of France',[42] nevertheless thought Gay-Lussac had been picking his brains.[43] Priority disputes have always been a feature of science and, when accompanied by national feeling, can be bitter. It was for Faraday an exciting introduction both to fundamental research and to scientific life. In a curious reversal of the story of Lavoisier and Priestley, Davy leaped to the new theoretical conclusion while his rival was cautiously experimenting. On 13 December Davy was elected a Corresponding Member of the Institut.

Davy was never introduced to Napoleon, home from Moscow, but Faraday reported seeing Napoleon 'sitting in one corner of his carriage, covered and almost hidden from sight by an enormous robe of ermine, and his face overshadowed by an tremendous plume of feathers that descended from a velvet hat';[44] they were some of the few Britons to see Boney and get away with it. They did visit the Empress Josephine at Malmaison, where a piece of porcelain was presented to Jane, and they saw the conservatories. Davy stalked out of an anti-English play and showed a distaste for the courtly etiquette of the French, sufficient for Napoleon apparently to remark to the leading members of the Institut that the young English chemist held them all in low esteem. This comes from Underwood, an unreliable informant; nevertheless, there seems no doubt that Davy carried the war into the very capital of France, but was certainly welcomed and admired by some groups among the cliques of men of science, notably those less committed to the regime. He delighted in finally overthrowing Lavoisier's theory of acidity in Paris, a Coriolanus embattled alone in the hostile city.

On 29 December the Davy Party left the city and headed south towards Italy. At Fontainbleau Davy wrote a poem, beginning with delight in nature and passing on to modern politics:

> Nearer I behold
> The palace of a race of mighty kings;
> But now another tenants. On these walls,
> Where erst the silver lily spread her leaves –
> The graceful symbol of a brilliant court –
> The golden eagle shines, the bird of prey, –
> Emblem of rapine and of lawless power:
> Such is the fitful change of human things:
> An empire rises, like a cloud in heaven,
> Red in the morning sun, spreading its tints
> Of golden hue along the feverish sky,
> And filling the horizon; – soon its tints,
> Are darken'd, and it brings the thunder storm, –
> Lightning, and hail, and desolation comes;
> But in destroying it dissolves, and falls
> Never to rise![45]

At the end of 1813 it did not perhaps indicate profound insight to foresee the fall of Napoleon, but his abdication and exile to Elba a few months later must still have seemed to Davy and his party something extraordinary, for which they had scarcely dared to hope.

They went to Montpellier, to Nice, Turin, Genoa and then Florence, where Davy used the great burning glass of the Grand Dukes of Tuscany to burn a diamond under very carefully controlled conditions. He had believed that there must be some chemical difference between diamonds and charcoal, but his researches in Florence and in Rome confirmed the idea of Smithson Tennant of Cambridge that there was none.[46] This astonishing result supported the notion that the material components of things were not crucial in determining their properties, since a single element could take such very different forms.

The Davys enjoyed Italy. In May they witnessed the return of the Pope, Pius VII, to Rome from his exile in France. Davy met him and from this time seems to have viewed the Roman Catholic Church with a respect unusual among his English contemporaries. The party went on to Naples, where they visited Vesuvius and collected samples of crystals around the crater. Faraday recorded their scrambles on the mountain in his journal; vulcanism was a subject of intense interest among geologists at this time, and few Englishmen had had the opportunity of seeing a volcano. They also went to Pompeii, where Davy saw some of the ancient paintings. As summer came on, they turned north, to Switzerland, Bavaria and Austria, coming down to Venice in October 1814. They spent some time in Geneva, where Faraday's talents were recognized as well as Davy's. At Milan Davy called upon Volta, who had apparently dressed up to receive his great disciple in the field of electrochemistry and was surprised to find Davy in scruffy travelling clothes. Davy and Faraday analysed gas from various natural vents in the mountains, discovering that it was the same as marsh gas, an observation that was to be of importance to them on their return to England.

During the winter of 1814/15 Davy did further experiments in Rome on iodine and chlorine and their compounds with oxygen;[47] and also on the colours used in painting by the ancients. This was an extremely important piece of work because it was the first research by a chemist on artists' pigments; the whole science of picture restoration in its modern forms stems from it. It is difficult to believe that it could be the work of a philistine, which confirms our political interpretation of his behaviour at the Louvre. Davy began by calling the attention of his readers to the importance of ancient statues and paintings as models for artists, and continued:

> There remains, however, another use to which they may be applied, that of making us acquainted with the *nature* and *chemical composition* of the colours used by the Greek and Roman artists. The works of Theophrastus, Dioscorides, Vitruvius, and Pliny, contain descriptions of the substances used by the ancients as pigments but hitherto, I believe, no experimental attempt has been made to identify them, or to imitate such of them as are peculiar . . . My experiments have been made upon colours found in the baths of Titus, and in the ruins called the baths of

> Livia, and in the remains of other palaces and baths of ancient Rome, and in the ruins of Pompeii.[48]

Through Canova and others, he had access to works of art and chose samples from them,[49] and took care to 'work upon mere atoms of the colour, taken from a place where the loss was imperceptible: and without having injured any of the precious remains of antiquity, I flatter myself, I shall be able to give some information not without interest to scientific men as well as to artists, and not wholly devoid of practical applications'.[50]

Davy's researches enabled him to cast light upon the way colours were applied in fresco painting, suggesting that ancient artists, like those of the Venetian and Roman schools, were probably 'sparing in the use of the more florid tints in historical and moral painting, and produced their effects rather by the contrasts of colouring in those parts of the picture where deep and uniform tints might be used, than by brilliant drapery'.[51] He compared the greatest artists of antiquity to Raphael and Titian, concluding on a note of sadness that many recent masterpieces were done in oil on canvas, vegetable materials liable to decomposition, so that, unlike ancient frescos done with mineral colours, they could not be passed 'down to posterity as eternal monuments of genius, taste and industry'.[52]

He wrote a tribute to Canova, the great sculptor of the day, which begins:

> Thou wast a light of brightness in an age
> When Italy was in the night of art: –
> She was thy country, but the world thy stage,
> On which thou acted thy creative part.[53]

It continues in this rather conventional vein. He also wrote poems about Mont Blanc, Rome and Pæstum, all a sign that he was enjoying himself. That on Rome is concerned with religion:

> Thy faith, O Roman! was a natural faith,
> Well suited to an age in which the light
> Ineffable gleamed thro' obscuring clouds
> Of objects sensible, – not yet revealed
> In noontide brightness on Syrian mount.
> For thee, the Eternal Majesty of heaven
> In all things lived and moved, – and to its power
> And attributes poetic fancy gave
> The forms of human beauty, strength, and grace.
> The Naiad murmur'd in the silver stream,
> The Dryad whisper'd in the nodding wood,

(Her voice the music of the Zephyr's breath);
On the blue wave the sportive Nereid moved,
Or blew her conch amidst the echoing rocks.
I wonder not, that, moved by such a faith,
Thou raisedst the Sybil's temple in this vale,
For such a scene was suited well to raise
The mind to high devotion, – to create
Those thoughts indefinite which seem above
Our sense and reason, and the hallowed dream
Prophetic. – In the sympathy sublime,
With natural forms and sounds, the mind forgets
Its present being, – images arise
Which seem not earthly, – 'midst the awful rocks
And caverns bursting with the living stream, –
In force descending from the precipice, –
Sparkling in sunshine, nurturing with dews
A thousand odorous plants and fragrant flowers,
In the sweet music of the vernal woods.
From winged minstrels, and the louder sounds
Of mountain storms, and thundering cataracts,
The voice of inspiration well might come![54]

Davy's romantic feeling for nature was still intact. He had visited the Wordsworths at Dove Cottage and had tramped among the mountains of Wales, Scotland and Ireland; but the Alps and the Appenines were for him among the great discoveries of this tour.

They returned to Naples, where Vesuvius was erupting, and then heard of Napoleon's return from Elba. On 21 March they left Naples and made a rapid return to London by way of Austria, where Davy treated a Tyrolean freedom fighter's rheumatism and was rewarded with a gun with which thirty Bavarians had been shot (he duly passed it on to Walter Scott); and the Rhineland and Ostend, getting home on 23 April 1815, which was, appropriately, St George's Day. In June the Battle of Waterloo put an end to all Napoleon's ambitions, and Davy wrote a letter to Lord Liverpool urging that the French be treated with severity in any subsequent peace treaty. To his mother he wrote: 'We have had a very agreeable and instructive journey, and Lady Davy agrees with me in thinking that England is the only country to *live* in, however interesting it may be to *see* other countries. I yesterday bought a good house in Grosvenor Street, and we shall sit down in this happy land.'[55] It turned out that a serious problem in applied science was awaiting him, and that in solving it he was to achieve even greater fame.

8

The Safety Lamp

In France and Italy Davy had been able to keep up with science at home to some extent. On 18 March 1814 he had written to his brother: 'There is now full communication between Italy and England: and tell me all the news – what you have done, what you have published, and what you are doing.'[1] Before and after this, he had managed to send home papers to the Royal Society, and in January 1815 in a letter to his mother he welcomed the news that Penzance was founding its own Geological Society: 'I am very happy to hear of a disposition to scientific activity in my native town, and shall be happy if I can do any thing to be useful to the museum. I will send to it some specimens from the Continent; and if there are subscriptions, pray get my name put down for £20.'[2] But the important news that awaited him was that at the beginning of October 1813 a Society for Preventing Accidents in Coal-Mines had been founded at Bishopwearmouth, now a part of Sunderland. The Rector (in succession to William Paley, the natural theologian) and chairman, Dr Robert Gray, who later became Bishop of Bristol, knew Davy a little and on his return to England 'judged it expedient to direct his attention to a subject, upon which, of all men of science, he appeared to be the best calculated to bring his extensive stores of chemical knowledge to a practical bearing'.[3] This was something Davy could not but welcome.

Sadi Carnot, son of Napoleon's Chief of Staff, in his little book propounding the Second Law of Thermodynamics, wrote that British power depended much less on the Royal Navy than on industry, and especially steam engines.[4] The demands of power technology, and the domestic needs of a rapidly expanding population flooding into the great industrial centres, had led to an enormous demand for coal. At the end of the seventeenth century the mines had got deep enough to run into water; the first steam engines of Newcomen and his successors had pumped them dry and dealt with that problem. A hundred years later they ran into explosive gas, the 'firedamp', and its aftermath, the 'choke-damp'.

The Established Church comes well out this story, in that the Reverend John Hodgson, of the parish of Jarrow, was, like Gray, prominent among those who urged that something should be done to avert the catastrophes caused by explosions. One of the most famous and shocking of these, vividly reported, happened at Felling Colliery in May 1812, when ninety-two men were killed in what had been believed to be an up-to-date pit.[5] Explosions made what had always been a dangerous business into one that seemed desperately unsafe, and threw large numbers of widows and orphans upon the generosity of the public, as commentators often noted. The sudden death in explosions was also feared by a generation that took the settling of affairs at the end of life very seriously.

Miners normally used candles to provide light, and stopped doing so only after a pit had proved itself to be unsafe; after all, working in the dark is nobody's idea of fun. They would use gunpowder to bring down the undercut masses of coal, igniting it with a slow match. If there were some reason to believe that gas was a problem, a 'fireman' would go down first, swathed in damp sacking as protective clothing and with a candle on a long pole to push cautiously round corners. Choke-damp was detected using a bird in a cage, which would collapse before men did. In fiery pits, and their number was steadily increasing, the only safe proceeding seemed to be the use of flint and steel, giving a feeble shower of sparks. This gave piteously little light, but insufficient heat to explode firedamp as a rule.

It was clear that a major industry could not be carried on that way. The first hope was better ventilation, but this did not by itself prevent explosions. The answer seemed to be to isolate the candle or oil flame from the surrounding atmosphere by enclosing it in glass. But since the flame could continue in a lantern only if it had access to air, an ordinary lamp was just as dangerous as a candle. William Reid Clanny, an Irishman trained at Edinburgh, was a doctor at Bishopwearmouth. He thought of forcing the air in and out of the lamp (with bellows) through water, the great antagonist of fire.

Clanny described his lamp in a paper communicated to the Royal

Society by William Allen, read in May 1813 and then published with an illustration.[6] He referred to the damp as 'inflammable air' (the old name for hydrogen) and as hydrogen, but also identified it as 'carburetted hydrogen', which is marsh gas, or methane. He remarked that 'in the short space of of seven years, upwards of two hundred pitmen were deprived, most suddenly, of their mortal existence, besides a great many wounded; and upwards of three hundred women and children were left in a state of the greatest poverty and distress.'[7] Steel mills were neither adequate nor safe to work by, but he had solved the problem, and 'the idea of insulating the light, and also the plan which I have adopted for carrying the idea into effect, by the construction of the apparatus or lamp, are perfectly original'.[8] He believed the lamp was strong and trouble free; exploding air inside the lamp could not ignite damp outside it. The lamp was successfully tried in Harrington Mill, where there had been a disastrous explosion in October 1812. Clanny later succeeded in reducing the weight of his cumbersome machine to thirty-four ounces, and in 1816–17 received medals from the [Royal] Society of Arts (a practical body, not to be confused with the Royal Society) for his invention.

Clanny's device was developed in the usual way in the history of technology, by evolution[9] from lamps and lanterns that were familiar; it drew upon no recent science. Although it was used in some collieries, it was not generally felt to have solved the problem, which is why Gray wrote to Davy. On 3 August 1815 Davy replied from Scotland, where he and Jane were on holiday:

> It will give me great satisfaction if my chemical knowledge can be of any use in an enquiry so interesting to humanity, and I beg you will assure the Committee of my readiness to co-operate with them in any experiments or investigations on the subject. If you think my visiting the mines can be of any use, I will cherrfully do so. There appears to me to be several modes of destroying the fire damp without danger; but the difficulty is to ascertain when it is present, without introducing lights which may inflame it. I have thought of two species of lights which have no power of inflaming the gas which is the cause of the fire-damp . . .[10]

A crucial point was indeed the detection of the damp, as well as a safe light when it was present. Gray put Davy in contact with John Buddle, 'viewer' of the colliery at Wallsend, near Newcastle, from which some of the best coal came, and on the 14 August Davy stopped in Newcastle on his way home and met Buddle and Hodgson.

Buddle's background was not unlike Davy's. His self-educated father, good at mathematics, had been schoolmaster at Kyo, near

Tanfield in County Duham. In 1781 he had been chosen as viewer, or manager, for the projected Wallsend pit, where, on his death in 1806, John succeeded him. John Buddle made many innovations in the ventilation of mines and in the methods of extraction of coal. Though over eight hundred men were said to have died in accidents in mines under his charge, he became both trusted by the miners and also expert in mining engineering. He was sufficient of a geologist to acquire a copy of a translation of papers from the French *Annales des Mines*,[11] edited by the future first Director of the Geological Survey, and to be External Examiner in Engineering at the infant University of Durham. He became known as the King of the Coal Trade. He was a warmly sympathetic person, in whom both practice and science were joined.

At first Davy seems to have hoped for some kind of low-temperature flame that would not ignite the damp, and some way of destroying the damp when it was present. Buddle reported.

> He made particular enquiries into the nature of the danger arising from the discharge of the inflammable gas in our mines. I shall supply him with a quantity of the gas to analyze; and he has given me reason to expect that a substitute may be found for the steel mill, which will not fire the gas. He seems also to think it possible to generate a gas, at a moderate expense, which, by mixing with the atmospheric current, will so far neutralize the gas, as to prevent it firing at the candles of the workmen.[12]

Davy was about to embark on the rather new process of going to the chemistry laboratory and studying the properties of gases and flames in the hope that these 'experiments of light' would lead, as Bacon had said, to 'experiments of fruit'. His success was a real triumph of applied science, making his tanning and agricultural work look no more than organized common sense. Unlike those researches, this one had the shock of surprise that comes with fundamental science. He hurried to London (missing a visit to Sir Joseph Banks at his country house, Revesby Abbey) for another of his autumn bursts of creativity.

Samples of the firedamp were sent by Buddle and by Hodgson, and Davy began to experiment with it, sometimes at considerable risk to himself. Exploding mixtures of gases was a long-standing game for chemists, used to setting off oxygen–hydrogen mixtures in eudiometers for example, and Davy had also worked on nitrogen trichloride and other dramatic substances. He took few safety precautions. He found that the firedamp was methane, the gas that he and Faraday had investigated in Italy and that, as 'natural gas', we use for cooking. He then made the crucial discovery that the mixture of methane and air is explosive only between certain limits, and will ignite only at a high temperature. On 30 October he sent a report to Gray:

> The fire-damp I find, by chemical analysis, to be (as it has been always supposed) a hydro-carbonate. It is a chemical combination of hydrogen gas and carbon, in the proportion of 4 by weight of hydrogen gas, and $11^{1}/_{2}$ of charcoal. I find it will not explode, if mixed with less then six times, or more than fourteen times its volume of atmospheric air. Air, when rendered impure by the combustion of a candle, but in which the candle will still burn, will not explode the gas from the mines; and when a lamp or candle is made to burn in a close vessel having apertures only above and below, an *explosive mixture* of gas admitted *merely enlarges* the light, and then gradually extinguishes it without explosion. Again, – the gas mixed in any proportion with common air, I have discovered, *will not explode* in a *small tube*, the diameter of which is less than $^{1}/_{8}$th of an inch, or even a larger tube, if there is a mechanical force urging the gas through this tube. Explosive mixtures of this gas with air require much stronger heat for their explosion than mixtures of common inflammable gas [ethylene, carbon monoxide]. Red-hot charcoal, made so as not to flame, if blown up by a mixture of mine gas and common air, does not explode it, but gives light in it; and iron, to cause the explosion of mixtures of this gas with air, must be made *white*-hot. The discovery of these curious and unexpected properties of the gas, leads to several practical methods of lighting the mines without any danger of explosion.[13]

The critical observation was that small enough tubes would prevent the flame in a lamp igniting firedamp, and Davy's first lamps had narrow ventilating tubes at the top and bottom. He described four kinds of lamp to Gray and promised to send models of them for testing.

He also wrote to Banks, who replied in a magnificent letter:

> My dear Sir Humphry
> Many thanks for your kind letter, which has given me unspeakable pleasure. Much as, by the more brilliant discoveries you have made, the reputation of the Royal Society has been exalted in the opinion of the scientific world, I am of the opinion that the solid and effective reputation of that body will be more advanced among our contemporaries of all ranks by your present discovery, than it has been by all the rest. To have come forward when called upon, because no one else could discover means of defending society from a tremendous scourge of humanity, and to have, by the application of enlightened philosophy, found the means of providing a certain precautionary measure effectual to guard mankind for the future against this alarming and increasing evil, cannot

> fail to recommend the discoverer to much public gratitude, and to place the Royal Society in a more popular point of view than all the abstruse discoveries beyond the understanding of unlearned people. I shall certainly direct your paper at the very first day of our meeting. We should have been happy to have seen you here; but I am still happier in the recollection of the excellent fruit which was ripened and perfected by the very means of my disappointment, your early return to London. I trust I shall arrive there on the morning of our first meeting, and that it will not be long after before I have the pleasure of seeing you.
> I am, my dear Sir H., always faithfully yours.[14]

Davy had intended his letter to Gray to be private, partly because it represented work still in progress rather than completed[15] and partly no doubt because he wanted his triumph to be reported first to the Royal Society, which was jealous in these matters. In the event, the news was first made public at a meeting in Newcastle on 3 November, and the formal paper presented to the Royal Society on the ninth;[16] for it Davy was awarded the Society's Rumford Medal, for work that represented the first fruits of modern science. His tone was modest: 'I was so fortunate as to discover some properties belonging to [fire damp], which appear to lead to very simple methods of lighting the mines, without danger to the miners, and which, I hope, will supply the desideratum so anxiously required by humanity.'[17]

He had worked out why narrow metal tubes were effective, noticing that if made of glass they did not prevent explosions:

> This phenomenon probably depends upon the heat lost during the explosion in contact with so great a cooling surface, which brings the temperature of the first portions exploded below that required for the firing of the other portions. Metal is a better conductor than glass, and it has already been shown that the fire-damp requires a very strong heat for its inflammation.[18]

He added that 'explosions likewise I found would not pass through very fine wire sieves or wire gauze', and some of the lamps illustrated do include gauze.[19] None of them look very like the typical Davy lamp or its descendants still in use today. Davy's suggestion was that 'it is only necessary to use air-tight lanterns, supplied with air by tubes or canals of small diameter, or from apertures covered with wire gauze placed below the flame, through which explosions cannot be communicated, and having a chimney at the upper part, on a similar system for carrying off the foul air';[20] common lanterns might easily be adapted for the purpose. By then he had already given up small tubes in favour of concentric 'canals' or gauze, though all the lamps but one shown have the flame enclosed in glass.

On 25 January 1816 Davy read a further paper to the Royal Society describing his experiments with metallic gauzes.[21] These included the now-familiar experiment of holding a gauze in a flame, which would not pass through it. He was also able to report that 'cylinder lamps have been tried in two of the most dangerous mines near Newcastle, with perfect success'.[22] The classic Davy lamp was described and illustrated in early 1816 in a new general scientific journal published for the Royal Institution, basking in its renewed fame, by John Murray, the most eminent publisher of the day. Davy wrote that his 'invention has the advantage of requiring no machinery, no philosophical knowledge to direct its use, and is made at a very cheap rate'.[23] He was delighted to report that, contrary to the expectations of those who undervalued science and expected that devices made in the laboratory would fail in real life, the lamps had proved a great success in the most dangerous pits in Newcastle and Whitehaven.

In June the same journal published a letter from Buddle, who wrote that the lamps had 'answered to my entire satisfaction' and that 'our colliers have adopted them with the greatest eagerness': 'In the practical application of the lamps, scarcely any difficulty has occurred. Those of the ordinary working size, when prepared with common cotton wick and the Greenland whale oil, burn during the collier's *shift*, or day's work of six hours, without requiring to be replenished.'[24] In mining, the long hours worked in factories were impossible in the cramped and fuggy conditions at the coal face, and, clearly, having to open up a lamp down the pit would be fatal.

> The only inconvenience experienced arises from the great quantity of dust, produced in some situations by working the coal, closing up the meshes of the wire-gauze, and obscuring the light; but the workmen very soon removed this inconvenience by the application of a small brush. We have frequently used the lamps where the explosive mixture was so high as to heat the wire-gauze red hot; but on examining a lamp which has been in constant use for three months, and occasionally subject to this degree of heat, I cannot perceive that the gauze cylinder of iron wire is at all impaired. I have not, however, thought it prudent, in our present state of experience, to persist in using the lamps under such circumstances, because I have observed, that in such situations the particles of coal dust, floating in the air, fire at the gas burning within the cylinder, and fly off in small luminous sparks.[25]

Coal dust was indeed to remain a problem in mines – especially where the viewer was less prudent than Buddle – and could give rise to explosions. Buddle also noted that the lamp was a wonderful detector of firedamp, for which purpose it continues to be used even in our day when mines have electric lights: 'Instead of creeping inch by inch with a candle, as is usual, along the galleries of a mine suspected to contain fire-damp, in order to ascertain its presence, we walk firmly on with the safe lamps, and with the utmost confidence prove the actual state of the mine.'[26] The appearance of the flame disclosed the presence of gas and showed how distant sources of firedamp might produce explosive mixtures at different places within the mine. The lamps were not only safe lights, but also scientific instruments.

Davy had (without Jane) visited Newcastle in March 1816, where he received a deputation from the coal owners, with a letter saying:

> They are most powerfully impressed with admiration and gratitude towards the splendid talents and brilliant acquirements that have achieved so important a discovery, unparalleled in the history of mining, and not surpassed by any discovery of the present day; and they hope, that whilst the tribute of applause and glory is showered down upon those who invent the weapons of destruction, that this great and unrivalled discovery for preserving the lives of our fellow-creatures, will be rewarded by some mark of national distinction and honour.[27]

It may surprise us to find such modern liberal sentiments at the end of the letter, but we have perforce again become aware of the role of science in weapons of destruction, and in 1816 honours of all kinds had been given to the officers of the army and navy that had defeated Napoleon. In the event, Davy was given a baronetcy in January 1819; since he had no children, this hereditary honour was an empty one and he seems to have made little of it. Nevertheless, it was then the highest honour ever conferred upon a man of science in Britain; Newton had been only a knight, and scientific peerages were far distant in the future.[28]

Davy refused to take out a patent for the lamp, magnanimously making it a gift to his country. He was duly rewarded by the coal owners with the gift of a set of plate, presented to him by Lambton, whom he had known with Beddoes. Davy had written that:

> as the Committee express themselves satisfied as to the utility of the Safety-lamp, I can only desire that their present, as it is highly honourable to me, should be likewise useful to my friends, and a small social circle, which it would be as a dinner-service for ten or twelve persons. I wish that even the plate from which I eat should awaken my remembrance of their liberality, and put me in mind of an event which marks one of the happiest periods in my life.[29]

On 13 September 1817 a dinner was held at the Queen's Hotel in Newcastle at which Lambton presented Davy with his set of plate, over £1500 having been collected towards it. 'Your brilliant genius,' he declared, 'which has been so long employed in an unparalleled manner, in extending the boundaries of chemical knowledge, never accomplished a higher object, nor obtained a nobler triumph.'[30] The absolute security of the lamp had been demonstrated in nearly two year's experience of its use.

Davy's modest reply was followed by cheers, 'drinking his health in three times three', at which he replied again, this time forcefully emphasizing the difference between science and mere rule of thumb:

> I am overwhelmed by these reiterated proofs of your approbation. You have overrated my merits. My success in your cause must be attributed to my having followed the path of experiment and induction discovered by philosophers who have preceded me: . . . it was in pursuing these methods of analogy and experiment, by which mystery had become science, that I was fortunately led to the invention of the Safety-lamp.[31]

The birth of modern science is generally credited to the seventeenth century, to Bacon, Descartes, Galileo, Kepler and their contemporaries, and its exemplar was astronomy; but it could be placed in the nineteenth century, with chemistry as the paradigm science and Davy as its greatest exponent. His great contemporaries, like Berzelius and Gay-Lussac, were, perhaps, chemists' chemists; with his work on the lamp he brought experimental science into the public gaze, indicating its power (in this case) for good.

However, attributing credit to Davy alone seemed to some to be unfair to Clanny, whose lamps had been first in the field and had been used in a few pits, and to George Stephenson, then unknown but later to be famous for his railway locomotives, who had invented what came to be called the Geordie lamp. Stephenson was a local man in the collieries, born in a little cottage (now preserved as a museum) beside the Tyne to the west of Newcastle, and with little education. His lamp was devised by trial and error, essentially, it seems, on the principle of restricting the access of air; and in October 1815 he came up with a lamp in which air was admitted through narrow tubes, later modified to a metal plate with perforations. Both Davy and Stephenson were self-made men whose reputation and fortune depended upon their discoveries, and who were thus sensitive about their intellectual property; but Davy was already famous and his heavyweight reaction might seem overdone.

Stephenson did not know explicitely the principle upon which his lamp depended. His achievement was, as his locomotives were to be, in the older tradition of technology, depending on intuition and common sense rather than abstruse science. Devotees of practical common sense, and those imbued with local patriotism, made a party of Stephensonians. Both sides accused the other of stealing ideas. It was then suggested that Davy's lamp depended upon his knowledge of unpublished work by Smithson Tennant,[32] who had found that explosive mixtures of gases will not ignite in narrow tubes. In his first paper on the lamp Davy had acknowledged that Wollaston had told him of these experiments, but only after he had done his own;[33] their context was the investigation in 1813 (when Davy was in France) of an explosion of a gas-holder in Westminster. This involved coal gas, mostly carbon monoxide, rather than methane. A committee of the Royal Society, including Banks, Wollaston, Hatchett and Brande, declared that:

> Sir Humphry Davy not only discovered, independently of all others, and without any knowledge of the unpublished experiments of the late Mr Tennant on flame, the principle of the non-communication of explosions through small apertures, but that he also has the sole merit of having first applied it to the very important purpose of a safety lamp, which has evidently been imitated in the latest lamps of Mr George Stephenson.[34]

Tennant was a man many of whose interests were closely parallel to those of Davy: he became Professor of Chemistry at Cambridge (a post which at the time did not involve many duties), worked on platinum metals in concert with Wollaston, was interested in iodine and in the pigments used in painting, and came to work on gas explosions. When Davy isolated potassium and sodium, Tennant wrote to a friend: 'I need not say how prodigious these discoveries are. *It is something to have lived to know them.*'[35] It should not surprise us that contemporaries sharing a world-view and investigating similar questions should come up with the same answers: the history of the sciences is full of simultaneous discoveries and priority disputes, and such things still go on. In the small world of Regency chemistry hints dropped in conversation might be half forgotten, but there seems no reason to dissent from the committee's judgement that Davy did not know of Tennant's work. Because all science is a matter of climbing on to the shoulders of giants, we would be foolish to expect complete originality in any case: discoveries always involve the previous work of others and there is no suggestion that Tennant ever thought of making a safety lamp.

We may regret that Davy regarded Stephenson as no more than a pirate, whose glass exploding-machine became a bit safer only as he adopted Davy's principles. But we can see the unpleasantness of Davy's critics in the contributions by one Mr John B. Longmire to Thomas Thomson's journal, *Annals of Philosophy*, in polemic, in which he was later joined by one J. H. H. Holmes. J. G. Children replied on Davy's behalf, only to be soundly abused for his pains. Longmire's first attack was based upon insinuations that the lamp was not really safe; his return to the attack, in company with Holmes (a Clanny supporter) includes some entertaining invective:

> Would Sir H. Davy be the *Grand Turk* of science? And has he arranged the media by which he would render into mutes, and strangle the efforts of, all who will not echo his opinions, or write in his praise? Sir H. Davy may succeed in such intentions when the King of Great Britain shall tremble at the Pope's Bull; or when Sir H. has brought his own wire-gauze lamp to perfection – but no sooner need he expect to succeed.[36]

Davy's angry domineering certainly later attracted unfavourable notice when he was President of the Royal Society, but hardly here, though all sorts of social tensions come out in the controversy. Davy might have secured a more peaceful life for himself had he patented his device, the philanthropist on his high moral plane being perhaps a particularly tempting target for a bit of muck, some of which sticks. The language of scientific periodicals in Davy's time was much more accessible than in ours, but also much more robust.

Thomson was no friend of the Royal Institution, and the same issue of his *Annals* also contained an article by Dr John Murray (not the publisher, but a Scottish chemist) – who had had a long controversy about chlorine with Davy's brother John – describing a lamp of his own devising for lighting coal mines.[37] His basic principle was that the explosive gas collects in the roof of the mine and that the lamp should therefore draw in air from floor level. It is to be hoped that miners got their ideas on lighting, and chemists their views on chlorine, from Davy rather than from Murray.

Although Davy's lamp was indeed safe, as Buddle and others demonstrated, if it was badly made or poorly maintained so that the gauze was broken, it could set off an explosion. There were also problems, as we saw, with coal dust in the air, and with strong draughts, which became a feature of the better ventilation of mines in succeeding years: in these circumstances lamps could set off explosions. Partly for these reasons, the number of disasters did not diminish in the years after the lamp was invented, though in proportion to the amount of coal won they did decrease. More important probably was the reluctance of owners and miners to bother with such safety precautions (the lamps giving a less good light than candles) in mines believed free of gas; they lived in a fool's paradise, from which they were jolted explosively in due course.

That *Annals* contained so much on safety lamps indicates the enormous interest aroused in what we may perhaps call the scientific community. Davy, having invented the device, went on to use it to investigate flames and gaseous combinations generally. This allowed him to infer something like a reaction mechanism. By now he could take his lamp as a standard piece of apparatus, in need of no further description or discussion. He noted that:

> When a wire-gauze safe lamp is made to burn in a very explosive mixture of coal gas and air, the light is feeble, and of a pale colour; whereas the flame of a current of coal gas burnt in the atmosphere, as is well known in the phenomena of the gas lights, is extremely brilliant . . . I have endeavoured to shew, that in all cases flame is a continued combustion of explosive mixtures; it became, therefore, a problem of some interest, 'Why the combustion of explosive mixtures, should produce such different appearances?'[38]

The critical thing seemed to be that outside the lamp there was a '*decomposition* of a part of the gas towards the interior of the flame where the air was in smallest quantity, and the deposition of solid charcoal, which, first by its *ignition*, and afterwards by its *combustion*, increased in high degree the intensity of the light'.[39] In the paler flame the heat was much greater.

Gas lighting, we should note, was a new development, in which Frederick Accum, who had worked at the Royal Institution in Davy's early days, was a pioneer.[40] Gas mantles were not yet invented, and the lights were smoky luminous flames, like those of a Bunsen burner with the air hole shut. As Davy saw it, and experiments with gauzes seemed to confirm his view, the reaction went in two stages when the gas was in a stream. He also remarked that his 'principle explains readily the appearances of the different parts of the flames of burning bodies',[41] something Faraday was later to enlarge on in his famous set of lectures on candle flames.[42] Davy investigated flames with chlorine instead of oxygen, and concluded that 'whenever a flame is remarkably brilliant and dense, it may be always concluded that some solid matter is produced in it', noting that 'the heat of flames may be actually diminished by increasing their light'.[43] He found that the flame from a mixture of oxygen and hydrogen under pressure in a blow-pipe apparatus was hardly visible yet so intensely heating that solid matters exposed to it gave light 'so vivid as to be painful to the eye'.[44] This was the principle of limelight and, later, of the incandescent mantle.

In the course of some of these experiments Davy had put platinum wires into the flames to see how hot they were. Platinum was a newly available metal, Wollaston having made his fortune and retired from doctoring after developing a process for getting it into malleable metallic form. It was expensive but not ruinously so, and any serious chemist had crucibles, wires and spatulas of this very refractory metal. It occurred to Davy that the temperature of flames was higher than was necessary for the ignition of solid bodies and that slow combustions might make a platinum wire glow. He found indeed that a fine platinum wire heated and put into cool explosive mixtures continued to glow red hot; if removed, it cooled down, but when put back into the mixture it instantly became red hot again. Very fine wires might become so hot that they ignited the mixture. Only platinum and palladium worked; copper, silver, iron, gold and zinc produced no effect. A slight coating prevented the effect. Davy introduced a little cage of platinum wire into a safe lamp in a receiver in which the proportions of coal gas and air could be varied. He found that as soon as any gas was present the platinum became ignited; when the concentration was such that the lamp flame went out, the 'platinum became white hot, and presented a most brilliant light'.[45]

Davy thought that lamps with such a cage would be useful in very fiery pits and duly sent some to be tried. A problem was that a wire of platinum happening to obtrude through the gauze could make the lamp unsafe, but he hoped that they would work in 'atmospheres in which no other permanent light can be produced by combustion': one would hope they would have been used only to assist in the evacuation of the miners. The practical value of this research lay far in the future, but what Davy had done was to investigate what was later called (by Berzelius) *catalysis;* in this case heterogeneous catalysis, because a solid is facilitating a chemical reaction in the gaseous phase. The catalyst (one of the rather few terms to come from technical chemistry into common use) initiates the reaction without itself being used up. Platinum catalysts are particularly valuable in chemistry, notably in cracking oil in refineries. Scientific investigation had led to a technological innovation and this had led to more science.

In October 1816 the Davys had gone to Bath because Jane was ill again. Davy wrote to Poole: 'Bath does not suit me much, nor should I remain here, but my wife has been indisposed, and the waters seem to benefit her, and promise to render her permanent service, and if that happens, I shall be pleased even with this uninteresting city.'[46] He went on to mention the safety lamp: 'It has pleased Providence to make me an instrument for preserving the lives of some of my fellow-creatures. You, I know, are of that complexion of mind that the civic crown will please you more than even the victor's laurel wreath.'[47] Writing over a year later, he hoped soon to show Poole the service of plate, and hoped to go to Stowey for woodcocks and to pursue an idea of buying an estate there. He added: 'I go on always labouring in my vocation. I am now at work on a subject almost as interesting as the last which I undertook. It is too much to hope for the same success; at least I will deserve it.'[48]

We can see what this project was from a letter to his mother, dated 25 May 1818 and curiously stilted:

> We are just going upon a very interesting journey. I am first to visit the coal miners of Flanders, who have sent me a very kind letter of invitation, and of thanks for saving their lives. We are then going to Austria, where I shall show Vienna to Lady Davy, and then visit the mines, and, lastly, before I return, we are going to Naples. I have the commands of his royal highness, the Prince Regent, to make experiments upon some very interesting ancient manuscripts, which I hope to unfold. I had yesterday the honour of an audience from his royal highness, and he commissioned me to pursue this object in the most gracious and kind manner. We shall be absent some months. With kindest love to my sisters and aunts, in which Lady Davy joins me, I am, my dear mother,
> Your most affectionate son,
> H. Davy.[49]

The Davy's harmonious relations with the Prince Regent (from 1820, King George IV) later cooled somewhat because in Italy they became friendly with his errant wife, Queen Caroline; but in 1818 he was warmly supportive of Davy's project to study Vesuvius and to use chemical techniques to unroll papyri found in Herculaneum. Davy believed that chlorine or iodine might do the trick and would not affect the writing, which was in carbon-based ink; small-scale experiments done in London were promising.

At first, all went well. The triumphal tour of the mining districts of Belgium was followed by travel along the Rhine, examining the formation of mist over water. Davy-concluded that this happens only when the water is warmer than the air. They visited Carniola, now Slovenia, which was to be his favourite Alpine retreat, on the way to Italy. There they had a happy time, looking at misty lakes and rivers, doing geology, fishing and being tourists. This time they could enjoy the Continent, free from wars and rumours of wars. In Rome they met and became friends, with Byron. Eventually they reached Naples, where letters from the Prince Regent and the Foreign Office smoothed Davy's way at first. Results looked promising, and on 25 February 1819 he wrote to his mother: 'We have been at Naples, and I have been perfectly successful in the object of my journey.'[50] The summer was spent in the Alps, fishing and delighting in the scenery; Davy belonged to the first generation of English intellectuals to love the Alps, in which he was followed by many men of science, including his successors at the Royal Institution, Faraday and John Tyndall.

Returning to Naples on 1 December 1819, Davy turned again to the experiments, but this time they went badly. Davy blamed the Italians at the museum for being no longer helpful but obstructive; he wrote up his work for the Royal Society in a handsomely illustrated paper that included a complaint about jealousy and obstacles.[51] He established that the papyri had not suffered through heat in the main, but through slow chemical changes that were turning them into something a bit like peat. A few fragments were separated and made legible with his chlorine process, but that was all.

On 13 March 1820 he wrote to his mother from Rome:

> I have written three times since I heard of your accident at Marazion, but I am doubtful, from Grace's last letter, whether you have received either of my letters. Her letter, like most human things, contained a mixture of good and bad; and the pleasure I received from hearing of your recovery was mixed with pain from the continuance of my aunt's illness ... I hope in the autumn we shall meet from different quarters of the world at Penzance. I have finished with success, and much sooner than I expected, the objects for which I came abroad. Lady Davy is not very well, and we are obliged to travel slowly, but I hope we shall be in London in the end of May or the beginning of June.[52]

In the event, Davy 'came over in consequence of the illness of Sir Joseph Banks, about the affairs of the Royal Society, rather before the time I intended', leaving Jane behind to follow at her own pace.[53] On 16 June he wrote that he had 'been very graciously received by the King;[54] on the nineteenth Banks died, having been President since 1778, and Wollaston took over as caretaker until 30 November, when Banks's successor would be elected.

Banks had been Davy's patron, and of him Davy wrote:

> He was a good-humoured and liberal man, free and various in conversational power, a tolerable botanist, and generally acquainted with natural history. He had not much reading, and no profound information. He was always ready to promote the objects of men of science; but he required to be regarded as a patron, and readily swallowed gross flattery. When he gave anecdotes of his voyages, he was very entertaining and unaffected. A courtier in character, he was a warm friend to a good

> King [George III]. In his relations to the Royal Society he was too personal, and made his house a circle too like a court.[55]

Banks was thus in Davy's eyes a gifted amateur; in the eyes of others he was an autocrat, who had tried to suppress specialized societies and had been unsympathetic to those working in the physical sciences. In later years his temper had been affected by the gout. His long reign made apparent the need for a new broom.

Davy was an obvious candidate, which was why he had hurried home. It was believed that Banks would have preferred Wollaston, a Cambridge man and a member of a distinguished intellectual family, but the safety lamp made Davy unstoppable if he chose to stand. He wrote to Poole in June:

> I regret very much that you could not join me at dinner this day. Tomorrow and the following day I shall be occupied by pressing affairs; but I shall be at home to-morrow till half-past eleven, and be most happy to see you. I am not very anxious to remove 'mists' for I feel that the President's chair, after Sir Joseph, will be no light matter; and unless there is a strong feeling in the majority of the body that I am the proper person, I shall not sacrifice my tranquillity for what cannot add to my reputation, though it my increase my power of being useful. I feel it a duty that I owe to the Society to offer myself; but if they do not feel that they want me, (and the most effective members, I believe, do) I shall not force myself upon them.
> I am, my dear Poole, very sincerely yours,
> H. Davy.[56]

This letter was prescient, but Davy decided to stand, Wollaston declined to make a contest of it and Davy's progress to the highest position in British science, as the nineteenth century's Newton, was assured. The first casualty was his relationship with Faraday, which we must next explore.

9

A Son in Science: Davy and Faraday

It seems that Davy's most famous pupil was Frankenstein.[1] The young Victor Frankenstein, after reading various alchemical writings, went to the University of Ingolstadt. Bored by natural philosophy, 'partly from curiosity and partly from idleness' he heard the attractive Professor Waldman lecture on chemistry:

> He concluded with a panegyric upon modern chemistry, the terms of which I shall never forget . . . *'The modern masters promise very little: they know that metals cannot be transmuted, and that the elixir of life is a chimera. But these philosophers . . . have acquired new and almost unlimited powers: they can command the thunders of heaven, mimic the earthquake, and even mock the invisible world with its own shadows.'*[2]

Frankenstein was duly moved to achieve even more, 'treading in the steps already marked'. He called upon Waldman, who welcomed him as a disciple.

Mary Wollstonecraft Shelley's father, William Godwin, knew Davy, who had visited their house when Mary was a young girl, and Waldman's lecture follows closely Davy's famous Inaugural Lecture of 1802, looking forward to the Bright Day, full of sexy rhetoric about the chemist:

> Not contented with what is found upon the surface of the earth, he has penetrated into her bosom, and has even searched the bottom of the ocean for the purpose of allaying the restlessness of his desires, or of extending and increasing his power . . . And who would not be ambitious of becoming acquainted with the most profound secrets of nature, of ascertaining her hidden operations, and of exhibiting to men that system of knowledge which relates so intimately to their own physical and moral constitution?[3]

Davy went on to compare the modern chemist, guided by the steady light of truth, with the deluded alchemist, adding that (in Galvanic electricity) 'a new influence has been discovered, which has enabled man to produce from combinations of dead matter effects which were formerly occasioned only by animal organs'.[4]

It was also a part of Davy's rhetoric that we could now expect to build upon foundations laid, and to proceed swiftly, but in steps rather than by speculative leaps. Shelley's book draws not only upon Davy, but also upon Captain Cook, whose 'ambition leads me not only farther than any other man has been before me, but as far as I think it possible to go';[5] but it ends with Victor defeated and dying, pursued by his nameless monster, hardly the modern Prometheus he had hoped to be. The fictional Waldman had the advantage over the real Davy that he taught at a university, if not exactly a research school, and therefore had students.

Although Davy and his generation were in at the beginning of new and powerful science, different in scale and excitement from the natural philosophy of the previous century, his pupils had to learn from a kind of informal apprenticeship. They had to be, in effect, what Berthollet called Gay-Lussac: his son in science. Berthollet really had a son, who committed suicide, recording notes of his asphyxiation until the pen dropped from his hand. Berthollet also taught in a modern institution, the École Polytechnique, so that Gay-Lussac was more pupil than apprentice. Gay-Lussac became a professor and Academician. In due course Berthollet's ceremonial sword came to him, as the mantle of Elijah came to Elisha, but as a good son he found it hard to disagree with Berthollet's scientific convictions, notably about chlorine. Perhaps he never fully escaped. Strong bonds between teacher and pupil became a feature of nineteenth-century chemistry. The childless Davy had four scientific sons: his cousin Edmund Davy, his brother John, his godson, James Tobin, and Michael Faraday. The important ones were John Davy and Faraday. All had the prolonged close intimate contact with Davy characteristic of apprentices and of sons.

It is often said that Faraday was Davy's greatest discovery, and, as often happens with discoveries, Davy did not see the full implications. This has much truth by in it, but the remark is curiously backhanded and can seem reductive. We should not really group potassium, chlorine, Faraday and the safety lamp together and ask which was the greatest. It is often used as part of a comparison of the two men, usually to Davy's disadvantage. Thus Bence Jones in his biography to Faraday wrote that Davy was hurt by success, that he had little self-control, method or order and that 'he gave Faraday every opportunity of studying the example which was set before him during the journey abroad, and during their constant intercourse in the laboratory of the Royal Institution, and Faraday has been known to say that the greatest of all his great advantages was that he had a model to teach him what he should avoid.'[6] This seems a catty remark, surprising in one whose contemporaries praised him as possessing all the virtues.

One great difference between the two men was that Faraday lacked the ambition of Davy or Cook. He believed that worldly ambition was wrong[7] and refused honours and official positions in learned societies; he knew his place, and those who did could be praised by their social superiors as nature's gentlemen. Davy, on the other hand, was ready to accept responsibilities, to believe that power ought to go with merit, and to look for greater usefulness. We may be queasy about ambition that goes beyond the urge to know, but it would be odd to condemn Davy for wanting to be President of the Royal Society and being pleased about his knighthood. In science, as elsewhere, saints are a bonus. Undoubtedly Faraday's life was happier than Davy's, but people are needed who are prepared to get their hands dirty in public life.

Faraday may have meant only that worldly success was corrupting, but the remark sounds more general; its tone may be understood as that of a son grumbling about an overbearing father. Bence Jones himself was uneasy about his simple view of Davy's relationship with Faraday, and later wrote in his book on the Royal Institution:

> Whenever a true comparison between these two Nobles of the Institution can be made, it will probably be seen that the genius of Davy has been hid by the perfection of Faraday. Incomparably superior as Faraday was in unselfishness, exactness and perseverance, and in many other respects also, yet certainly in originality and eloquence he was inferior to Davy, and in love of research he was by no means his superior. Davy, from his earliest energy to his latest feebleness, loved research, and notwithstanding his marriage, his temper, and his early death, he first gained for the Royal Institution that great reputation for original discovery which has been and is the foundation of its success.[8]

Our liking, and that of biographers, for comparison and for ranking means that Davy has too often been seen through the unflattering eyes of Faraday, who changed from hero-worshipper to sceptic, as sons are apt to do.

What Faraday needed to avoid is perhaps indicated in a chatty column in *Fraser's Magazine* of February 1836, which is devoted to him as number sixty-nine in a 'Gallery of literary characters'. Davy, wrote the columnist,

> that great and good man (so abominably caricatured by the ass Paris) rushed to the rescue of kindred genius [Faraday]. Sir Humphry immediately appointed him as assistant in the laboratory, and after two or three years had passed, he found Faraday qualified to act as his secretary . . . and he is now what Davy was when he first saw Davy . . . in all but *money* . . . The future Baronet is a very good little fellow . . . a Christian, though, we regret to add, a *Sandemanian* (whatever that may signify) – a Tory (as might have been inferred from Rat Lamb's [the Prime Minister, Lord Melbourne] hostility).[9]

Like many sons, Faraday did not want to walk exactly in his father's footsteps or to live in his shadow: he married a *hausfrau* rather than an heiress, and refused public honours. Nevertheless, he did work in the institution and the tradition established by Davy, who made his career possible both by taking him on and by making the Royal Institution a place where Faraday could live his public and private life in the way he needed, with research and lecturing.[10]

There was also a steely side to Faraday, as in any saint worthy of the name. John Tyndall, his admiring successor (and a famous agnostic) wrote of him:

> We have heard much of Faraday's gentleness, and sweetness and tenderness. It is all true, but it is very incomplete. You cannot resolve a powerful nature into these elements, and Faraday's character would have been less admirable then it was had it not embraced forces and tendencies to which the silky adjectives 'gentle' and 'tender' would by no means apply. Underneath his sweetness and gentleness was the heat of a volcano.[11]

As Tyndall knew him, Faraday was not slow to anger, but was highly self-disciplined. He showed a curious mixture, or perhaps we should say compound, of humility and proper pride.

While it is fun to try to get inside the psyches of the dead and distinguished, and that indeed is part of the duty of the biographer, there is more reward in looking at the context in which Davy and Faraday inosculated, to their mutual pleasure and pain. In the early nineteenth century there was no obvious way into science as there was into the great professions, the church, the law and medicine. Entry into science, as into the oldest profession, was uncontrolled, and unless one had wealth or could practise another profession, one ran the risk of prostituting one's talents, devoting too much time to low-grade activities in the effort to make ends meet. There have been some dynasties in science, like the Wollastons, the Becquerels and the Huxleys, but it is less of a hereditary activity than the old professions.

By the late nineteenth century it had become something like a profession: there were degree courses, graduate schools and specialized learned societies open only to graduates. Posts in higher education and in some industries were available only to the suitably qualified. But in Davy and Faraday's day, science worked by patronage. Davy's meteoric career to the professorship at the Royal Institution was made possible through Gregory Watt, Davies Giddy, Beddoes, Rumford and others. Beddoes was, perhaps, the nearest to a father in science, but from the start Davy's brilliance was apparent, and Beddoes was converted to his protégé's theory of heat and light at the outset of their relationship. Davy's move to London after the publication of his book on the oxides of nitrogen brought him out of Beddoes' shadow, if indeed he had ever been in it.

Davy saw in Beddoes an example of what to avoid: in this case, the dissipation of energy. On Beddoes death, he wrote to Coleridge on 27 December 1808:

> Alas! poor Beddoes is dead! He died on Christmas eve. He wrote me two letters, on two successive days, 22nd and 23rd. From the first, which was full of affection and new feeling, I anticipated his state. He is gone at the moment when his mind was purified and exalted for noble affections and great works. My heart is heavy. I would talk to you of your own plans, which I endeavour in every way to promote; I would talk to you of my own labours, which have been incessant since I saw you, and not without result; but I am interrupted by very melancholy feelings, which, when you see this, I know you will partake of.
> Ever, my dear Coleridge, Very affectionately yours,
> H. Davy.[12]

Davy was then thirty and world-famous; nobody would ever have thought of comparing him with Beddoes. Beddoes and Coleridge were people with whose hard times he sympathized. Faraday's position at thirty, in 1821, was very different.

Faraday's background was not unlike Davy's, but urban: his father was a blacksmith who moved from Westmorland to London just before Michael was born.[13] Bence Jones, in the first edition of his biography, wrote of the family receiving a handout of bread each week in the hard times of 1801 – correcting this soon after: 'I was too easily led into this error by my wish to show the height of the rise of Faraday by contrasting it with the lowliness of his starting point.'[14] Faraday's biographers, like Davy's, emphasized his social mobility, but he was from the respectable poor, who did not have to rely upon charity. This was, no doubt, why he later found beggars so offensive: he did not think, as some recent commentators would have had him do, like a student of the class of 1968;[15] but nor did Davy. They were prickly men who had made their way, with a strong sense of what was their intellectual property.

Davy's career reached its height when he filled the niche vacated by Sir Joseph Banks, a landed grandee and Oxford graduate: he was thus presiding over his betters in a highly stratified society. Not everybody likes being governed by their social inferiors, however brilliant, and Davy's reputation was to suffer from the judgements *de haut en bas* of contemporaries, the 'gentlemen of science' who came to occupy positions of power by the 1830s.[16] In such company Davy and Faraday were both in danger of being fish out of water; but Faraday knew where social mobility had better stop.

Science, like everything else in the Regency period, worked by patronage. Faraday's method of obtaining it by writing to a great man was not unique. Nor was his first choice: Banks. Banks had helped a number of young men, notably sending them as naturalists on voyages and then finding them jobs at Kew and other botanic gardens round the world. To young men of Faraday's station, Banks could never have been a father: a patron, a fairy godfather perhaps, putting up with grumbles on occasion, but never anything like an equal. Patronage is a hit-and-miss business. On applying to Banks with his letter, Faraday was rebuffed: no answer. At much the same time Gay-Lussac's associate Thenard threw out an under-age laboratory assistant called Boussingault, who went to South America.[17] Returning ten years later with a reputation as a scientific explorer, Boussingault became an eminent agricultural chemist and won the support of Thenard. That story had a happy ending, but Thenard's help was almost too late, like Lord Chesterfield's patronage of Johnson. It is a great gift to be able to spot youthful talent in unexpected places.

As Tyndall put it, 'Davy was helpful to the young man, and this should never be forgotten.'[18] Even in Faraday's time it was unusual to learn chemistry entirely through apprenticeship or pupillage; most chemists had had some kind of medical training. It was also unusual to train with the most brilliant chemist of the day. Berzelius took in pupils one at a time, but they had had some training and were not with him very long; he was more like a supervisor for a higher degree. Faraday was lucky. He broke into what had been Davy's system for helping friends and relations.

Edmund Davy was born in Penzance in 1785. In January 1807 Davy was 'obliged to dismiss George Farrant from the Royal Institution, in consequence of idleness and general neglect of duty – I may say, very general ill conduct'.[19] On the twelfth Davy recommended Edmund to the managers of the Royal Institution as 'a young man of good conduct and some promise . . . and willing to devote himself wholly to the business of the Institution at a salary of £60 per annum and the usual advantages attached to a residence in the House'.[20] Edmund was duly appointed; a room on the Attic Floor was found for him, and a bedstead and hangings ordered. Six months later he got a better room; and in January 1809, on Davy's proposal, his salary was raised to £70 in view of his new duties in looking after and showing the mineral collection, some of it collected by Davy on field trips and holidays. It was his duty to look after the laboratory. This was not an easy job: on 13 September 1809 Davy wrote a memorandum.

> Objects much wanted in the laboratory of the Royal Institution.
>
> Cleanliness
> Neatness
> &
> Regularity
>
> The laboratory must be cleaned every morning when operations are going on before ten o'clock.
> It is the business of Wm. Payne to do this and it is the duty of Mr. E. Davy to see that it is done & to take care of and keep in order the apparatus.
> There must be in the laboratory, *pen, Ink & Paper & wafers* and these must not be kept in the slovenly manner in which they usually are kept. I am now writing with a pen & ink such as was never used in any other place.
> There are wanting, small graduated glass tubes blown here & measured to ten grains of mercury.
> There are wanting four new stop cocks fitted to our Air pump.
> There are wanting 12 green glass *retorts*.
> There are wanting, most of the common metallic and saline Solutions, such as – Acetate of copper, Nitrate of *silver*, Nitrate of Barytes. Most of these made in the laboratory.
> All the wine glasses should be cleaned.
> And as operations cease at 6.0 Clock in the evening, there is plenty of time for getting things in order before night, but if they are not got into order the same night they must be by ten O Clock the next day.
> The laboratory is constantly in a state of dirt & confusion.

> There must be a *roller* with a coarse towel for washing the hands – & a basin of water, & soap & every week at least, a whole morning must be devoted to the inspection & ordering of the Voltaic Battery.[21]

Edmund must have managed to meet these standards, more or less, for his salary was raised to £100 and he continued there until the end of 1812, when he resigned:[22] changes following Humphry's marriage and departure were unacceptable to him. Humphry had built up enthusiasm for chemistry in Ireland, and Edmund was to benefit from it; after six years at the Royal Institution, and in his late twenties, he launched out on his own. In 1813 he was appointed Professor at the Royal Cork Institution, and subsequently at the Royal Dublin Society.

John Davy was born in 1790, the youngest child in the family and eleven years younger than Humphry.[23] Their father had died when John was very young, and Humphry had supervised his education: writing to his old headmaster:

> The little boy who brings you this letter is my brother. It is my desire & it is my mother's desire that he should become your pupil. I fear his mind at present is in a very uncultivated state; but he seems to possess sensibility; which I have been accustomed to consider as the foundation of all power and activity. – Under your tuition at all events, he must be improved & if he is not capable of becoming learned, he will at least become virtuous. His disposition at present is good; but his labours are irregular. These habits you will easily correct without pain & at a future period he will thank you – and if indeed the benefits he may derive from you are at all analogous to those his brother has derived, he will never forget his instructor.[24]

On 24 October 1808 at the Royal Institution 'Mr Davy stated that his Brother, Mr John Davy, is coming to London, and requested permission of the Managers that he might reside in the House of the Institution, and that he might attend the Lectures and the Laboratory. *Resolved*, that this request be complied with.'[25] By now Davy had determined that his brother should have the formal training in medicine that he had never had, and had formed the plan that John should spend a few years at the Royal Institution before going to the flourishing University of Edinburgh, where the medical course was the best in Britain. Unlike Edmund, John was not on the payroll of the Institution.

John Davy's time at the Royal Institution, he used to say, was one of the happiest and best-employed periods of his life. In 1810 he went to Edinburgh. He graduated MD in 1814, upon which he was elected a Fellow of the Royal Society and joined the Royal Institution as a Subscriber. Obtaining a higher degree is an important rite of passage, crossing a Rubicon, and marks the end of anything like apprenticeship. In 1815 he joined the Army and had a very successful career, ending up as Inspector General of Army Hospitals and retiring to Ambleside in the Lake District after various tours of duty abroad.[26] He was an able man, publishing numerous papers on anatomy and physiology, some later collected in book form,[27] and travel books. He made no astonishing discoveries; he did well in his profession, and was a well-trained and efficient practitioner of what is nowadays called normal science.

His obituarist wrote: 'It is much to the credit of Dr Davy's moral nature that no shadow of mortification or jealousy ever darkened his meditations on his brother's achievements, into comparison with which he was so constantly forced to bring his own.'[28] He also remarked more darkly on 'the great reputation which, in spite of all efforts to the contrary, has settled round the name of Sir Humphry Davy',[29] and attributed the protection of Davy's reputation in large part to John. Blood is thicker than water, and readers of *The Selfish Gene* will know that it is as rational to help a brother as a son.[30] John Davy was a distinguished scientist, genial, energetic and methodical, but, like most scientists, he lacked that volcanic quality necessary to the great discoverer or revolutionary: he was not driven on by a demon.

The Davy brothers were both keen fishermen, and Humphry seems to have really relaxed only with rod (or sometimes gun) in hand; his closest friends were his fishing companions. Faraday, although he went along on fishing trips (and loaded Davy's gun for him too), was not really interested.[31] We learn from George Barnard that 'Faraday did not fish at all during these country trips, but just rambled about geologising or botanising'.[32] He lacked a key to Davy's heart.

John Davy and Faraday died within six months of each other and nearly forty years after Humphry, so it is important to remember that they did not belong to a different generation, being only eleven and twelve years younger than he was. In different ways they behaved like sons. It was the good son, John, who rushed to attend Humphry in his last illness in Italy and Switzerland. He may have been tempted to see Faraday as the bad son, but that misses various interesting ambiguities. Faraday wrote that 'these polemics of the scientific world are very unfortunate things; they form the great stain to which the beautiful edifice of scientific truth is subject. *Are they inevitable?* They surely cannot belong to science itself, but to something in our fallen natures.'[33] And again, on family rows: 'We may well regret such incidents. It is not that they are not to be expected, for they belong to our nature, but they ought to be repented.'[34] One of the last acts of his life was to subscribe 'most liberally to the fund for raising a monument to Sir H. Davy at Penzance'.[35]

In 1805 Faraday went as an apprentice to the fatherly bookbinder George Ribeau, and in 1810 his father died. In 1812, at the age of twenty-one, he finished his term and went to work with Henri De la Roche, a 'very passionate man', who gave him a great deal of trouble, but who said: 'I have no child, and if you will stay with me you shall have all I have when I am gone.'[36] Faraday's choice of Davy for father instead meant that he got a position rather than possessions. Although John was in Edinburgh, Davy does not seem to have felt any need for a successor.

During his apprenticeship Faraday had read some of the books he was binding, and had resolved upon self-improvement. With other young men, he went to science lectures; and one of Ribeau's customers, William Dance, invited him to come to what turned out to be Davy's last four lectures at the Royal Institution in March and April 1812. Faraday made a fair copy of his notes and presented them to Davy with a covering letter. Apparently Davy showed it to one of his close friends, saying:

> 'Pepys, what am I *to do*, here is a letter from a young man named Faraday; he has been attending my lectures and wants me to give him employment at the Royal Institution, *what can I do?' 'Do'*, replied Pepys, 'put him to wash bottles; if he is good for anything, he will do it directly; if he refuses, he is good for nothing.' 'No, no,' replied Davy, 'we must try him with something better then that.'[37]

In the event, Davy (by now, Sir Humphry) saw Faraday, told him that science was a harsh mistress in the pecuniary point of view, smiled at his notion of the superior moral feelings of scientific men and advised him to 'attend to the book binding'.[38] Publishers did not issue cased, 'hardback', books until around 1830: books came in paper wrappers, often with cardboard front and back, and most owners had them bound; it was a flourishing trade.

Faraday nevertheless persisted in his wish to enter science. He may have acted briefly as amanuensis when Davy was disabled by an explosion of nitrogen trichloride in October 1812, but his big chance came when on 19 February 1813 Payne, whose duty we saw was to clean up the laboratory, hit Newman, the instrument maker, and was fired on the twenty-second. Davy at once got in touch with Faraday – the message arriving at bedtime – and on 1 March he was appointed, the Managers hearing from Davy that 'his habits seem good, his disposition active and cheerful, and his manner intelligent'.[39] The cheerfulness was to be tested, but the activity and intelligence were never to be in doubt. Henceforward neat copies of Davy's writings were prepared by Faraday.

There is a fascinating source for insight into Faraday's time with Davy beyond the letters and journals he wrote, and that is Faraday's copy of Paris's biography of Davy, a handsome large quarto, for which he was one of the informants. Into his interleaved copy Faraday bound a whole series of manuscripts. These include a doodle, remarks on space and time, a geological sketch, drafts of scientific papers and letters from Davy. The volume is an extraordinary monument, which cannot be read without emotion.[40] A manuscript on iodine is annotated by Faraday on January 1832: '*An original paper by Sir H. Davy*. It was my business to copy these papers. Sir H. Davy was in the habit of destroying the originals but on my begging to have them he allowed it and I now have two volumes[41] of such manuscripts in his handwriting.'[42] The iodine MS is on paper of different sizes, but shows Davy in excellent form, reasoning from analogy and no prey to the sterile virtue of tidiness.

Faraday went to France with the Davys as assistant and amanuensis, but also found himself acting as valet, and, as the French tell us, nobody is a hero to his valet. Nevertheless, although like John, Faraday came rapidly to loathe Jane, his respect for Humphry remained profound. In the letter in which he grumbled about her, he wrote about iodine: 'The French chemists were not aware ⟨of⟩ the importance of the subject until it was shewn to them and now they are in haste to reap all the honours attached to it but their haste opposes their aim they reason theoretically without demonstrating experimentally, and errors are the results [.]'[43] He was never to lose this love and respect for experiment. Much later, the eminent chemist J. B. Dumas wrote: 'we admired Davy, we loved Faraday',[44] and this may have been true at the time.

Faraday had never been more than twelve miles from London and was clearly homesick on these travels: he would have sympathized with Davy's poem about London, dated 1814, celebrating (prematurely) victory over Napoleon:

> Such art thou! mighty in thy power and pride;
> No city of the earth with thee can vie;
> Along thy streets still flows the unceasing tide
> Of busy thousands. E'en thy misty sky
> Breathes life and motion, and the subject waves,
> That wash thy lofty arches, bear the wings
> Of earthly commerce, where the winds, thy slaves,
> Speed the rich tribute to the ocean kings. –

Thy graves and temples filled with mighty dead
Are awful things.
Here in the dust the noble and the proud,
The conquerors of nature and of man, –
Those for whom Fame her clarion sounded loud,
Who triumphed o'er the ocean, earth, and air,
All now are found beneath a few carved stones, –
Conquerors and sages, deep beneath the sod. –
Shall future mightier piles e'er hide such bones
As these high worthies were? Allied to God,
Gifted with noble hopes and aspirations, –
And perfecting their will, – and rising high,
(The wonder and the blessing of the nations,)
To the true source of immortality,
Showing a virtue which can never die.[45]

Faraday, certainly, and Davy, possibly, were relieved when the project of going via Greece (still under Turkish rule) to Constantinople was abandoned and the party returned hastily home.

Faraday was at first without a job on their return, but Davy lost little time in getting him reappointed at the Royal Institution.[46] He assisted with the safety lamp, declaring it in a lecture to be: 'the result of experimental deduction. It originated in no accident, nor was it forwarded by any, but was the consequence of regular scientific investigation ... an instance for Bacon's spirit to behold. Every philosopher must view it as a mark of subjection set by science in the strongest holds of nature.'[47]

His copy of Paris's biography contains manuscripts connected with the lamp, and in his published papers Davy acknowledged Faraday's assistance: such candour was and is not always found in eminent men of science, and we can see Davy bringing forward a promising student. In 1816 Davy gave him an analysis to do on his own:

> Sir Humphry Davy gave me the analysis to make as a first attempt to make in chemistry at a time when my fear was greater than my confidence, and both far greater than my knowledge; at a time also when I had no thought of ever writing an original paper on science. The addition of his own comments and the publication of the paper encouraged me to go on making, from time to time, other slight communications . . .[48]

Faraday later acknowledged Sergeant Anderson's assistance, but he never encouraged him to publish a paper; Davy was treating Faraday as more than a mere assistant. In 1817 Davy recommended 'a gentleman' to take private tuition from Faraday, in the manner again of a supervisor giving an opportunity to a research student. Faraday wrote to his old friend Ben Abbott: 'I am engaged to give him lessons in Mineralogy & Chemistry thrice a week in the evening for a few Months . . . Our lessons do not commence till 8 o'clk and as my gentleman is in the immediate neighbourhood I am at liberty till that hour.'[49] Faraday was evidently teaching him at his house.

When the Davys went abroad again, Faraday was left to hold the fort, and letters bound into the *Life* show how things developed. On 5 June 1818 Davy wrote from Frankfurt; the letter has no beginning and ends formally: 'I am very sincerely yours, H. Davy'.[50] Humphry's letters to John, by contrast, always began 'My dear John', and ended in a variety of ways, including 'Tell me what you are doing and what you wish, and command me as your affectionate friend, and love me as your very affectionate brother' and the more usual 'Pray . . . believe me to be, My dear John, Your affectionate Brother and Friend, H. Davy'.[51]

The tone to Faraday began to change during the trip. Writing from Venice, Davy ended in what became his customary form, 'I am dear Mr Faraday Very sincerely your friend & wellwisher H. Davy'.[52] From Rome he told Faraday that 'Mr. Hatchetts letter contained praises of ⟨y⟩ou which were very gratifying to me & pr⟨ay⟩ believe me there is no one more interested in your success and welfare'.[53] It is always delightful to get independent confirmation of a student's capacities, and Charles Hatchett's opinion, like that of an external examiner, helped change the relationship again from that of supervisor and student to one of senior and junior colleague. Now Davy asked: 'Pray when you write me give me any scientific news – for neither journals nor transactions reach this place', and groused to Faraday about his successor as professor, W. T. Brande, who had published without authority what Davy had written about the Herculaneum MS: Davy was 'more vexed . . . than I can well express' and hoped he would not again be treated 'with so little ceremony', but asked that it be taken no further.[54] He added: 'It gives me great pleasure to hear that you are comfortable at the Royal Institution & I trust you will not only do something good and honorable for yourself; but likewise for Science.'[55] Now that Faraday was in his confidence, Davy could, by November 1819, discuss chemical theory as with an equal, in a letter that began 'Dear Mr Faraday':

> Be so good as to say to Mr Brande that I have received his letter & that I do not consider myself as *adopting* the opinion of Berzelius in regarding silica as acting the part of an acid. I think Mr Smithson is the first person who made this statement & long before Berzelius I pointed out

> the analogies between the silica & Boracic bases – Whatever perfectly neutralizes an alkali may be regarded as possessing the opposite chemical powers; but the term *acid* applies to physical properties such as sour &c & I certainly never meant to attribute any acid qualities to silica but merely to express that it acted the part of an acid in producing solution of an oxide &c.[56]

In the care he devoted to terminology Faraday again was to show that he had learned from Davy.

Then Davy came home and was elected President of the Royal Society. Great and impossible hopes of him were entertained. Newton had not always been amiable and straightforward in his dealings in that post, and nor was the would-be Newton of chemistry. He knew what he ought to do, as we know from the poem about eagles teaching their young to fly:

> Their memory left a type, and a desire;
> So should I wish towards the light to rise,
> Instructing younger spirits to aspire
> Where I could never reach amidst the skies,
> And joy below to see them lifted higher,
> Seeking the light of purest glory's prize.[57]

That was written in August 1821, and that year was to be an *annus mirabilis* for Faraday. He became a full member of the exclusive and fundamentalist Sandemanian Church and he got married. Davy wrote to him, in a letter bound in the biography, 'I hope you will continue quite well & do much during the summer & I wish your new state all that happiness which I am sure you deserve.' With Davy's backing, he was promoted to Superintendant of the House at the Royal Institution, and his first paper to be published by the Royal Society appeared in the *Philosophical Transactions*. He became thirty. It should have been evident that his period of tutelage was past. In 1810 Davy had written a memorandum for his assistants:

> No experiments are to be made or carried on in the laboratory without the consent and approbation of the Professor of Chemistry and the attempt at original experiments unless preceded by knowledge merely interferes with the process of discovery. There is a sufficient number of new and interesting objects which a modest student would wish to pursue and in which the path is marked and distinct.[58]

In *Consolations* Davy also ascribed a modest role to an assistant, who should have 'no preconceived ideas of his own'.[59]

Although Faraday was no longer just 'Chemical Assistant' at the Royal Institution, Davy did not find it easy to remember his status; he still expected to read over drafts of Faraday's papers and add explanatory notes to them in a manner that had been welcome in 1816 but was no longer. Faraday liquefied chlorine following a suggestion from Davy that heating it under pressure might be interesting,[60] it rankled with him that Davy, like a modern professor probably would, took much of the credit for the outcome. During Wollaston's brief tenure of the presidency of the Royal Society the Council (which included Davy) awarded the Copley Medal to H. C. Oersted for his discovery of electromagnetism. Davy and Wollaston were long-standing friends (Davy had made a fisherman of him) and rivals in electrochemistry, and had to be careful and correct in their dealings. Wollaston hoped to make a wire rotate on its axis when it carried a current in a magnetic field; he tried the experiment with Davy at the Royal Institution, with no result. In September 1821 Faraday produced electromagnetic rotation. There was a feeling that he had muscled in on Wollaston's territory (then and now staking claims is a major feature of science) and had not acknowledged Wollaston's work. He wrote an agonized letter to Wollaston, who responded coolly, and Davy seemed to sympathize more with Wollaston than with Faraday.[61] The young are always well advised to butter up very eminent elders who have made vaguely helpful suggestions, but Faraday was unworldly: he expected his credit to go where it was due.

In 1823 Faraday had his name go forward for the Royal Society.[62] The nomination was organized without Davy's involvement by Richard Phillips, but Davy's friends Babington and Children as well as Wollaston were signatories. The President did not by convention propose candidates, but Banks had expected to be consulted and had been free with his opinions. To Banks, and Davy, such vetting had seemed essential if unworthy candidates were to be excluded, and Davy felt affronted at what looked like a deliberate challenge to his authority. He dared not appear weak and acquiescent, especially as Faraday was his protégé. We need not assume extreme jealousy on his part; it is enough that the two protagonists were socially insecure and highly emotional. It should have been obvious that embarrassment was to be expected: the Royal Institution was felt by some to be too powerful an interest within the Royal Society, and Davy was trying to reduce admissions so as to increase the scientific character of the Society. This policy was not popular and he could not be involved in favouritism. On tactical grounds he was against Faraday's nomination because he would otherwise be assumed to be behind it. Faraday was nevertheless elected; Davy was the loser, as those who cut off their children always are. In the end he had betrayed a person for what he saw as the sake of an institution and his own position within it.

The last item in Faraday's copy of the *Life* is a draft of a note by Davy. In the handwriting of both Davy and Faraday (who drafted it), this clarified the position vis-à-vis Wollaston. Then comes Faraday's manuscript chronology of these events. Then nothing. Faraday wrote in 1835:

> I was by no means in the same relation as to scientific communication with Sir Humphry Davy after I became a Fellow of the Royal Society as before that period, but whenever I have ventured to follow in the path which Sir Humphry has trod, I have done so with respect and the highest admiration of his talents.[63]

On this chilly note, and with various ripostes to John Davy over the years, the relationship ended. The business does not vindicate the confidence in the 'superior moral feelings of scientific men' that Faraday expressed to Davy when they first met, but Tyndall's later analysis seems very convincing:

> A father is not always wise enough to see that his son has ceased to be a boy, and estrangement on this account is not rare ... It is now hard to avoid magnifying this error. But had Faraday died or ceased to work at this time, or had his subsequent life been devoted to money-getting, instead of to research, would anybody now dream of ascribing jealousy to Davy? Assuredly not. His reputation at this time was almost without a parallel: his glory was without a cloud.[64]

He adds that they were both proud men, but that, while very different, 'in one great particular they agreed. Each of them could have turned his science to immense commercial profit, but neither of them did so ... I commend them to the reverence which great gifts greatly exercised ought to inspire.'[65] Davy unfortunately at the critical moment gave way to 'irritation and anger',[66] but not to anything more base. Cantor suggests that Faraday was plagued by guilt at having put himself foward in an un-Sandemanian manner.[67]

Davy continued to be Faraday's patron, getting him promoted to Director of the Laboratory at the Royal Institution in 1825, and making him Secretary of the newly formed intellectual club, The Athenaeum, of which he was a founder; all these things were in Davy's interest as much as Faraday's. But Faraday perhaps learned to avoid the sort of entanglement and opportunity he himself had had with Davy. There is an anecdote recorded in 1872: 'Dr. Stenhouse says Faraday was selfish and narrow-minded. That a man once went to him, as he himself had gone to Davy, and that F. sent the young man to Graham, of which incident Graham made a standing joke.'[68] When the young Thomas Huxley went to see Faraday, he received courtesy but no fathering. Faraday was assisted by Anderson, who always did exactly as he was told. Faraday wrote that he could never work, as some professors did, through students, and he made it a rule never to write testimonials.[69] Although Tyndall was in many ways his heir, Faraday was never really close to him. He was a proud and humble and intensely private man – not unlike Davy. Sylvanus Thomson found magnanimity in Faraday's later attitude to Davy;[70] most of us might deplore its absence on both sides, regret the end of a story that had involved generosity and affection, and mark how inarticulate two of the greatest orators of science were in an emotional crisis. The clash of two temperaments led to no offer of compromise or truce.

Faraday's privacy was intimately connected with his happy, though childless, marriage and his Sandemanian membership. His faith deepened through the 1820s, and in 1832 he became a deacon in his church. Davy was by no means irreligious, but whereas Faraday's faith was deeply biblical and personal, Davy's owed more to natural religion, was undogmatic and ecumenical, absorbed in the great problems of death, freedom and immortality, and closer to what Victorians were to call theism. We shall return to this later, but end with a poem of about 1815:

The massy pillars of the earth,
The inert rocks, the solid stones,
Which give no power, no motion birth,
Which are to Nature lifeless bones,

Change slowly; but their dust remains,
And every atom, measured, weigh'd,
Is whirl'd by blasts along the plains,
Or in the fertile furrow laid.

The drops that from the transient shower
Fall in the noonday bright and clear,
Or kindle beauty in the flower,
Or waken freshness in the air:

Nothing is lost; the etherial fire,
Which from the farthest star descends,
Through the immensity of space
Its course by worlds attracted bends,

To reach the earth; the eternal laws
Preserve one glorious wise design;
Order amidst confusion flows,
And all the system is divine.

If *matter* cannot be destroy'd,
The *living mind* can *never* die;
If e'en creative when alloy'd,
How sure its immortality!

Then think that intellectual light
Thou loved'st on earth is burning still,
Its lustre purer and more bright,
Obscured no more by mortal will.

All things most glorious on the earth,
Tho' transient and short-lived they seem,
Have yet a source of heavenly birth
Immortal – not a fleeting dream.

The lovely changeful light of even,
The fading gleams of morning skies,
The evanescent tints of heaven,
From the eternal sun arise.[71]

10

President

The President of the Royal Society occupies a unique position in Britain even today. It is the oldest surviving scientific society in the world, after all, and it enjoys prestige and access to government funds that other societies cannot match. While it has some of the form of a club, it is now very comparable with the Academies of Sciences that exist in most countries outside the English-speaking world: Fellows are now expected to be immensely distinguished in order to be eligible, so that election marks a climax in a career and usually comes in middle age. The Society is run by and for scientists, though there is provision for electing a few members distinguished in public life. The president and other officers serve for a limited term.

In 1820 almost everything was very different. There was the Linnean Society, for natural history, and the recently founded Geological Society, but otherwise the Royal Society was the only national scientific body. Now there are presidents of numerous specialized societies and of the peripatetic British Association for the Advancement of Science to be spokespersons for science; then the President of the Royal Society and the Astronomer Royal (who embodied the scientific civil service) were the outstanding pair, and the former was in most spheres much the more important. He was consulted officially or unofficially on all sorts of topics; even before being actually elected, Davy was, for example, asked to advise on the appointment of a director of the Ordnance Survey. The position was one of great power and patronage: though there were no regular government grants to disburse, applications were made from time to time for funds for scientific expeditions or for experimental devices, such as Charles Babbage's 'calculating engines', the clockwork ancestors of computers.

There was no fixed limit to the reign of the president; though he came up for election each year, serious challenges to an incumbent were not really to be expected. Newton had been president for over twenty-three years, and Banks for over forty-one. In choosing Davy at the age of forty-one the Fellows were running the risk that he might well go on for over twenty years, and perhaps for forty. Newton and Banks had also been autocratic in the way they perceived their role within the Society, keeping down any who disagreed with them and not brooking rivals or challenges. The Society was sometimes turbulent: in 1783, after five years in office, Banks was faced with a formidable revolution led by mathematicians;[1] Davy was not alone in finding it hard to remain popular.

A good deal of administration was required; it has been written of Banks that 'few honest applicants ever appealed to him for help in vain [Faraday was one], and he was always ready to give the benefit of his experiences and statesmanship to great and small, from the King to deportees from his beloved Lincolnshire'.[2] It has been estimated that Banks wrote at least 50,000 letters and probably about 100,000, about fifty a week over the active period of his life:

> Banks never shirked his duties as a correspondent: he frequently replied at length to letters that seem to merit no answer, and he nearly always replied promptly to the many letters he received from strangers, adventurers and nonentities of all kinds, and often responded handsomely to the frequent appeals that were made to him for financial and other aid. In the hundreds of Banks' letters that have come down to us, we almost always find him courteous, polite and sympathetic . . . The numerous official, scientific, civic and philanthropic bodies with which he was associated brought him into contact with the active elements in almost every branch of human activity and with men of the most diverse views and opinions . . . Thus from his little room in Soho Square, Banks constantly performed single-handed duties which ministers and officials, with office organizations, secretaries and clerks to assist them, could not have so effectively performed.[3]

To play the part of Banks would not have been easy, especially as he had brought to it the easy confidence of the landed aristocrat.

In 1820 the Royal Society was essentially a club for those interested in the natural sciences.[4] There were dining clubs associated with it; and during the London season, from November round to the early summer, its meetings were one of the attractions of the metropolis for such folk, especially from among the gentry, the professions and the armed services. The meetings cannot have been very exciting affairs, since there were no questions or debates following the reading of papers, but the Society published the most prestigious journal in Britain, the *Philosophical Transactions*, which Davy, as a Secretary, had for a time edited. Would-be Fellows were proposed and seconded, after consultation with Banks; other Fellows might also sign the nomination paper, which was publicly displayed; and after a fixed period the candidate was voted on, and, as a rule, elected. There was no need to be eminent in science, though, increasingly, when proposing those who had published, their backers did refer to this. Davy tried to make it compulsory.

The Society was therefore large, and the subscriptions paid for the handsome publications of the scientifically active minority: papers in the *Philosophical Transactions* were well printed on excellent paper in an impressive quarto format, with beautifully engraved copper-plate illustrations. The Society was run by a Council, elected each 30 November; in Banks's time those who had published papers were in a minority on this body as well as in the Society as a whole. Almost all the Fellows were amateurs in the sense that their income did not come from their science, and by 1820 there was some feeling that the Society itself was amateurish.

Davy disliked France and seems to have avoided it on his various travels after 1813, but he and his contemporaries could not but be aware that Paris was the centre of matters scientific. They were aware that academicians at the Paris Academy of Sciences (briefly dissolved and then refounded as the First Class of the Institute in the Revolution) were salaried and were appointed to fill dead men's shoes; there were set numbers of chemists, mathematicians and so on, and the body was, in effect, a self-perpetuating oligarchy, though under government control. At first, it had the ideal of collective and utilitarian research, but it had become much more individualistic.[5]

There were other institutions in Paris admired by Britons, notably the École Polytechnique, a military school with emphasis upon research and teaching;[6] and the Museum associated with the botanical garden and the zoo, where Jussieu, Lamarck, Cuvier and others worked and taught. There seemed to be a career structure open to men of science in France, with a range of salaried positions. Associated with them was honour and respect; titles and positions seemed to go to chemists and physicists in gratifying recognition of their merits. The grass is always greener on the other side of the fence, and Frenchmen saw the frustrations of cliques, of pluralists accumulating positions and of

Parisian dominance; but when, after Waterloo, the military threat from France was no more, Britons began to see it as a more modern country than their own.[7]

In Paris the men of science governed themselves; in London the Royal Society was run for them. Power was in the hands of men of high social rank, who became the patrons of those, like Davy, whose extraordinary abilities they spotted. Pressure for reform and the ending of Old Corruption began to build up again. In the 1790s this had been kept down by the authorities, and the long wars had meant little change for a generation; by the 1820s the times were becoming ripe for change.[8]

The English universities – and England, like Aberdeen, had two – had begun to reform themselves at the turn of the century. At Oxford this meant that a humanistic course based upon Classics and philosophy (called 'Greats') became the norm; degrees in mathematics could also be taken, but they enjoyed much less prestige. Following the lead of Oriel College, the other colleges began awarding fellowships on the basis of merit. These developments squeezed out the science professors, whose lectures were optional extras in universities, which were part seminaries for the Church of England, part finishing schools for young gentlemen who probably did not actually take a degree. In Cambridge, mathematics became the prestigious subject; and because it included the applied (or 'mixed') branch, some of what we would call physics was a part of the course. After 1815 a group including John Herschel, William Whewell, Babbage and George Peacock succeeded in modernizing the syllabus to take account of the advances in mathematical analysis made by the French.[9]

Cambridge men like these formed a network, which has been called the 'gentlemen of science', that hoped to make science a self-governing enterprise in which power and patronage lay with those eminent for their research. The Royal Society had no laboratory of its own, but London had always been a centre of scientific activity, and the Royal Institution in Davy's time had become the greatest centre of research there. It had an aristocratic and fashionable feeling about it; like the Royal Society, it depended upon subscriptions from members who were passive scientifically to underwrite the research of its few professors. There was also the British Museum, which held antiquities and a great library as well as collections in natural history. The President of the Royal Society was *ex officio* a Trustee, and Banks had been assiduous. He had also played a large part in the transformation of Kew Gardens into a great centre for botany, and both these places had posts for zoologists and botanists. The Royal College of Surgeons had a fine museum of specimens, which included much comparative anatomy. It had begun as the teaching collection made by John Hunter, and was named after him: in Davy's time Everard Home was its curator.

For all these groups and institutions the President of the Royal Society was the representative, spokesman and figurehead. The hopes felt on Davy's elevation were unrealistic. Banks had taken up office in 1778 in the apparently secure days of the *ancien régime*; he had kept mathematicians in their place, promoted natural history and seemed to the young to be dictatorial and out of date. Many hoped for a transformation into something like the French had, with dilettantes excluded, but others felt that such a society would lose all its influence and importance; and the gentry interested in science liked things as they were. Men living in the provinces, like John Dalton, saw little in Banks' Royal Society for them: they wanted something less metropolitan and more democratic.[10] Davy was between hostile camps, unable to establish a consensus.[11] No president could have pleased everybody, but in the event he satisfied none of these constituencies.

Davy saw himself as a modernizer and reformer, but Fellows suspicious of the Royal Institution and fashionable popularizing perceived him as snobbish. He saw the Athenaeum as an intellectual club, taking over that function from the Royal Society, and sought to ensure that new Fellows of the Society were already distinguished in science, thus restricting the membership and slowly bringing about a change. During his presidency the Council came to have a majority of men on it who had published a scientific paper. With Sir Stamford Raffles, he was a founder of the London Zoo, hoping that it would be a centre for the acclimatization of foreign species that might flourish in Britain. Flocks or herds of ostriches and zebras did not come to grace this green and pleasant land, but London could at last match Paris in having a zoo with an associated scientific society rather than just an animal show.

In 1827 Davy published the addresses he had delivered to the Royal Society as President. There was a fuss about who bore the costs,[12] but Davy dedicated the work to the Council as 'intended to communicate general views . . . not minute information' and 'to endeavour to keep alive the spirit of philosophical inquiry and the love of scientific glory'.[13] In 1820 he had welcomed 'the separate and independent bodies [which] have arisen for registering observations and collecting facts, each in a different department . . . I trust that, with these new societies we shall always preserve the most amicable relations, and that we shall mutually assist each other.'[14] Davy was presiding over the dissolution of Banks's learned empire, as specialization became the order of the day. He hoped for a kind of commonwealth, where the new societies would feel 'respect and affection for the Royal Society . . . and co-operate in perfect harmony, for one great object, which, from its nature, ought to be a bond of union and peace'.[15] He referred to various current scientific researches, including the scientific voyages to the Canadian arctic, and concluded with an appeal:

> Gentlemen, the Society has a right to expect from those amongst its Fellows, gifted with adequate talents, who have not yet laboured for science, some proofs of their zeal in promoting its progress; and it will always consider the success of those who have already been contributors to our volumes, as a pledge of future labours. For myself, I can only say, that I shall be most happy to give in any way assistance, either by advice or experiments, in promoting the progress of discovery. And though your good opinion has, as it were, honoured me with a rank similar to that of a general, I shall be always happy to act as a private soldier in the ranks of science. Let us then labour together, and steadily endeavour to gain what are perhaps the noblest objects of ambition – acquisitions which may be useful to our fellow-creatures. Let it not be said, that, at a period when our empire was at its highest pitch of greatness, the sciences began to decline; let us rather hope that posterity will find, in the Philosophical Transactions of our days, proofs that we were not unworthy of the times in which we lived.[16]

Later annual discourses included brief obituaries of Fellows who had died, and addresses on the award of medals. One of Davy's successes, through Robert Peel, were the Royal Medals, endowed by King George IV, increasing the marks of honour and respect available to men of science. After it had been used by his wife and brother, Davy's presentation plate was, in accordance with his will, given to the Royal Society to endow another medal.

Our view of science, owing much to people like Davy, as the work of solitary geniuses, makes us tend to underrate those who administer it; and just as the best popes have not been saints, so the best presidents have not been geniuses. Important offices leave little time for foot-soldiering in science, and Davy's attempt to delegate his research to Faraday did not work out. As a general, Davy continued Banks's practice of holding scientific parties at his house every week during the season; he had hoped to invite women, but the Council vetoed this. Banks had held them on Sunday; Davy, perhaps in deference to sabbatarian Dissenters among the Fellows, changed the day to Saturday. These parties were paid for by Davy. After 1826, when they were held in the Royal Society's rooms, the bill for the food was still paid by the president; it was not an office for a poor man.

He also set up committees. One, which reported in 1823, was to look at the statutes of the Society and to recommend how its scientific character might be enhanced; it recommended reducing the intake of members. It was at this point that Faraday was nominated, but so were others much less qualified in terms of scientific reputation, and Davy's attempts to control admission through prior consultation were a complete failure. Other committees were concerned with practical questions; Faraday's work on optical glass and on steel was in response to two of these. Davy hoped strongly to build up a tradition of applied science beneath the auspices of the Royal Society and Royal Institution, with elements of collective discussion and individual experimentation on the French model.

One investigation he took upon himself from 1823 to 1825, in which Faraday, who assisted, was first exposed to electrochemistry. Since 1761 ships of the Royal Navy had been covered with copper plates below the water-line to protect them from attack by shipworms, but the process was expensive and it appeared that the copper bottoms dissolved slowly in sea water. Davy applied his knowledge of electrochemistry: he had found that when negatively charged, or in contact with a more positive metal, metals become more inert chemically. He duly fixed lumps of zinc or iron to copper sheets in tanks of sea water in the laboratory and the copper was preserved from attack. This principle is now called 'cathodic protection' and is involved in galvanized iron and in tin plate, but in Davy's case it failed to work in the complex world of ships at sea. The copper was protected, but, as a result, organisms adhered to it so firmly as to slow the ship's sailing significantly. Davy was much ridiculed as a result and his programme of applied science suffered.

During the researches the experiments were gradually scaled up, from plates to model ships and then to boats in Portsmouth harbour. In June 1824 the steamer HMS *Comet* was sent to Norway for survey purposes, and Davy sailed on her to experiment with protectors at sea. The ship encountered a gale that washed away the protectors, and Davy did some scientific tourism and some fishing in Scandinavia while new protectors were made and tried. The steamer caused some astonishment – people thought it had caught fire. Davy briefly visited Berzelius and Oersted, who had in 1820 made the astonishing discovery of electromagnetism, and the astronomers Olbers and Gauss. On the rough voyages he suffered from seasickness, writing 'the sea is a glorious dominion, but a wretched habitation'.[17] He decided philanthropically not to patent his invention.

Experiments on ships began in 1824 with every hope of success, but were abandoned in 1828. Davy was much mortified by the consequent criticism and ridicule, writing to Children: 'A mind of much sensibility might be disgusted, and one might be tempted to say, Why should I labour for public objects, merely to meet abuse? – I am irritated by them more than I ought to be; but I am getting wiser every day – recollecting Galileo, and the times when philosophers and public benefactors were burnt for their services.'[18] Applied science is a difficult business, and we have become much more familiar with promising ideas that fail in their first applications. Davy's friends felt that he was too much mortified by this failure: 'that while he became insensible to the voice of praise, every nerve was jarred by the slightest note of disapprobation'.[19] Public life demands a thick skin, and perhaps more in the early nineteenth century than at some other times. But it seems that in 1823 Davy had the first symptoms of the disease that was to kill him, and that 'the change of character which many ascribed to the mortification of wounded pride, ought in some measure to be referred to a declining state of bodily power, which had brought with it its usual infirmities of petulence and despondency.'[20]

As general in the army of science (it is not clear who they were fighting – rhetoric in science and medicine tends to be military) patronage came Davy's way. Here, as politicians soon come to know, you cannot please everybody, but in 1821 when William Leach had to retire from his post as Keeper of Zoology at the British Museum, Davy's conduct in the choice of a successor was noisily criticized. One candidate was William Swainson.[21] He was one of the great illustrators of the day, a pioneer in the new technique of lithography, which was to transform works of natural history; it was both cheaper and less inhibiting than copper-plate engraving. He had worked with Leach and felt himself to be well qualified for the post, but it went to Davy's friend Children, who needed a job because his banking business had failed. Swainson's correspondence is preserved at the Linnean Society, and from it we can see how one could earn a living from natural history and also how he reacted to not being appointed. Davy was not the only man of science to display anger; indeed, it is a pretty fair assumption that in nineteenth-century England leading men in any field would be barely on speaking terms; egotism was rife and great rows were commonplace.

In fact, Swainson was an eccentric candidate: he became an enthusiast for the 'quinarian' system of classifying organisms, and when after Davy's death he tried again for a post at the Museum, he did not get it, upon which he humphed himself off to New Zealand. But Children was not a strong candidate either: he was rather reluctant to be considered, because he was a chemist and mineralogist, not a zoologist. Government ministers are still appointed on the basis that they will soon learn about their department on the job; then science professors at Oxford, Cambridge and Yale were chosen on the same basis, and often did very well, and so in the event did Children. He had the advantage of being a gentleman, the social equal of the Trustees, who were wont to put men of humbler background (like Swainson) firmly in their place.

Nevertheless, the business was damaging because Swainson told his Liverpool friend Thomas Traill what he believed had happened, and Traill publicly assailed the British Museum as a centre of corrupt jobbery. Swainson believed that the Archbishop of Canterbury, another Trustee, had been responsible for his rejection; Davy did not put him right. The accusations discredited the Museum, which Banks by his labours and legacies had done much to promote as a great centre of natural history, and a consequence was that specimens collected on voyages, even those, like that of HMS *Beagle*, under government sponsorship, did not finish up in the national collection. Patronage ill-exercised had weakened an institution that Davy had especially meant to strengthen.

Davy courted the Cambridge network, making Herschel one of the secretaries of the Society and welcoming the Astronomical Society when it was founded. But in 1826, when the other secretaryship was vacant on the resignation of Brande, Davy again chose Children to fill it. This infuriated Babbage, who had hoped for the post, and his circle, but Babbage was notoriously irascible. From Davy's perspective, to have Cambridge mathematicians filling both secretaryships would have looked very ill-balanced, and although Cambridge men felt let down by this choice, it would seem that no sensible president would have opted for Babbage at that point, if indeed at all. In 1827 Davy was followed by Gilbert (Banks's preferred candidate in 1820); in 1830, Babbage was the campaign manager for Herschel in his bid for the presidency against the Duke of Sussex, one of Queen Victoria's wicked uncles (reformed in old age and genuinely interested in science), and the Duke won. The changes that both Davy and the Cambridge network wanted did come about, but slowly, and we should be careful not to see Davy's reign through the impatient eyes of firebrand reformers. Davy was a Whig, favouring evolutionary rather than revolutionary change in society and in the Society; but to some noisy Fellows his course seemed indirect and peevish.

Summing up, John Davy wrote of the presidency:

> As an honorary situation, wihout profit or emolument of any kind, but occasioning considerable expense to the individual, a stranger to the nature of its duties would suppose the office of President of the Royal Society, for a man of science, not only the most elevated but the most agreeable possible. It undoubtedly should be so; but it never can be so, as long as pretension to knowledge, vanity, and presumption, are more common (and they will always be more intrusive) than real knowledge, modesty, and diffidence. The pleasures of office, and especially of honorary office, are generally in anticipation and imaginary – the trials and troubles, real and incessant. These are the rocks and glaciers, the storms and torrents of the Alpine heights; the other, the rosy hues of reflected light, lost on near approach, – to be seen only in the distance, at which all asperities are invisible.[22]

We should not however see the whole period as one of gloom for Davy, although by 1824 he was beginning to feel his age and noticing the greyness of things:

> Whatever burns consumes, – ashes remain;
> And tho' in beauty and in loveliness,
> And infinite variety of forms,
> The primitive beings shone, their relics sad
> Have the same pale and melancholy hue.
> Such are the traits strong passions leave behind,
> Consumers of the mind and of the form.
> The auburn, flaxen, and the ebon hair,
> Take the same hoary hue; the blooming cheek
> Of beauty, the bronzed brow of manly strength,
> And the smooth front of wisdom, sadly show
> The same deep furrows: intellect alone
> Does not so quickly waste itself; but like
> The tranquil light which in the ocean springs,
> When living myriads in succession quick
> Sport on the wave, it lives, and in the storms
> And change of things appears more beautiful,
> Triumphant o'er the elements.[23]

Nevertheless, he did enjoy his holidays, now usually taken separately from Jane, going to his favourite Ireland and to Scotland, fishing and geologizing; to Cornwall, where in Penzance a civic banquet was laid on for him; and to Scandinavia. He seems to have avoided public events connected with George IV, setting out from London one day before the Coronation, 'that splendid ceremony, which all the world was crowding to see' (and Queen Caroline was disturbing), and leaving

Edinburgh just before the King got there: 'Scotland is all in commotion. I dined with Sir W. Scott the day before I left Edinburgh, who is, in fact, master of the royal revels; and I was very much amused to see the deep interest he took in the tailors, plumassiers, and show dressmakers, who are preparing this grand display of Scotch costume.'[24] His Majesty appeared in a kilt, which must have been a sight for sore eyes, but by then Davy was fishing.

In 1821 in reflective mood he entered in a notebook:

> It is now eleven years since I have written anything in this book; I take it up again, February 17, 1821. I have gained much since that period, and have lost something; yet I am thankful to Infinite Wisdom for blessings and benefits; and I bow with reverence beneath his chastisements, which have been always in mercy. May every year make me better, – more useful, – less selfish, – and more devoted to the cause of humanity and science![25]

He also managed to get some research done, notably stimulated by Oersted's work, of which he heard a vague report in October 1820. He investigated electromagnetism, estimated the conducting power of various metals at various temperatures, and did some experiments on passing electricity through evacuated tubes. All these were projects carried much further subsequently by Faraday; Davy did not manage in this area to experiment as tellingly as his erstwhile disciple was to do, or to quantify so successfully, but his pioneering work was interesting.

In 1826 he gave another Bakerian Lecture to the Royal Society, reverting to electrochemistry: he referred back to his work of twenty years before and described the researches on protectors as well as other recent experiments. This might seem like an old man summing up his life work (though Davy was under fifty), but even when he believed himself dying, in Italy in 1829, he was experimenting, trying to get enough current from an electric fish (the torpedo) to perform electrolysis. He also did some experiments on crystals that seemed to prove igneous action, confirming 'plutonism' as against those geologists ('Neptunists') who attributed everything to the action of water. It would therefore be wrong to think of his giving up research, though no doubt the interruptions and vexations of presidential life made it more difficult, and he did nothing as exciting as the work on potassium, chlorine or the lamp.

In 1825 he wrote a little poem about death, with which he was constantly trying to come to terms.

And when the light of life is flying,
  And darkness round us seems to close,
Nought do we truly know of dying,
  Save sinking into deep repose.

And as in sweetest, soundest slumber,
  The mind enjoys its happiest dreams,
And in the stillest night we number
  Thousands of worlds in starlight beams;

So may we hope the undying spirit,
  In quitting its decaying form,
Breaks forth new glory to inherit,
  As lightning from the gloomy storm.[26]

In the following year his health deteriorated[27] and his plan to visit Cornwall in October was forestalled by his mother's death in September. This was a great blow. It is hard to know exactly what was wrong with him. Diseases are in part real entities and in part social constructions, and medical theory and taxonomy in the early nineteenth century were very different from ours. We can identify mumps from a clinical description in the Hippocratic writings, but it is impossible to be confident about more complex illnesses like Davy's, Faraday's or Darwin's. His first two biographers were both medical men, but that does not help us much. His own father had died young, but his mother and brother both lived to a ripe old age. It looks as though he had heart trouble and then a stroke.

In 1826 he was re-elected President:

> When he delivered that discourse which was his last to the Royal Society, at the anniversary meeting on St. Andrew's day, 1826, it was done with such effort that drops of sweat flowed down his countenance, and those near him were apprehansive of his having an apoplectic seizure; and he was so much indisposed after, that he was unable to attend the dinner of the Society.[28]

John returning to England in December after four years abroad, found him well but stouter, complaining of numbness in his hand and arm, occasional 'inordinate action of the heart',[29] and weakness in the right leg. He had been recommended to follow a strengthening diet of 'animal food' – meat three or four times a day – rather than an abstemious regimen; this accorded with 'the convivial epicurean habits of London society',[30] but did not make him well.

On 10 December Davy wrote to Knight, regretfully turning down an invitation:

> I continue very much indisposed, and I am advised by my physicians [Babington and Holland] to try a journey to the south of Europe, which I trust I shall attempt. I suffer from almost constant pain in the region of the heart and chest . . . I have been twice at the British Museum, but I despair of anything being done there for natural history. The Trustees think of nothing but the arts, and money is only obtained for those objects. A thousand thanks for your kind invitation.[31]

Not long after, John was summoned to his bedside, and 'found him much worse than I expected, and labouring under a paralytic attack, affecting the right side. It had come on suddenly while shooting at Lord Gage's.'[32] Medical treatment seemed to be no real help: 'as he gained strength, however, the symptoms gradually diminished, and we were very sanguine that he would recover completely.'[33]

On 22 January, in company with John, he set out for the Continent, to recover 'in the absence of the many annoyances and causes of injurious excitement to which he was exposed at home',[34] and in the warmer climate of Italy. They had a cold and wet journey through France, aiming to travel forty or fifty miles a day in a post-chaise accompanied by postilion and courier. A bottle of leeches in the chaise was frozen all the way to Reggio. They crossed the Alps in snowstorms and then

> my brother accepted the offer of two stout mountaineers, and was drawn by them down to Susa in a small Alpine sledge with safety and rapidity, though in a manner not very agreeable, at least for an invalid, owing to the great steepness of many of the descents, and the heat from the sun in the very clear sky, and the reflection from the snow.[35]

By 20 February they were at Ravenna, 'lodged in the Apostolical palace, by the kindness of the Vice-Legate of Ravenna, a most amiable and enlightened prelate, who has done everything for me that could have been done for a brother'.[36] John was happy enough about Humphry's progress to return to work in Corfu, then under British rule, leaving him taking rides in the pine forest, reading Byron's poetry and enjoying the company of the Legate, Spada, whom he wrongly supposed to be a cleric and to whom he bequeathed the snuffbox that had come as a present from the Tsar. In order to avoid excessive heat, he was to go to the Alps in the summer. He brooded upon the transience of things, writing a number of poems, for example:

> Our life is like a cloudy sky 'midst mountains,
> When in the blast the watery vapours float.
> Now gleams of light pass o'er the lovely hills,
> And make the purple heath and russet bracken

Seem lovelier, and the grass of brighter green;
And now a giant shadow hides them all.
And thus it is, that in all *earthly* distance
On which the sight can fix, still fear and hope,
Gloom and alternate sunshine, each succeeds.
So of another and an unknown land
We see the radiance of the clouds reflected,
Which is the future life beyond the grave![37]

He also observed natural history, tried some electrical experiments on frogs and torpedos, and dissected fish he caught. He went to his beloved Julian Alps and on into Austria, but his health did not really improve. His eyes gave him trouble, and so he wore green spectacles and gave up his daily glass of wine.

On 30 June 1827 he wrote to Jane from Salzburg about the Royal Society:

> If I had perfectly recovered, I know not what I should have done with respect to the P., under the auspices of a new and more enlightened government [Liverpool was no longer Prime Minister]; but my state of health makes the resignation *absolutely* necessary. To attempt business this year would be to prepare for another attack. The vessels of the brain very slowly recover their tone; the slightest causes, even the writing of this letter, sends blood into them, and it is only by the strictest discipline, physical and mental, that I can hope to recover. I shall give no opinion to the Fellows of the Society respecting my successor, but I am exceedingly pleased with the idea of Mr. Peel becoming P.R.S. He has wealth and influence, and has no scientific glory to awaken jealousy, and may be useful by his parliamentary talents to men of science. The prosperity of the Royal Society will always be very dear to me, and there is no period of my life to which I look back with more real satisfaction than the six years of labour for the interests of that body. I never *was*, and never could be, unpopular with the active and leading members, as six unanimous elections proved; but because I did not choose the Society to be a tool of Mr. —'s journal jobs, and resisted the admission of improper members, I had some enemies, who were listened to and encouraged from Lady —'s chair. I shall not name them, but as Lord Byron has said, 'my curse shall be forgiveness'.[38]

At the same time he wrote to Gilbert, his patron and now Vice-President, soon to be his successor, describing his symptoms in more graphic detail and giving the same message:

> though I have had no new attack, and have regained to a certain extent the use of my limbs, yet the tendency of the system to accumulate blood in the head still continues, and I am obliged to counteract it by a most rigid vegetable diet, and by frequent bleedings with leeches and

> blisterings, which of course keep me very low. From my youth up to last year, I had suffered more or less from a slight hemorrhoidal affection; and the fulness of the vessels, there only a slight inconvenience, becomes a serious and dangerous evil in the head, to which it seems to have been transferred. I am far from despairing of an ultimate recovery; but it must be a work of time . . .[39]

Half a century later the Alps would have been full of professors, but Davy was a pioneer in this respect. Unlike Tyndall and most of us, he does not seem even when vigorous to have especially wanted to get to the tops; he admired the views from the summits of passes, and his great pleasure was to find the source of rivers, jotting down a 'thought which occurred to me last night: – 1, the river, like human life, has its origin from *infinity* (that of air), and is lost in immensity (that of ocean)'.[40] But he found that 'my spirits cannot bear this constant solitude, where there is no amusement and no books'.[41] In letters to Jane he rejoiced in her recovery: 'with health and the society of London, which you are so well fitted to ornament and enjoy, your 'viva la felicitá' is much more than any hope belonging to me'.[42] He was sometimes gloomy, and asked her to come and spend the winter in Italy:

> Your society would undoubtedly be a very great resource to me, but I am so very well aware of my own present unfitness for society, that I would not have you risk the chance of an uncomfortable moment on my account. I often read Lord Byron's *Euthanasia*; it is the only case, probably, where my feelings perfectly coincide with what his were. Though *solitary*, I am not *sad*, and now my eyes are better I can amuse myself with writing, and there are some thoughts and opinions which I wish to connect in composition, that I believe are original.[43]

She did not come, and at the end of September he resolved to return to England, writing to her:

> I think you will find me altered in many many things – with a heart still alive to value and reply to kindness, and a disposition to recur to the brighter moments of my existence of fifteen years ago, and with a feeling that though a burnt-out flame can never be rekindled, a smothered one may be. God bless you! From your affectionate, H. Davy. I hope it is a good omen that my paper by accident is *coleur de rose*.[44]

His constitution shattered, forced into early retirement, without much expectation that Jane would fan the flame, Davy returned to London, to find himself having to console Knight on the accidental death of his son, for whom Davy had felt great affection too. But he had found himself a future: writing dialogues to communicate his love of nature and of science.

11

Salmonia

Davy later described his return to London, 'the grand theatre of intellectual activity', which in youth and in the prime of manhood he 'never entered' without feelings of pleasure and hope':

> I now entered this great city in a very different tone of mind, one of settled melancholy, not merely produced by the mournful event which recalled me to my country, but owing likewise to an entire change in the condition of my physical, moral, and intellectual being. My health was gone, my ambition was satisfied, I was no longer excited by the desire of distinction; what I regarded most tenderly was in the grave ... my cup of life was no longer sparkling, sweet, and effervescent; – it had lost its sweetness without losing its power, and it had become bitter.[1]

Peter Medawar tells us that 'no working scientist ever thinks of himself as old, and so long as health, the rules of retirement, and fortune allow him to continue with research, he enjoys the young scientist's privilege of feeling himself born anew every morning'.[2] This does not seem quite true of Davy; nevertheless, he set down to work on some dialogues concerned with fishing, hoping to write a *Compleat Angler* for fly-fishers. He set the dialogues beside various streams and lakes where he had himself passed happy days. There are nine of them. Cuvier described Davy as a dying Plato, and certainly he used interlocutors with Greek names in dialogues organized into an ennead.[3] It is and was not quite clear what sort of book *Salmonia* is; it was clearly a new departure for Davy, surprising in a sick man. Illness and retirement enabled him to organize thoughts jotted down over his whole lifetime.

Salmonia appeared anonymously in 1828, but its authorship seems to have been a very open secret and its contemporary popularity is perhaps surprising. An unkind critic wrote that it seemed 'a patch-work composed of shreds of anniversary speeches before the Royal Society, articles in Philosophical Journals, and lectures on Natural History to Mechanical Institutions'.[4] We might perhaps then expect that when Mind was on the March, a sugared pill of popular science, containing information, edification and entertainment, by the most eminent natural philosopher of the day would be in demand. But, in fact, the book is very different from the *Discourses* given to the Royal Society, which all relate to particular occasions, seem to have lost money for the Society and must soon have been interesting only to historians of scientific institutions.

The critic was John Wilson, who was a major contributor to the obstreperous and fiercely Tory *Blackwood's*. He noted that in angling Davy was more than an amateur; that, indeed, he did it well, for a gentleman. Davy's tone is gentlemanly throughout,[5] whereas Walton had written of coarse fishing, a plebeian sport.[6] He took it for granted, or hoped to bring it about, that science will interest men of leisure in search of hobbies and subjects of conversation. That was the spirit in which he had recruited Roderick Murchison to science in 1823.[7] Murchison went on to become a most eminent geologist and geographer. And it looks as if Wilson was wrong: the dialogues of *Salmonia* may be not too far from actual conversations on moors and beside streams. Both fly fishing and shooting were impeccably gentlemanly pursuits; and we are reminded that in the 1820s Davy looked for an estate to purchase, to establish himself amongst the landed: not actually an easy thing to do for a *nouveau riche*, as Wedgwood experience showed, and he may have been wise not to try.

In his *eloge* of Davy, Cuvier referred to *Salmonia* and remarked that the many curious observations it contained would make it always an important work in ichthyology; but he added that science must regret that so powerful a genius had had to pursue such distractions and forget chemistry in trying to save his tottering health.[8] Such insignificant occupations as fishing were necessary to one whose accelerated career had prematurely brought on the infirmities of age. While Cuvier should have been a good authority on what was important in ichthyology, he was perhaps being generous; in William Yarrell's standard work on British fish of 1836 *Salmonia* hardly featured at all, though there is a reference to Davy's 'good history' of the grayling.[9]

Davy seems to have read the standard writings of Bloch and of Lacepède on fishes, who were the great authorities until superseded by Cuvier and Louis Agassiz, and when in 1824 there was a Select Committee of Parliament set up to consider the salmon fisheries of the United Kingdom, Davy delivered a brief paper on 8 May, which was published as an appendix to the report.[10] He made three proposals: that more fish should be allowed to spawn, of all ages and sizes; that fish should not be killed in rivers after spawning; and that fry, or young fish, should not be killed. These would entail further restrictions on netting, prohibition of fishing at night with lights, and a longer close season. He referred to experiments reported in Bloch, showing salmon eggs hatch only in running water saturated with air, but really there is little science in his remarks. In the old pattern he used science in support of common-sense proposals, which in fact favoured fly-fishing gentlemen against professional fishermen.

In Edward Jesse's *Gleanings of Natural History* of 1834 *Salmonia* is cited in discussion of the generation and habits of eels,[11] a moot point in contemporary ichthyology, and Jesse quotes 'poetical' passages on the swallow[12] and on the beauties of nature, expressing 'the gratification I have derived in viewing what is beautiful in nature – my pleasant walks by some clear and lively stream, and my strolls through woodland scenery'.[13] Such happy hours 'passed in contemplating the works of a beneficent Creator' should be looked upon 'as neither mis-spent nor unprofitable'.[14] Davy would probably have welcomed these reactions: *Salmonia* is not a work of original science or a textbook, but a fishing book suffused with natural theology.

Anybody coming to it in hopes of tips on fishing would have been rather disappointed. We learn much more about how fish were actually caught, and get a much stronger feeling of a passion for fishing, in William Scrope's *Days and Nights of Salmon Fishing on the Tweed* of 1843,[15] which was, like Davy's and Jesse's books, published by John Murray II, also born in 1778. Scrope was the father-in-law of Poulett Scrope, the eminent geologist, who took his name, and was, like Davy, a friend of Scott. He was also wittier than Davy. In *Salmonia* there is a rather ponderous discussion of the ethics of fishing, where Davy argues that fish feel little or no pain when hooked. Scrope's case is that he simply makes an artificial fly, which the fish lawlessly attempts to murder, thus wantonly intruding himself on the hook in the attempt to eat the fisherman's property.

If *Salmonia* is neither simply a work of natural history (and like Yarrell's these could be discursive) nor a manual of fishing, perhaps it should be classified as literature, and, indeed, the references to 'poetry' by Cuvier, Jesse and Scott indicate that it was so perceived by some readers. Wilson in *Blackwood's* found it badly written, in a review that began by blowing Davy's cover: 'This is a book on a very delightful subject, by a very distinguished man. But although it is occasionally rather a pleasant book than otherwise, it is not by any means worthy either of the subject or the man – the one being angling, and the other Sir Humphry Davy.'[16] Scott, reviewing it enthusiastically in the dignified *Quarterly*, declared that 'we are indebted to the most illustrious and successful investigator of inductive philosophy which this age has produced',[17] which at the time would have been universally recognized as a circumlocution for Davy. Foes and friends, in what Coleridge called 'the Age of Personality', knew who they were writing about, even when reviewing anonymous works.

Wilson continued with gusto. In the preface to *Salmonia* we are told that it was composed during some months of severe and dangerous illness; Wilson felt that the author must have been comatose and his recovery miraculous. He was an expert writer of comic dialogue, famous for his episodes involving the Ettrick Shepherd, originally modelled upon the novelist James Hogg but soon acquiring a life of his own. He found Davy's dialogues 'stupid' and 'drawling', the interlocutors being 'introduced without dramatic skill'[18] so that they have no character – one never knows who is speaking. He noted that one of them is called Poietes and was said to be modelled upon Wordsworth, though a poor prosy person; another was called Ornither, though in a discussion of eagles he said nothing.

Wilson complained that we were never told how the four persons got to their various exotic locations or what they did when not fishing; they seemed to constitute an 'Exclusive Angling Club', so much so that no fifth person could ever join in and make their society less boring. What some readers found poetical were for Wilson mere purple passages, 'sad common-place stuff – very very trashy indeed'.[19] At one point a large fish was caught and was to be sent to a prince; here Davy's Whig and aristocratic views had led him into writing a fairy-tale. Some of the anecdotes seemed to Wilson evidence that Davy was ignorant of natural history, and as a good Scotsman he was indignant at patronizing remarks about sabbath observance in North Britain.

'Trashy' was a term usually used in criticism of novels. Like 'flashy', which was applied by stern readers to Macaulay's *History*,[20] it meant that the author had disregarded the advice given by Coleridge to aim at 'an austere purity of language both grammatically and logically; in short a perfect appropriateness of the words to the meaning'.[21] Wilson was said to be robust but not malicious in his writing, and a public man who was a Whig was fair game, even if ailing. Many of Wilson's criticisms are appropriate: we cannot tell who is speaking, and those who tried to find consistently real people in the dialogues must have had a hard time. The sentiments expressed are sometimes those associated with Davy's friends, but the voices are really all his own, speaking through transparent masks. We read the book not to meet interesting fictional characters, but to meet Davy himself in his leisure moments. John Davy wrote of his brother's companions in fishing and shooting that 'they, indeed, I will now say, were almost his only true friends who were his associates in these sports', and that even when ill 'there he recovered the hilarity natural to his disposition, and appeared in his true character most cheerful, amiable, and entertaining, and the delight of his friends'.[22] It is sad that in his last years so much of his angling was solitary.

Scott perceived *Salmonia*, reviewing it at Jane Davy's request, as an informal work revealing its author. Davy was avid to see the review. 'When great men condescend to trifle', Scott began in the grand manner, 'they desire that those who witness their frolics should have kindred sympathy with the subject which these regard.'[23] Fishing was the male equivalent of needlework or knitting, half business and half idleness, but in the writing of the leading natural philosopher of the day 'we are led to discover the sage even in his lightest amusements'.[24]

Where Wilson had deplored taking the antiquated, Cockney and stiff *Compleat Angler* as a model, Scott saw this as very proper, and was delighted to see Davy go beyond Walton in his geographical and social range. The tone of the dialogue, with its references to the Creator who has planted a religious instinct in mankind, was just 'what a great and good man's mind might be expected to exhibit'[25] on slow recovery from serious illness. Scott noted how wary Davy was of drawing conclusions in natural history and how he avoided dogmatic scepticism; this last reference is to some remarks about omens, which might go with Davy's alleged superstition. Scott concluded with the fear that over-fishing and the draining of the country would soon extirpate the salmon, so that future generations would read *Salmonia* as we read of the chase of deer, wild boar or wild cattle.

By the time *Salmonia* was published, Davy was again abroad. He believed the very climate of England kept 'the nervous system in a constant state of disturbance', making his countrymen 'preeminently active' so that few 'very distinguished men' lived to a ripe old age.[26] Jane did not accompany him, but this time he had a valet, George, and also a young companion and amanuensis, John James Tobin, a medical student and the son of the old friend for whose play Davy had written the prologue. Tobin was the last of Davy's scientific sons and was not a success. In his will Davy wrote 'I beg Lady Davy to be so good as to fulfill my engagements with the persons who are travelling with me but without any favour as I have no reason to praise either their attention or civilities.'[27] When he set out on 29 March 1828 there can have been little hope that Davy would ever recover, or indeed that he would return; and in the event he never saw England again. In *Consolations* he described his setting forth for Slovenia, where:

> I had sought for and found consolation, and partly recovered my health after a dangerous illness, the consequence of labour and mental agitation; there I had found the spirit of my early vision. I was desirous therefore of spending some time in these scenes, in the hope of re-establishing a broken constitution . . . Nature never deceives us, the rocks, the mountains, the streams, always speak the same language . . . And, nature affords no continued trains of misfortunes and miseries such as depend upon the constitution of humanity, no hopes are ever blighted in the bud, no beings full of life, beauty and promise taken from us in the prime of youth. Her fruits are all balmy, bright and sweet; she affords none of those blighted ones so common in the life of man and so like the fabled apples of the Dead Sea, fresh and beautiful to the sight, but when tasted full of bitterness and ashes.[28]

He was delighted by what he heard of the reception of *Salmonia*, writing to Jane that 'it has almost rekindled my love of praise',[29] and planning a second edition. No doubt she did not pass on Wilson's review. He wrote to John in September: 'You will have seen *Salmonia* by this time. I have made the second edition twice as large, and I hope twice as amusing. It contains many of my philosophical views, and some new and I hope true opinions in natural history. I send the copy for the second edition to Murray by the next opportunity.'[30] The terms Scott applied to him in the review, 'poet', 'philosopher' and 'sage' were, gratifyingly, just those Davy had used of himself in a poem about his love of nature.

An important episode in *Salmonia* also featured in a poem, that about the eagles.[31] The party saw an eagle fishing on a highland loch, and one of them described how he had seen a pair of eagles teaching their young to fly. For Scott, this scene united the sublimity of Salvator Rosa to the accuracy of Gilbert White, whereas for Wilson it perpetuated ornithological blunders, for the kind of eagle described would not fish as Davy said it did, dropping from a height into the water. Davy described the bird (Poietes used the word 'animal', to Wilson's indignation) as the 'gray or silver eagle', and distinguished it from the osprey; in seems likely that it was the white-tailed eagle, *Haliaetus albicilla*, now seen only as a vagrant in Scotland but in Davy's time not uncommon there, and regarded as a pest.

Haliaetus is derived from a poetical Greek form of *Halieus*, a fisherman; this is the name given to the leader of the fishing party, or Exclusive Angling Club, who 'is supposed to be an accomplished fly fisher' and who, like the parent eagles, is instructing the others. They are: Ornither, a gentleman fond of field sports; Poietes, an enthusiastic lover of nature; and Physicus, a person fond of natural history and philosophy, but a novice in angling. Apart from their different expertise in fishing, all the characters fit Davy himself: the president encouraging younger spirits perhaps even to surpass his achievements; the keen shooter of partridges and snipe; the poet who loved nature as never mortal man before had done; and the inquirer. It is no wonder we cannot be sure which is which.

Davy himself wrote that 'these personages are of course imaginary, though the sentiments attributed to them, the Author may sometimes have gained from recollections of real conversations with friends, from whose society much of the happiness of his early life has been derived'.[32] No doubt those who knew him got some pleasure from guessing who said what. In the last dialogue we are given the clue that 'a likeness . . . will not fail to be recognised to that of the character of a most estimable Physician, ardently loved by his friends, and esteemed and venerated by the public';[33] this was William Babington, to whom *Salmonia* was dedicated 'in remembrance of some delightful days passed in his society, and in gratitude for an uninterrupted friendship of a quarter of a century'.[34] John Davy warned against identifying Halieus elsewhere with Babington, and also refers to Davy describing his own state of mind and sentiment 'in the character of Physicus' at the end of the book;[35] but Physicus, a man of science taking up fishing in middle life, was probably originally based on Wollaston, as is implied in a note. The fourth edition of *Salmonia* has Davy's name on the title page, and a composite portrait of Babington, Davy and Wollaston on the first page.

Wilson seized upon an old-maidish passage where Halieus forbids another bottle of claret: 'A half pint[36] of wine for young men in perfect health is enough, and you will be able to take your exercise better, and feel better for this abstinence. How few people calculate upon the effects of constantly renewed fever, in our luxurious system of living in England!' The consequence of too rich living was that 'the heart is made to act too powerfully, the blood is thrown upon the nobler parts', ending up in the head. 'Free livers' must expect, especially if they are in the habit of wading while fishing or shooting, to be killed off by apoplexy or made miserable by palsy; such folk were those who consumed 'as much animal food as they could eat, with a pint or perhaps a bottle of wine per day'. There were indeed some old men who had both drunk and waded freely, but they were *'devil's decoys* to the unwary, and ten suffer for one that escapes'.[37]

This sound medical advice came from Halieus, who continued:

> I could quote to you an instance from this very country, in one of the strongest men I have ever known. He was not intemperate, but he lived luxuriously, and waded as a salmon fisher for many years in this very river; but before he was fifty [Davy was forty-nine], palsy deprived him of the use of his limbs and he is still a living example of the system which you are ambitious of adopting.[38]

Davy is clearly referring to himself, though his case was not quite as bad as that, and goes on to produce a maxim that sounds as though it comes from a nanny, but is attributed to Boerhaave: 'Keep the feet warm, the head cool, and the body open.'[39] When Davy was taken ill, he was at first recommended to eat meat, as we saw, and then put on to an abstemious regimen, with leeches to remove excessive blood, on the hypothesis that too much blood was going to the head.

In Rome on his fiftieth birthday Davy heard of Wollaston's illness and impending death, writing to John on 21 December: 'Poor Dr. Wollaston has had an attack of paralysis, and I am sorry to hear is without hopes. His severe and ascetic life has not preserved him. This complaint is certainly becoming more common in England. I have heard of two or three other friends who have likewise suffered, – spare, abstemious men.'[40] Connoisseurs of funerals will know the survivor's triumphant tone, especially when all would have expected him to be the one to go first. To Jane, he wrote: 'He at least has not suffered from indulgence. I cannot help thinking that a certain quantity of nervous or vital power is given to man, which, when consumed, cannot be replaced, and which limits the period of activity and existence.'[41] Not only had Wollaston taken great care of his health, but he had amassed a fortune from platinum, and Davy wrote rather self-righteously to Jane about it, and about his disease: 'It was not worth his while to have died so rich; but I suppose there is pleasure in accumulating. So will W. die! . . . You speak of his malady being a family one. This likewise is my case: my grandfather and six or seven of my great uncles died of apoplexy.'[42]

Davy was always concerned that the wealthy should promote science, closely concerned as it was in his mind with usefulness, and thus with philanthropy. What is more striking is that whereas when writing *Salmonia* he had attributed his illness to gluttony and wading, a year later he had come to see it either as the consequence of burning up his capital of nervous energy in the pursuit of science or of hereditary predisposition. Neither of these are morally objectionable. He was coming to terms with his disease.

Salmonia sold well and the second edition came out in 1829. At the Royal Institution there is an annotated copy of the first edition, with the amendments written in. To say that the book became twice as large was an exaggeration: it had 335 pages against 273. Many changes are minor, involving changing a word, and the excessive commas characteristic of Davy's style are trimmed. There is little change to the specifically fishy parts, though we do find more information about the *hucho*, presumably because Davy had learned more about it while fishing in the Alps.

The purple passage with which we began this biography became a good deal purpler, and it seems to call out for declamation; it is good of its kind, but we cannot imagine anybody saying it in coversation between friends. A passage on Home's dissection of an eel, which he supposed hermaphrodite, said bluntly: 'this circumstance demands confirmation'; this was changed to the more courtly 'I hope this great comparative anatomist will be able to confirm his views by new dissections, and some chemical researches upon the nature of the fringes and the supposed melt.'[43] Davy did not really hope this at all, disagreeing with Home; and, as John Davy pointed out in the 1851 edition, Davy was right. The two had been associated in the Animal Chemistry Club, but Home seems to have lost Davy's respect by 1823, when he wrote to John: 'Sir Everard has published his Lectures – a magnificent book as to engravings, chiefly in consequence of the liberality of the Royal Society; and it contains certainly a great mass of valuable matter, with much loose speculation and *microscopic* physiology';[44] 'microscopic' here clearly meaning petty. Babbage later attacked the Royal Society's 'liberality' in this case, and Home's failure to acknowledge it.

Salmonia was illustrated. The first edition had woodcuts of fish, and plates of flies. Plates were engraved for the second edition, but replaced by woodcuts for the fourth (copper wears less well than wood).[45] One of the engravings shows the eagles, low in the foreground; the others are picturesque views of the rivers. Paris tells us that, according to Jane Davy, the wood engravings of fish scattered through the text were from drawings by Davy himself. This seems likely, because, although John does not mention it, he refers to Humphry's skill in drawing geological features in a landscape and to his journals with 'figures drawn with pen or pencil of any forms, especially of fish, which were new and interesting to him'.[46] The woodcuts are attractive. They face either way, where the usual convention is that fish point to the right in zoological illustrations.[47] In Davy's annotated copy the second woodcut of the grayling has an amendment made to the tail, making it more sharply forked; and this change seems to have been carried out in later editions.

The biggest changes were in the general discussions, where much was added, making it less of a fishing book. Davy wrote to Poole: 'A *second edition* will soon be out, which will be in every respect worthy of your perusal, being, I think, twice (not saying very much for it) as entertaining and philosophical.'[48] If we are to meet Davy through *Salmonia*, it is the revised version we should use. In looking within the fisherman for the philosopher we should not entirely forget the exterior: with his 'curious and elaborate' tackle, his broad-brimmed white hat garnished with flies, breeches with knee-caps made from old hat, and jacket with numerous pockets – an outfit, as John remarked, 'not unoriginal and considerably picturesque'.[49] He was apparently a very successful fisherman. Scott said that *Salmonia* failed to indicate the inequality between anglers; Davy was famous for some of his good catches – a salmon of forty-two pounds, caught in the Tweed, for example, and not mentioned in the book. There were some that got away, and some unlucky days: John writes of spending a long June day with Humphry from dawn to dusk and not raising a fish.

The dialogues begin in London, with a discussion of the morality of fishing. Anecdotes of fish being caught very soon after breaking a line, and therefore feeling no great pain from a hook, help to reassure the squeamish. The party then meets at Denham, on the Colne in Buckinghamshire, and this day, unlike any other, is dated 'May 1810', when Davy was at the height of his chemical work. John wrote of this period:

> When his pursuits did not keep him in town, he often made short visits to friends residing in the neighbourhood of London, or went to some trout stream, of which there are so many good ones within twenty or thirty miles of the metropolis, and breathed the fresh air by the river side, and enjoyed the country and his favourite exercise and amusement of fishing together.[50]

The passage in *Salmonia* describes such a visit, with fishing in the afternoon, in the evening (after dinner at five) and on the following morning.

A fair amount of informal entomology is taught in the dialogues, with a mention of Kirby and Spence's recently completed classic work (written in the form of letters).[51] There is also discussion, not in the first edition, of development and inheritance, referring to David Hartley, Buffon and Erasmus Darwin. Of the three, Hartley (much admired by Priestley) is praised for his 'profound ideas', the others being speculative and given to wild fancies. Having thus distanced himself from materialism, Davy proposed a theory of the inheritance of acquired characteristics, using evidence from domestic selection, including horses, merino sheep,[52] poultry and domestic pigeons; and went on to apply this to trout – suggesting that cross-breeding commonly takes place even between supposed different species. This does not mean that had he been spared for a few more years, Davy might have written the *Origin of Species*, but it indicates that his mind was as restless as ever, and that nobody with a professional interest in agriculture and a leisure interest in ichthyology could avoid such questions.

We also find a mention of trade secrets being purchased, but in general in *Salmonia* we are not in the world of the Industrial Revolution, except in a negative way, as when we find the happy and religious peasantry of the Habsburg lands contrasted with the idle and conceited lower classes of England, who have had education forced upon them (by some of Davy's patrons and friends, like Sir Thomas Bernard and William Allen, if the truth be told) and 'have been become such tumultuous subjects of *King Press*, whom I consider the most capricious, depraved, and unprincipled tyrant, that ever existed in England'.[53] Davy had had trouble with the media, and Poietes, who makes the foregoing remark, is supposed to be a Roman Catholic. It is one of the surprising things about Davy that, in a century when men of science in particular were hostile, he should have been sympathetic to this Church. He was delighted when Catholic Emancipation was carried, writing to Jane: 'I have always considered this point as essential to the welfare of England as a great country, and connected with her glory as a liberal, philosophical, and Christian country.'[54] Experience in the Tyrol and Slovenia had reinforced feelings developed in Ireland, and on his last journeys, in Ravenna and in Rome, he had met intellectual Roman Catholics also, one of whom, Dr Morichini, was to benefit from his will 'in remembrance and memory of his great kindness to me'.[55]

The next dialogue is set on Loch Maree, where Davy had been in 1821, and then, on 12 August, shooting grouse on the moors of Sir George Mackenzie, well known for his interests in chemistry, geology and agriculture.[56] As well as fishing and advice on the cooking of salmon, we find remarks on Sunday observance and the introduction of a poacher; which give some local colour to the dialogues. There is also an interesting discussion, not in the first edition, of the process of dying. The peaceful deaths of William Cullen, Joseph Black and Charles Blagden are described, and suffocation is said to be rather agreeable than otherwise, from Davy's own experiments with gases at Clifton and the suicide of young Berthollet.

Also in Scotland, we find a discussion of instinct. This was a topic of current scientific debate, on which Kirby and Spence had much to say and were in disagreement.[57] Davy considered that instincts might be referred to the immediate of God, but seems to have preferred to think of them as the result of general laws governing the universe. He did not believe, with Paley, that there was a good analogy between the world and a machine, seeing organisms and machines as essentially different, because the former embodied a principle of conservation as well as perfection – we continue through the flux of the particles that compose us. Mankind, having reason, does not need ordinary instincts, but Davy believed that we have a religious instinct. This leads to natural religion, which, when guided aright by revelation, could produce that most 'pleasurable state of the human mind . . . when, with intense belief, it looks forward to another world and to a better state of existence, or is absorbed in the adoration of the supreme and eternal Intelligence'.[58] In rejecting Paley's 'watchmaker'[59] analogy, Davy urged the inscrutability of God, 'the one incomprehensible Cause of all being';[60] in this, as in his chemistry, he was in reaction against the received opinions of the late eighteenth century. The dialogue form was useful for this kind of exploration of ideas.

The next place where the anglers met was Downton, near Ludlow, the home of Knight, the plant physiologist who supplied notes for the third edition of *Agricultural Chemistry* and to whom the fourth edition was dedicated; Davy wrote that his experiments (notably on peas) were 'not merely curious, but useful'.[61] The house and grounds had been laid out by Richard Payne Knight, the aesthete, in the picturesque manner of which he was a pioneer, and on his death had passed to his younger brother. Davy often fished there and was a friend of the family. Davy was desolated, as we saw, by the death of Knight's son in a shooting accident, and in turn, so Knight's daughters wrote after Davy's death, there was nobody in whose society their father 'so delighted, and whom he could so ill at this time have spared; there were many points in which the feelings of both were peculiarly in accordance'.[62]

Here they were after grayling, and the general discussion is first about eels and then entomology, with descriptions and illustrations of artificial flies. The conversation again turns to religion, with the insect as the type of man, rising from his chrysalis-coffin to a new and higher life, and the happiness of death under the vision of celestial glory and the pure, intense love of God. When the party find themselves later in Austria, it is the design and wisdom evident in the correlation of the parts of creatures that they discuss. This principle made much of by Cuvier, enabled him to reconstruct extinct creatures from jumbles of fossil bones; it enabled the angling party to reject stories of mermaids (Davy tells a story of being mistaken for one), krakens and other such fabulous creatures. It also contained all that was of any value in phrenology, 'the science of the bumps' popular in the 1820s as a mechanistic psychology, where the shape of the skull was supposed to indicate the nature of the mind within it. Davy took a less sceptical view of omens and superstitions: 'The deep philosopher sees chains of causes and effects so wonderfully and strangely linked together, that he is usually the last person to decide upon the impossibility of any two series of events being independent of each other.'[63] Nevertheless, he did explain the basis of some superstitions, while feeling unable to account for 'the more mysterious relations of moral events and intellectual natures'.[64]

Particularly with the revised *Salmonia*, Davy had found a new role: he could express his religious and philosophical views in the form of dialogues, which, despite Wilson, got a generally favourable response. Although he could no longer be the laboratory researcher he had been, his intellectual life could go on under different circumstances; he could see himself as a sage. Moving between Italy and the Alps, he dictated the new passages of this book and then the dialogues that became *Consolations*, to Tobin, and they became more and more important to him, giving his life a new urgency. Tobin as companion earned no legacy, but as the (unnamed) amanuensis he was left £100 in the final codicil to the will, which then deals with the publication of *Consolations* and the copyright of *Salmonia*.

The dialogues formed a kind of substitute for real conversations with real friends, for Davy seems a curiously modern figure cut off by his meteoric rise from early friends and family, he was rootless; his attitude to politics, as reported by John Davy, was uncommitted – 'free from party bias, and who regarded the then perplexing and anxious state of things very much in the light of a problem to be solved',[65] no doubt by a kind of social engineer; he hoped for love from his fellow-countrymen, but all he got was recognition. Although there are gloomy letters from his solitary trips abroad, on the whole

Davy seems to have been pretty satisfied with his own company and that of a few foreigners, boring though this was to Tobin. He delighted in the works of God rather than those of man, though getting a delight from ruins like a true Romantic. In *Salmonia* we do not really meet the chemical researcher, but we do see almost all other aspects of the man – even the founder of the zoo, hoping to breed exotic fish there to introduce on fishing rivers.[66] In the final chapter of his life we shall see that he brought in the chemistry too, coming back to those wrestlings with materialism that had first brought him into science, and trying to find the meaning of life.

12

Consolations

On 28 May 1829 Davy died in a hotel room in Geneva, where on the following Monday he was given a public funeral and buried at the cemetary of Plain-Palais outside the walls. He had wished to be buried where he died, but wanted burial delayed in case he should be merely comatose (for similar reasons, he had refused to allow an autopsy). However, the laws of Geneva did not permit it and decay was beginning. Jane had a memorial tablet put up in Westminster Abbey soon afterwards, at a cost to her of £142 to the disgust of the *Athenaeum*.

At the Royal Society on St Andrew's Day Davies Gilbert had to refer to the deaths of Wollaston and Young as well as Davy, but he spoke particularly of his predecessor and protégé and 'his high character of an inventive philosopher, by which he has added to the credit, to the honour, and to the fame of this most distinguished Society, by which he has diffused a lustre on the province which gave him birth, and on the entire nation to which he belonged'.[1] As the only Fellow who had known the young Davy, he recounted some anecdotes of his early years. It was only three years since Davy himself had delivered the Presidential Address, but there does not seem to have been any great feeling that he had been prematurely cut off. He had fitted a great deal of science into a relatively short life and he had also left no school of disciples to carry on his work. Faraday and J. F. Daniell worked in electrochemistry, but by 1829 Davy's researches seemed something in the distant past: organic chemistry was becoming the most exciting sphere, and Germany (particularly through Gay-Lussac's student,

Justus Liebig)[2] the centre of things. Davy did not leave the gap one might have expected in the scientific community.

Davy's legacy was the *Consolations*. Young Tobin found his travels with Davy a great disappointment, for he was stopped from having a good time by the demands of the morose and reclusive man of science, whose only pleasures were apparently in wild and secluded spots, where he fished and shot, and who wanted to be read to each evening and to dictate dialogues. What might have turned out as something like the young Darwin's voyage on HMS *Beagle* as companion to the captain went sour; perhaps science was not really Tobin's forte, competent doctor though he may have become. Indeed, he was a desirable companion because he spoke German; despite Davy's love for the Austrian and Slovenian alps and people (he preferred both to those of Switzerland), he never learned to speak either language.

In the early nineteenth century there were two recommended ways of learning a foreign language. One was Macaulay's: one acquired a Bible in the language (this was the heyday of the British and Foreign Bible Society) and read it, preferably on a long sea voyage. The other was Byron's: one acquired a girl-friend who spoke it. In Italy, Davy met Teresa Guiccioli, pensioned off by her aggrieved husband as a result of this system, and made friends with the daughter of an innkeeper in Ljubljana. Tobin gloomily noted that it was the third-best inn and the attractions of Josephine Detela or Dettela, whose pet name was Papina, were all that recommended it.

We meet her in *Consolations* as a guardian angel (or archetype) dreamed of at the crisis of Davy's gaol fever:

PHIL: There was always before me the form of a beautiful woman with whom I was engaged in the most interesting and intellectual conversation.

AMB: The figure of a lady with whom you were in love?

PHIL: No such thing; I was passionately in love at the time, but the object of my admiration was a lady with black hair, dark eyes and pale complexion; this spirit of my vision on the contrary had brown hair, blue eyes and a bright rosy complexion, and was, as far as I can recollect, unlike any of the amatory forms which in early youth had so often haunted my imagination. Her figure for many days was so distinct in my mind as to form almost a visual image; as I gained strength the visits of my good angel, for so I called it, became less frequent, and when I was restored to health they were altogether discontinued.

ONU: I see nothing very strange in this, a mere reaction of the mind after severe pain, and, to a young man of twenty-five, there are few more pleasurable images than that of a beautiful maiden with blue eyes, blooming cheeks and long nut brown hair.

PHIL: But all my feelings and all my conversations with this visionary maiden were of an intellectual and refined nature.

ONU: Yes, I suppose, as long as you were ill.

PHIL: I will not allow you to treat me with ridicule on this point till you have heard the second part of my tale. Ten years after I had recovered from the fever, and when I had almost lost the recollection of the vision, it was recalled to my memory by a very blooming and graceful maiden fourteen or fifteen years old, that I accidently met during my travels in Illyria; but I cannot say that the impression made upon my mind by this female was very strong. Now comes the extraordinary part of the narrative; ten years after, twenty years after my first illness, at a time when I was exceedingly weak from a severe and dangerous malady, which for many weeks threatened my life, and when my mind was almost in a desponding state, being in the course of travels ordered by my medical advisers, I again met the person who was the representative of my visionary female; and to her kindness and care I believe I owe what remains to me of existence. My despondency gradually disappeared, and though my health continued weak, life began to possess charms for me which I had thought were for ever gone; and I could not help identifying the living angel with the vision which appeared as my guardian genius during the vision during the illness of my youth.[3]

Although the sceptical character Onuphrio goes on to account for the dream as mere fancy and imagination excited by disease, there seems no doubt that for Davy Papina was the fulfilment of a dream. He duly bequeathed to his 'kind and affectionate nurse' £100 or 1000 florins, and then added another codicil adding £50; she would, therefore, have had for a dowry what was then quite a large sum, from one of the most distinguished natural philosophers of her time.

From what Tobin says of Davy, it seems as though he needed cheering up as much as nursing:

> Sir Humphry's health was in so shattered a state, that it often rendered his inclinations and feelings sensitive and variable to a painful degree. Frequently he preferred being left alone at his meals; and in his rides, or fishing and shooting excursions, to be attended only by his servant. Sometimes he would pass hours together, when travelling, without exchanging a word, and often appeared exhausted by his mental exertions.[4]

In other words, as is clear elsewhere, he was a curmudgeon, open only to the charms of scenery and Josephine. On arriving in Rome for the winter, he wrote to the dark and pale complexioned Jane on 18 November 1828 (the day he wrote the first bequest to Papina into his will):

> I hope I shall find some shooting here, and I trust some literary amusement, in adding to my vision of the Colosseum, otherwise I shall vegetate – for my spirits are too feeble to bear general society, and I fear I shall find no Illyrian nurse here, such as the spirit that dispelled my melancholy at Laybach; but I live always in hope, that I may still be useful to others, and that my existence is continued for some useful purpose. God bless you, my dear Jane!
> Your most affectionate H. Davy.[5]

On 20 January 1829 he wrote again:

> I feel my vital powers (quoad the body) are not restoring; and my stomach is feebler than ever in this climate. I cannot go into society, for even a conversation of half an hour exhausts me, and my voice is almost inaudible. My mind, however, I hope, has all its vigour, and too much discriminating sensibility. Should I survive the winter, I shall like to return to Illyria, and the Save and the Traun in spring; and shall be most happy to meet you there. You talk of some nymph Bettina. Alas! my Bettinas are with the years of the old Romans, and amongst the ruins of the eternal city, images which cannot return. But if you mean my little nurse and friend of Laybach, I shall be very glad to make you acquainted with her. She has made some days of my life more agreeable than I had any right to hope. Her name is Josephine or Papina. I am at this moment much interested in an inquiry of science, which has opened itself unexpectedly in the functions of the torpedo; and I regret more than ever the stiffness and weakness of my left hand, which limit my powers of dissection and manipulation.[6]

It does not seem likely that if his relationship to Papina had been adulterous, Davy would have mentioned it to Jane in this way, so perhaps here again he was a dying Plato – he does not seem to have been very highly sexed even as a young man. Bettina Brentano, later von Arnim, was notorious for her hero-worship of Goethe, hence the joke;[7] and whether or not Davy and Papina's conversation was always intellectual and refined, she was clearly no Bettina.

On 19 February Davy added the further codicil to his will, and on the twentieth 'most unexpectedly, for there had been no premonitory symptoms, and quite suddenly, he had that severe attack which ultimately proved fatal'.[8] After dictating an addition to the sixth and last dialogue of *Consolations*,[9] he had another stroke, and a lowering, or 'antiphlogistic', treatment had no effect. John and Jane were summoned to his bedside. To John he dictated a letter of *nunc dimittis*:

'I bless God that I have been able to finish all my philosophical labours.'[10] In a later letter he reported on his experiments on the electricity of the torpedo.

Another letter went to Jane, and that player of many roles now knew her part.[11] She wrote back:

> I have received, my beloved Sir Humphry, the letter signed by your hand, with its precious wish of tenderness, bearing date the 1st of March. I start to-morrow, having been detained here by Drs. Babington and Clarke, till to-day. I shall travel with all the expedition I can, to arrive not quite useless. I trust still to embrace you, for so clear and beautiful expressions and sentiments cannot be the inhabitants of decay, however of feeble limbs and frame. I shall to the extremest point hold your wishes sacred, and obey in ready willingness the spirit even more than the letter of your order. God still preserve you, and know that the lofty and noble tone of your letter deepens all love and faith I have ever borne to you, and believe the words of kind effort will be a shield to me through life. I cannot add more than that your fame is a deposit, and your memory a glory, your life still a hope.
> Your ever faithful and affectionate, Jane Davy.[12]

The 'orders' related to her position as sole executrix. She got to Rome before he died, as did John, and together they were bringing him home when he expired.

Thus in his last days Davy had the intermittent consolations of two women, but most of the time neither of them were there and, like Boethius, he turned to the consolations of philosophy in his dialogues. Years before, in Clifton, he had written: 'Philosophy has warmed me through life on the bed of death she does not desert her disciple. The frost of the grave can never chill these burning energies connected with the thoughts of future existence.'[13] *Consolations* begins with a vision in the Colosseum in Rome, and Davy claimed that, like the story of the guardian angel, this began as a real dream. Tobin, who took it down from his dictation, says that it had been Davy's

> pleasure and delight during his mornings at Ischl, and when he was not engaged in his favourite pursuit of fishing, to work upon this foundation, and to build up a tale, alike redundant with highly beautiful imagery, fine thoughts, and philosophical ideas; and the hours thus passed with Sir Humphry have afforded me high mental gratification and advantage, for I have then marked his mind wandering, as it were, with the associates of his early days; those days in which he was evidently, by the exercise of his extraordinary powers and quick perception, exciting not only his own mind to dive into, and to unfold to clearer view, the mysteries of creation, but that too of other congenial spirits; thus

> most naturally collecting around him a constellation of shining lights, the remembrance of whom often awakens vivid thoughts of the past, and rouses his whole soul to action.[14]

At this stage, the vision was an isolated dialogue, but Davy soon determined to add to it. On his fiftieth birthday he wrote to Jane:

> I lead the life of a solitary. I go into the Campagna to look for game, and work at home at my dialogues the alternate days. I hope I shall finish something worth publishing before the winter is over. This day, my birth-day, I finish my half century. Whether the work I am now employed on will be my last, I know not; but I am sure, in one respect it will be *my best*; for its object is to display and vindicate *the instinct or feeling of religion*. No philosopher, I am sure, *will* quarrel with it; and no Christian *ought* to quarrel with it.[15]

The vision presents a progressive view of the history of mankind, from savagery to civilization, and a view of people as immortal souls, which after a time wear out and outgrow the mortal machinery in which we meet them. After death, they can expect reincarnation as higher and more spiritual beings, with more ethereal bodies, on other planets. John Davy tells us that a dialogue about the hellish fate of the wicked was also planned. One of the great lessons of the book is that death is necessary for birth or rebirth; and another that we are not mere matter, and that the laws of chemistry and physics do not alone govern living beings.

In the vision an important message was that progress results from the work of men of genius, demi-gods, which is never lost:

> Look at these groups of men who are escaped from the state of infancy: they owe their improvement to a few superior minds still amongst them. That aged man whom you see with a crowd around him taught them to build cottages; from that other they learnt to domesticate cattle; from others to collect and sow corn and seeds of fruit. And these arts will never be lost; another generation will see them more perfect. ... in general, it is neither amongst sovereigns nor the higher classes of society, that the great improvers or benefactors of mankind are to be found. The works of the most illustrious names were little valued at the times when they were produced, and their authors either despised or neglected; and great, indeed, must have been the pure and abstract pleasure resulting from the exertion of intellectual superiority and the discovery of truth and the bestowing benefits and blessings upon society, which induced men to sacrifice all their common enjoyments and all their privileges as citizens, to these exertions.[16]

We should not take this too seriously, because although there have been some martyrs to science, Davy was one of the most famous men in Europe, a Colossus among natural philosophers, and could hardly be said to have had to sacrifice enjoyments or privileges.

Higher beings on other planets could understand the natural and moral government of the world much better than mankind could, and

> as their highest pleasures depend upon intellectual pursuits, so you may conclude that those modes of life bear the strictest analogy to that which on earth you would call exalted virtue. I will tell you however that they have no wars, and that the objects of their ambition are entirely those of intellectual greatness, and that the only passion that they feel in which comparisons with each other can be instituted are those dependant upon a love of glory of the purest kind.[17]

The vision is discussed upon the summit of Vesuvius. There are three participants, whom we have met talking about Papina: Philalethes, representing Davy himself; Ambrosio a liberal Roman Catholic; and Onuphrio, a sceptic. The focus is the tension between a naturalistic and a specifically Christian account of mankind and its history. While supporting the latter, Ambrosio considers 'all the miraculous parts of our religion as effected by changes in the sensations or ideas of the human mind and not by physical changes in the order of nature'.[18] His

> first principle is, that religion has nothing to do with the common order of events; it is a pure and divine instinct intended to give results to man which he cannot obtain by the common use of his reason and which at first view appear contradictory to it, but which when examined by the most refined tests, and considered in the most extensive and profound relations are in fact in conformity with the most exalted intellectual knowledge, so that indeed the results of pure reason ultimately become the same with those of faith, – the tree of knowledge is grafted upon the tree of life and that fruit which brought the fear of death into the world budding on an immortal stock becomes the fruit of the promise of immortality.[19]

Immortality was clearly extremely important for Davy. He was uncertain about man's fall or rise, and impatient with the kind of natural theology that William Paley had made popular, based on the idea that the world was a great clock showing evidence of a Designer.[20] Davy believed that 'we can hardly comprehend the cause of a simple atmospheric phenomenon, such as the fall of a heavy body from a meteor; we cannot even embrace in one view the millionth part of the objects surrounding us, and yet, we have the presumption to reason upon the infinite universe and the eternal mind by which it was created and is governed'.[21] Metaphysical speculations 'must begin with a foundation of faith'.[22] Whereas Paley sometimes seems to congratulate

God upon a particularly clever bit of designing, Davy's position is closer to that of Coleridge: 'Assume the existence of God, – and then the harmony and fitness of the physical creation may be shown to correspond with and support such an assumption; – but to set about *proving* the existence of God by such means is a mere circle, a delusion.'[23]

Although Davy recognized a connection between the happy and friendly state of the Austrian peasantry, as he saw it, and the Roman Catholic Church, his religion is clearly an individual rather than a social matter; like most Englishmen, he was deeply Protestant, with a strongly personal religion. Accordingly, he belonged to no particular Church, 'excepting, in an enlarged sense, to the "Church of Christ"'.[24] Probably on Sundays he preferred to worship God with fishing rod in hand, finding sermons in stones and good in everything, rather than in a pew. That it was private and personal does not mean that it was unimportant to him, but it was clearly very different from Faraday's faith,[25] for example, intertwined with the strong community life of a small sect. In sickness and age, especially, 'submission in faith and humble trust in the divine will . . . creates powers which were believed to be extinct, and gives a freshness to the mind, which was supposed to have passed away for ever, but which is now renovated as an immortal hope; then it is the Pharos, guiding the wave-tost mariner to his home'.[26] To that ancient mariner at least it brought relief, although clearly it did not always work, as we can see from the grumpy Davy portrayed in Tobin's *Journal*.

We might expect that in the 1820s it would be the account of Creation in Genesis that would raise problems for a man of science. Davy had visited Kirkstone Cave in North Yorkshire with William Buckland, who had interpreted the fossils there in his *Reliquiae Diluvianae* (1823) as the remains of a hyenas' den destroyed in Noah's Flood. Buckland first published his discoveries in the Royal Society's *Philosophical Transactions*, Davy being President, after describing them at a meeting of the Society. He was not a literalist and was denounced by those calling themselves 'Scriptural Geologists'; but he believed that recent research established the essential accuracy of Genesis. He was at the time Professor of Geology at Oxford, and went on to become Dean of Westminster in a successful academic and ecclesiastical career. Later, many of the effects he had attributed to the Deluge came to seem more plausibly due to glaciers, like those Davy had seen in the Alps, and Buckland was one of the first to recognize and popularize the idea of an Ice Age. He also found the first dinosaur fossils, though these were of marine species and did not arouse the excitement of later finds. He was very excited by the first finds of dinosaur footprints, and electrified scientific audiences by hopping around as dinosaurs must have done.

Davy had all his life been interested in rocks and minerals, and in the geological processes that had shaped the Earth. He gave various geology lectures at the Royal Institution, incorporated parts of the science into his *Agricultural Chemistry*, and went on study tours or field trips in Cornwall, Wales, Scotland and Ireland. Later, on the Continent, he investigated Vesuvius, as well as petrifying streams, caverns, waterfalls and other phenomena described in *Consolations*. He had to withdraw from the infant Geological Society because of Banks's opposition to a potential rival, but he claimed to have given the first public lectures on geology in London, and he also grumbled that some of his geological ideas had been pirated. One of the great controversies of the day was that between so-called Neptunists and Plutonists, arguing about the role of fire and water. Most significant was that the former school, of which James Hutton of Edinburgh had been a founder, allowed for enormous tracts of time in the past rather than for the 6000 years or so of the Genesis story taken literally, as by the seventeenth-century Archbishop Ussher of Armagh with his famous date of 4004 BC.

To Davy's generation, Hutton's geology was alarming in its apparent atheism, characteristic of Enlightenment science. Hutton ended his paper with the sentence: 'The result, therefore, of our present enquiry is, that we find no vestige of a beginning, – no prospect of an end.'[27] This was written in 1785 and also appeared in his book, which turned out to be almost unreadable, until popularized by John Playfair in 1802.[28] At the same period Sir James Hall at Edinburgh made experiments with melts trying to synthesize igneous rocks. Playfair and Hall made Hutton's ideas seem more plausible, but those final words resonated through the educated community; it appeared that he had no need (like Laplace) of the hypothesis of God, though, in fact, natural theology is prominent in his writing.

Buckland seemed to have shown that the Great Flood was a real event, but in 1830 his pupil Charles Lyell began to bring out his classic *Principles of Geology*, an 'attempt to explain the former changes of the Earth's surface, by reference to causes now in operation.'[29] This meant a return from the short time-scales of Buckland and Cuvier, involving series of violent catastrophes, to the long times of Hutton and Playfair, in which processes we actually see at work have gradually modelled the Earth. Lyell quoted from Davy's recently published *Consolations*, from the same publisher. Lyell approved of Davy's emphasis upon the recent appearance of man, but deplored his notion that 'in the successive groups of strata, from the oldest to the most recent, there is a progressive development of organic life, from the simplest to the most complicated forms'. 'No geologists', he added, 'who are in possession of all the data now established respecting fossil remains, will for a moment contend for the doctrine in all its detail, as laid down by the great chemist to whose opinions we have referred.'[30]

Lyell is now chiefly famous because his protégé and friend, Charles Darwin, drew heavily upon his ideas in evolving his theory of natural selection, so that late in life Lyell changed his mind about progressive development, which he had disliked in Davy and especially in Lamarck. Why he should pick on *Consolations* is not quite clear: it enabled him to make the point about geologists and a chemist, that science was by now specialized and a distinguished practitioner of one branch was an amateur in others; and his target was safely dead and would not take up the gauntlet. Lyell came by the 1860s to be recognized as the greatest theorist of his day, but in the 1830s he had seemed an extremist, interesting but unsound, to his geological colleagues, such as Davy's friend Murchison. Davy's views were, in fact, close to those of the mainstream of geology in Britain, appropriate for what was meant to be a popular work of natural philosophy by one opposed to narrow specialization.

Ambrosio sets off the discussion: 'You mistake my view, Onuphrio, if you imagine I am desirous of raising a system of geology on the book of Genesis.'[31] In the early books of the Bible we find divine truths, 'yet clothed in a language and suited to the ideas of a rude and uninstructed people. And, when I state my satisfaction in finding that they are not contradicted by the refined researches of modern geologists, I do not mean to deduce from them a system of science.'[32] By now a fourth character has joined in, a mysterious figure, the Unknown, who also seems to represent Davy. He has no objection to 'the *refined plutonic view*' for explaining many phenomena, but believes that there must have been world-wide catastrophes producing extinctions and discontinuities, 'a succession of destructions and creations preparatory to the existence of man'.[33] Like the vision of souls progressing through more and more ethereal bodies, this was a progressive view of the Earth's development: *Consolations* is, after all, a hopeful book.

We meet further dreams and visions, and then the action shifts to the Falls of the Traun in Austria, where Philalethes is carried over the falls in a boating accident that could have ended badly had not the Unknown happened to be at the bottom and hauled him out with his fishing tackle. Davy himself experienced no such disaster abroad, but in 1804 had nearly drowned in a fishing accident, about which he wrote a poem; and, Tobin fell from a rickety staircase in a limestone cave called 'the Waterfall' from its stalagtites: 'Not being able to recover myself, I slipped from rock to rock, turning twice head over heels, but without injury, and with perfect presence of mind, although I expected every instant to be dashed over the edge of a precipice.'[34] Like Davy, who also in the dialogue retained his presence of mind, Tobin had glimpsed the abyss.

Davy wrote a poem about these falls in July 1827:

From the high rock thy lovely waters burst,
As if a new creation from the wand
Of Israel's mighty prophet, sprung to life
To save his people! But the dreamy thought
Of that most blessed, tho' but scanty rill,
Gives but faint image of thy might, and power,
And awful force, and fulness: as if a spirit
Imprison'd by magic art, and now released,
Thou thunderest on, determined to destroy;
And thy mild functions to produce and cheer
Are changed for attributes more terrible,
Saddening, destructive, wildly carrying on
Rocks, trees, before thee, e'en the mighty pine,
Rending the mountain, through a new-torn vale,
Opening thyself a passage to the plain.
But in thy wayward and most perilous leaps,
Thou still art pure, and still might image well
The innate mind of poet or of sage.
In thy bright azure depths, and when thy foam
Sinks into quietness, I seem to view
That season of our life when pleasure fades,
And sober reason with its heavenly light
Fills the deep cool of th'unimpassion'd mind,
Escaped from turbulent and fretful youth,
Its troubles, passions, bubbles, noise, and foam,
Which are well imaged in the falling stream.
E'en as I look upon thy mighty flood,
Absorb'd in thought, it seems that I become
A part of thee, and in thy thundering waves
My thoughts are lost, and pass to future time,
Seeking the infinite, and rolling on
Towards the sea eternal and unbounded
Of the all-powerful, omnipresent mind![35]

The dialogue that begins at the falls is 'The Proteus, or Immortality'. In the caves of Slovenia curious newt-like creatures called *Proteus* were found; they were white and blind, and had gills and lungs. It was not clear where they came from – perhaps they were thrown up from a sunless sea – or whether they were larvae or perfect animals. They were therefore a good symbol for mankind, the Great Amphibium, whose souls have sight of that immortal sea that brought us hither, and whose birth is but a sleep and a forgetting – Davy's *Consolations* is not so far from Wordsworth's great Immortality Ode, though much more prosy. We are ambiguous creatures, the larvae of what we shall become after death: Davy used Proteus to revive the old image of man as a caterpillar who will die into his chrysalis-coffin and be transformed into a butterfly.

Then come discussions that must recall what went on in the Animal Chemistry Club. We are assured that life is independent of its material basis, as Davy had shown that chemical properties do not depend upon components; and that biology cannot be completely reduced to chemistry or physics. The dreams of those who supposed that living creatures were machines or laboratories have passed away and will pass away: matter in living organisms obeys new laws, and the methods of the physical sciences are inappropriate for studying our minds:

> The doctrine of the *materialists* was always, even in my youth, a cold, heavy, dull and insupportable doctrine to me, and necessarily tending to atheism. When I had heard with disgust, in the dissecting rooms, the plan of the physiologist, of the gradual accretion of matter and its becoming endowed with irritability, ripening into sensibility and acquiring such organs as were necessary, by its own inherent forces, and at last rising into intellectual existence, a walk into the green fields or woods by the banks of rivers brought back my feelings from nature to God; I saw in all the powers of matter the instruments of the deity . . . I saw *love* as the creative principle in the material world, and this love only as a divine attribute.[36]

We should not see this as just another dying man babbling of green fields. Davy's generation turned against even the Christian materialism of Priestley (associated with the resurrection of the body) because of its closeness to that of Holbach and other French philosophes. William Lawrence's lectures, published in 1819 as *The Natural History of Man*, had caused a great scandal and were regarded as blasphemous because they were too close to recent French work and reductionist in tone. The scientific mind must also be both creative and sceptical; scepticism about reductive theories is usually in order, and Davy's empiricism was not a consequence of belated religious fervour. Davy may seem modern in seeing progress in the fossil record, and ancient in opposing recent physiology, but we do not have to give him good or bad marks, and should note that his cautious views were very typical of established men of science.

Consolations ends with two dialogues, on chemistry and on time, which were worked up from earlier writings. The last stresses the relationship between decay and birth, and the cycles going on in the world. Birth and death are necessarily connected. The discussion of chemistry less obviously fits in with the drift of the book, but it forms a justification of a life spent in the science. Chemistry is not just the province of the apothecary and the cook; it is a progressive science wrestling with the fundamental nature of matter and, moreover (both actually and potentially), the most useful of sciences. It could now be elegantly carried on even in a drawing room, using small quantities of reagents, though the chances of a chemist retaining a good eye and a steady hand through life are slim, for laboratory life is full of dangers.

The chemist's life is therefore not merely worthy, but risky and adventurous, and scientific ambition is the highest kind, unlike that of politicians or lawyers desiring mere worldly success and prepared to sacrifice everything for it. What was no doubt drafted as a recruiting document for chemists became in its new context in *Consolations* an *apologia pro vita sua*: the chemist both comes to understand God's world and becomes the benefactor of his fellow men, in a realization of Bacon's vision. What is interesting is that Davy's recommendation for a chemical education is very different from what he had had: he suggests a sound liberal education, as at Oxford or Cambridge, followed by training in the science. Conservative thinkers, alarmed at the prospect of two cultures, echoed such ideas for another generation.

We can see how the book was Davy's legacy, and how he had come to terms with his coming death and with his loss of worldly position. He did love aristocrats, and rejoiced in letters from crowned heads, but fundamentally his pleasures were simpler, and in his retirement and travels he found this out again for himself. In a sense he recovered his youthful vision and his life came full circle. In Podkoren, on the Wurzen Pass and just into Slovenia, a house that he rented for fishing bears a blue plaque in English and Slovenian. The village is still very small; the house is one of the best in it, but it is by no means large – rather like what a country doctor near Penzance might have lived in. In front are the fields, and behind the beech woods and streams where he must have walked. Nearby is the Sava, or Save River, where he loved to fish.[37] The view of the Julian Alps must be still much as he saw it, and perhaps even more than at the Royal Institution (where so much has happened since his day) we can look through his eyes there and meet the distinguished fisherman in his curious outfit. His solitary travels had their gloomy times, but also their pleasures:

> The mountains here are like the needles of Chamouni, of limestone, but of the noblest forms that limestone ever assumes. They surround the village on all sides, and rise with their breast of snow and crests of pointed rock into the middle sky. The source of the Save is a clear blue lake surrounded by woods, and the meadows are as green [in August 1827] as those of Italy in April, or of England in May. Two or three miles nearer the Italian frontier, a fine clear torrent, one of the sources of the Gailen, rises from the lake of Weeper-see, fed by the snows of Mount Manhart, a noble mountain; and two posts below, the Isonzo boils up almost like a sea from a limestone cavern, clear, blue, and cold.[38]

His last days did not wholly lack consolation.

Consolations was dedicated to his old friend Poole, who wrote 'to bear testimony of his general intellectual elevation, and to the warmth, sincerity and simplicity of his heart'.[39] A very independent person, he believed that

> neither the importance of his discoveries, nor the attentions of the exalted in rank or science, whether as individuals or public bodies, nor the honour conferred upon him by his sovereign, made the least alteration in his personal demeanour, or in the tone of his correspondence. No man was ever less spoiled by the world . . . to be useful to science and mankind was that in which he *gloried*, to use a favourite word of his. He was enthusiastically attached to science, and to men of science; and his heart yearned to be useful to mankind, and particularly to the humblest of mankind.[40]

In their tanning and agricultural work, what had struck Poole was 'the quickness and truth of his apprehension. It was a power of reasoning so rapid, when applied to any subject, that he could hardly himself be conscious of the process; and it must, I think, have been felt by him, as it appeared to me, pure intuition. I used to say to him, "You understand me before I half understand myself!" '[41]

Poole had set out 'not to speak of Sir Humphry Davy's discoveries in science, his various literary productions, or his able and upright conduct as a member of public bodies – these are before the public, and evince his greatness – but . . . to show that he was not only one of the greatest, but one of the most amiable of men.'[42] We are not on oath when writing about our dead friends, but Poole's remarks should be set against those of others who reacted otherwise to Davy. Not everything about Davy's character was admirable: he overreacted to criticism, he did have worldly ambitions, he did not always treat others well, he liked to shine, and while protesting belief in a great republic of science above national quarrels, he was keen to outdo the French.

Poole points to two of the key factors in Davy's life: his brilliant mind turned towards understanding nature, and his urge to make his knowledge useful. These are characteristics likely to be found in anybody doing really well in science, though not all societies and times propel the able into science rather than other useful occupations. A combination of fortunate circumstances and an ability to take advantage of them meant that, very unusually, he excelled both in discovery and its useful application. He made real Bacon's vision of knowledge as power in the newly industrializing society of his day. His very success perhaps promoted the idea of the man of science as the solitary trouble-shooter or miracle worker, a consultant to be called in when things went wrong rather than someone to be employed to keep things going right; and his reputation as a genius perhaps did not promote the careers of ordinary works chemists, of whom British industry did not have many in the nineteenth century.

What Poole only hinted at was the social mobility involved in Davy's life. He made a career in chemistry and showed how science could be, as the medieval Church had been, a vehicle for an able man to rise. Faraday, Huxley, Tyndall and others were to follow him along this slippery slope, though Faraday's religion stopped him going all the way. Davy went through the world of patronage and became a patron himself; and though he was unable to realize his vision of the Royal Society, it came to contain more people like him. Whether Davy's life was happy it is difficult to say. There is no doubt that his science gave him times of immense joy, still visible in his scribbled notes; that he also derived great pleasure from nature, and from the fishing and shooting that then went easily with it; and that other parts of his life were associated with unhappiness and frustration. Through the emotional ups and downs of his life came eventually, at least as he saw it, wisdom to be passed on. He was undoubtedly a great man.

We can let Tobin have the last word. He reports that Davy had 'often taken large doses of laudanum and acetate of morphine (of the latter in one day upwards of twenty grains) even more than his physicians approved'; and after his stroke in Rome it seemed 'impossible for him to exist without being read to, and on one day I read Shakespeare to him for *nine* hours'.[43] On what turned out to be the last evening of Davy's life Tobin had read Smollet's *Humphry Clinker* to him until about ten o'clock in the evening. At six the next morning Tobin was told that Davy had died:

I had so often, whilst at Rome, seen Sir Humphry lie for hours together in a state of torpor, and to all appearance dead, that it was difficult for me to persuade myself of the truth; but the delusion at length vanished, and it became too evident that all that remained before me of this great philosopher, was merely the cold and senseless frame with which he had worked. The animating spirit had fled to its oft self-imagined planetary world, there to join the rejoicing souls of the great and good of past ages, soaring from system to system, and with them still to do good in a higher and less bounded sphere, and I knew that it was freed from many a wearisome and painful toil: yet I could not look upon Sir Humphry as he was, without remembering that which he had been, and my tears would fall, spite of my effort to restrain them.[44]

Notes

Preface

1 An excellent recent discussion of this is Wolf Lepenies' *Between Literature and Science: the Rise of Sociology* (Cambridge, CUP, 1988).

Chapter 1 *Beginning: the Meaning of Life*

1 [H. Davy], *Salmonia*, pp. 325ff; the passage was less purple in the first edition. For a bibliography of Davy's writings, see June Z. Fullmer, *Davy's Published Works.*
2 H. Mayhew, *Wonders of Science.*
3 A. Treneer, *Mercurial Chemist*. See also M. Neve, 'The Young Humphry Davy, or John Tonkin's Lament' in S. Forgan (ed.), *Sons of Genius*, pp. 1–32.
4 J. J. Tobin, *Journal*, pp. 210ff, 241.
5 Letter at the Royal Institution.
6 M. Berman, *Social Change and Scientific Organization* is a social history of the early Royal Institution.
7 J. A. Paris, *Humphry Davy*; see esp. p. vi; it was published in one volume quarto, and also two octavo.
8 John Davy appears in *DNB* and in *DSB*.
9 Anne Treneer in her *Mercurial Chemist* is one of the few biographers who tries to be fair to Jane Davy, who was eminent enough in her own right to get into the *Dictionary of National Biography*.
10 J. Davy, *Fragmentary Remains*.

11 H. Davy, *Collected Works*.
12 See my paper 'Davy and Faraday: Fathers and Sons', in D. Gooding and F. James (eds.), *Faraday Rediscovered*.
13 J. Davy, *Memoirs*, Vol. 1, p. 377; Vol. 2, p. 157.
14 H. Hartley, *Humphry Davy*.
15 The classic work on the emergence of professionals is J. Morrell and A. Thackray, *Gentlemen of Science*.
16 H. Hartley, letters of 24 May and 18 December 1967.
17 See my *The Age of Science*. On 'scientist', Coleridge and Davy, see M. Fisch and S. Schaffer (eds.), *William Whewell: a Composite Portrait*, (Oxford, OUP, 1991), p. 178.
18 C. A. Russell, G. K. Roberts and N. Coley, *Chemists by Profession*, pp. 24–6.
19 D. Hartley, *Observations on Man*, (London, Leake, 1749), Vol. 2, p. 243.
20 T. S. Kuhn *The Structure of Scientific Revolutions*, 2nd edn (Chicago, University of Chicago Press, 1972).
21 H. Davy, *Consolations*.
22 S. T. Coleridge, *Confessions of an Enquiring Spirit*, H. St J. Hart, (London, Black, 1956); K. Coburn, *Inquiring Spirit*.
23 J. Davy, *Memoirs* Vol. 2, p. 385; R. Floud, K. Wachter and A. Gregory, *Height, Health and History*, Cambridge, CUP, 1990.
24 R. Siegfried, 'Davy's "Intellectual Delight" and his Lectures at the Royal Institution', in Forgan (ed.), *Sons of Genius*, pp. 177–99. Siegfried has usefully stressed Davy's empiricism.
25 J. Z. Fullmer, 'Humphry Davy, Reformer' in Forgan (ed.), *Sons of Genius*, pp. 59–94.
26 D. P. Miller, 'Between Hostile Camps: Sir Humphry Davy's Presidency of the Royal Society of London, 1820–1827', *British Journal for the History of Science*, 16 (1983) pp. 1–47.
27 See my paper 'Davy's *Salmonia*' in Forgan (ed.), *Sons of Genius*, pp. 201–30.
28 A. Geikie, *The Life of Sir Roderick I. Murchison*, 2 vols, (London, Murray, 1875), Vol. 1, p. 94.
29 J. Davy, *Fragmentary Remains*, p. 310.
30 M. P. Crosland, 'Explicit Qualifications', *Notes and Records of the Royal Society*, 37 (1983), 167–87; and on the Society in the nineteenth century, M. B. Hall, *All Scientists Now*.
31 H. Hartley, *Humphry Davy*, p. 148.
32 J. Davy, *Memoirs*, Vol. 2, pp. 281ff.
33 J. Davy, *Fragnentary Remains*, p. 288.
34 M. Neve, 'The Young Humphry Davy, or John Tonkin's Lament', in Forgan (ed.), *Sons of Genius*, p. 2.

35
> The progress of the human race, its destiny, the fate of each of us, the course of thousands of globes of which astronomers can see only very few; these formed the subject of dialogues where the poet shines no less than the

philosopher, and where, among the various stories, a powerful mind applies itself to the most serious of questions. We might say that once escaped from the laboratory, he found again those sweet dreams, those sublime thoughts which had enchanted his youth; it was in a way the work of a dying Plato.

G. Cuvier, *Eloges Historiques* (Paris, n.d.), p. 354.

36 J. Davy, *Fragmentary Remains*, p. 14.
37 J. Davy, *Memoirs*, Vol. 2, p. 345.
38 ibid., p. 384.
39 H. Davy, *Collected Works*, Vol. 9, pp. 383–8.
40 M. Rudwick, *The Great Devonian Controversy* (Chicago, 1985).
41 [J. Marcet], *Conversations on Chemistry*, Vol. 1, pp. v–vi.
42 G. Chilton and N. G. Coley, 'The Laboratories of the Royal Institution', *Ambix*, 27 (1980), pp. 173–203.
43 M. Faraday, *Chemical Manipulation*; A. E. Jeffreys, *Faraday: Lectures and Published Writings*.

CHAPTER 2 *Growing Up*

1 A. Briggs, *Iron Bridge to Crystal Palace: Impact and Images of the Industrial Revolution* (London, 1979).
2 For a brief account of Watt, see D. S. L. Cardwell's essay in R. Porter (ed.), *Man Masters Nature* (London, BBC, 1988).
3 J. Farey, *A Treatise on the Steam Engine*, (Newton Abbott, David & Charles, 1971) Vol. 2 [written 1851], is chiefly concerned with Cornish engines.
4 W. Borlase, *The Natural History of Cornwall*, [Oxford, 1758], reprint intr. F. A. Turk and P. A. S. Pool, (London, E & W, 1970); *Antiquities of Cornwall*, [2nd edn, London, 1769] reprint intr. P. A. S. Pool and C. Thomas, (Wakefield, E. P., 1973).
5 For a slightly later time, see R. Barton (ed.), *Life in Cornwall in the Early Nineteenth Century: Extracts from the West Briton Newspaper, 1810–1835* (Truro, Bradford Barton, 1970).
6 R. Southey, *The Life of Wesley and the Rise and Progress of Methodism* [1820], new edn (London, 1890), pp. 283ff, 287ff, 294.
7 J. Davy, *Memoirs*, Vol. 1, p. 14.
8 F. Galton, 'On Men of Science, their Nature and their Nurture', *Proceedings of the Royal Institution*, 7 (1874), pp. 227–36, esp. pp. 231, 234.
9 J. A. Paris, *Humphry Davy*, p. 5.
10 J. Davy, *Memoirs*, Vol. 1, p. 17.
11 ibid., p. 20.
12 On these men and on Davy's medicine, see M. Neve, 'The Young Humphry Davy, or John Tonkin's Lament', in S. Forgan (ed.), *Sons of Genius*, pp. 1–32.
13 J. Davy, *Memoirs*, Vol. 1, pp. 22ff.

14 H. Hartley, *Humphry Davy*, p. 148.
15 Paris, *Humphry Davy*, p. 14.
16 Ibid, pp. 16–20; A. Pritchard (ed.), *Poetry*, pp. 2ff.
17 E. Burke, *A Philosophical Enquiry into the Origin of Our Ideas of the Sublime and Beautiful*, 2nd edn (London, 1759).
18 A. Cunningham and N. Jardine (eds.), *Romanticism and the Sciences*.
19 Paris, *Humphry Davy*, pp. 30ff.
20 ibid., p. 10.
21 R. G. Neville and W. A. Smeaton, 'Macquer's *Dictionnaire de Chymie*: a Bibliographical Study', *Annals of Science*, 38 (1981), pp. 613–62.
22 See my 'Revolutions in Chemistry', in W. Shea (ed.), *Revolutions in Science*, pp. 49–69.
23 A. D. C. Simpson (ed.), *Joseph Black, 1728–1799* (Edinburgh, Royal Museum Scotland, 1982).
24 E. Robinson and D. McKie (eds.), *Partners in Science*.
25 R. E. Schofield, *The Lunar Society of Birmingham*, pp. 420ff.
26 ibid.
27 Paris, *Humphry Davy*, p. 34.
28 Hartley, *Humphry Davy*, p. 14.
29 Cunningham and Jardine (eds.), *Romanticism and the Sciences*, pp. 105ff, 215.
30 For a biography, see D. A. Stansfield, *Thomas Beddoes MD*; see also T. H. Levere, 'Dr Thomas Beddoes: Science and Medicine in Politics and Society', *British Journal for the History of Science*, 17 (1984), pp. 187–204.
31 H. Davy *Works*, Vol. 2, p. 3.
32 ibid., p. 24.
33 See G. Cantor, *Optics after Newton: Theories of Light in Britain and Ireland, 1704–1840*, (Manchester, Manchester University Press, 1983).
34 H. Davy, *Works*, Vol. 2, p. 36.
35 M. Pera, 'Radical Theory Change and Empirical Equivalence: the Galvani–Volta Controversy', in W. Shea, *Revolutions in Science*, pp. 133–56.
36 H. Davy, *Works*, Vol. 2, pp. 82ff.
37 ibid.
38 ibid.
39 Neve, 'The Young Humphry Davy', p. 9.

Chapter 3 *Clifton*

1 Royal Institution, Davy notebook 13e, p. 23.
2 J. Davy, *Memoirs*, Vol. 1, p. 65.
3 J. Cottle, *Reminiscences*, p. 263.
4 Davy notebook 13e, p. 33.
5 J. A. Paris, *Humphry Davy*, p. 56.
6 H. Davy, *Researches*, p. 454.
7 Cottle, *Reminiscences*, pp. 269–70.
8 J. Davy, *Memoirs*, Vol. 1, p. 104.

9 H. Davy, *Consolations*, p. 252.
10 H. Davy, *Researches*, p. 458.
11 ibid., p. 487.
12 A. Hayter, *Opium and the Romantic Imagination*, London, 1968.
13 Cottle, *Reminiscences*, pp. 463ff.
14 H. Davy, *Researches*, p. 462.
15 ibid.
16 See the paper by M. Lefebure in R. Gravil and M. Lefebure (eds.), *The Coleridge Connection*, (London, 1990); J. Cottle, *Reminiscences*, p. 274; R. Holmes, *Coleridge*, p. 245.
17 Paris, *Humphry Davy*, p. 51.
18 See Simon Schaffer's essay in D. Gooding, T. Pinch and S. Schaffer (eds.), *The Uses of Experiment*.
19 My book *The Nature of Science* explored it under these three aspects.
20 J. Ziman, *Public Knowledge* (Cambridge, CUP, 1968); but see M. Polanyi, *Personal Knowledge* (London, 1958). In my *Ideas in Chemistry* there is a chapter concerned with the occult science of alchemy.
21 J. Davy, *Memoirs*, Vol. 1, p. 120.
22 Cottle, *Reminiscences*, p. 267. On English radicals, see J. Epstein, 'Understanding the Cap of Liberty', *Past and Present*, 122 (1989) pp. 75–118.
23 H. Davy, *Researches*, pp. 497ff.
24 ibid.
25 ibid.
26 ibid.
27 ibid., p. 536.
28 J. Davy, *Memoirs*, Vol. 1, p. 102; H. Davy, *Researches*, p. 557.
29 H. Davy, *Researches*, pp. 537ff.
30 Ibid.
31 Cottle, *Reminiscences*, p. 270.
32 A. Cunningham and N. Jardine, *Romanticism and the Sciences*.
33 Davy notebook 13e, p. 24.
34 Paris, *Humphry Davy*, pp. 54ff.
35 ibid.
36 ibid.
37 H. Davy, *Consolations*, p. 233.
38 J. Davy, *Memoirs*, Vol. 1, p. 119; Royal Institution, Davy notebooks 13d, pp. 9–10, and 13f, pp. 3ff, 6ff.
39 ibid.
40 J. Davy, *Memoirs*, Vol. 1, p. 117.
41 J. Davy, *Memoirs*, Vol. 1, p. 97; Davy notebook 13c, p. 1.
42 Cottle, *Reminiscences*, pp. 279ff.
43 R. Sharrock, 'The Chemist and the Poet: Sir Humphry Davy and the Preface to the *Lyrical Ballads*', *Notes and Records of the Royal Society*, 17 (1962), pp. 57ff.
44 H. Hartley, *Humphry Davy*, p. 24.
45 Treneer, *Mercurial Chemist*, pp. 63–4; A. Pritchard (ed.), *Poetry*, p. 14.

46 J. Priestley, *Experiments and Observations*.
47 I owe this contrast to J. J. Thomson, comparing himself to Crookes, in his *Recollections and Reflections* (London, Bell, 1936) p. 379.
48 H. Davy, *Researches*, p. 558.
49 See M. Pera, 'Radical Theory Change and Empirical Equivalence: the Galvani–Volta Controversy', in W. Shea (ed.), *Revolutions in Science*.
50 A. Volta, 'On the Electricity excited by the mere Contact of Conducting Substances of different kinds'. *Philosophical Transactions*, 90 (1800), pp. 403ff.
51 J. Priestley, *Electricity*, Vol. 1, p. xiv.
52 H. Davy, 'Experiments on Galvanic Electricity', [*Nicholson*'s] *Journal of Natural Philosophy, Chemistry and the Arts*, 4 (1800), p. 275.
53 H. Davy, 'An Account of some Galvanic Combinations', *Philosophical Transactions*, 91 (1801), pp. 397–402.
54 Paris, *Humphry Davy*, p. 109.
55 H. Davy, 'Observations on the Causes of the Galvanic Phenomena' [*Nicholson's*] *Journal of Natural Philosophy, Chemistry and the Arts*, 4 (1800), p. 326.
56 J. Davy, *Memoirs*, Vol. 1, p. 131.
57 ibid., p. 132.

CHAPTER 4 *The Bright Day*

1 J. A. Paris, *Humphry Davy*, p. 89.
2 J. Morrell and A. Thackray, *Gentlemen of Science* describes the emergence of the scientific community in Britain.
3 B. Thompson, *The Collected Works of Count Rumford*; Vol. 1 deals with the nature of heat.
4 M. Berman, *Social Change and Scientific Organization*.
5 F. Greenaway (ed.), *Archives*, reprints these documents in facsimile.
6 H. Davy, *Consolations*, p. 234.
7 T. H. Huxley, *Lay Sermons, Addresses, and Reviews*, 6th edn, (London, 1877), p. 77; prints an address delivered in 1854.
8 T. Young, *A Course of Lectures on Natural Philosophy and the Mechanical Arts* (London, Johnson, 1807; reprinted New York, Johnson, 1971).
9 Paris, *Humphry Davy*, pp. 79–80.
10 R. Hudson (ed.) *Coleridge among the Lakes and Mountains* (London, Folio Society, 1991), p. 119.
11 Berman, *Social Change and Scientific Organization*, p. 49.
12 ibid.
13 E. Meteyard, *A Group of Englishmen (1795–1815): being Records of the Younger Wedgwoods and their Friends* (London, 1871).
14 H. Davy, *Consolations*.
15 H. Davy, 'The constituent parts of certain astringent Vegetables', *Philosophical Transactions*, 93 (1803), pp. 233–73.
16 ibid.

17 ibid., p. 272.
18 C. L. Berthollet, *Essai de Statique Chimique* (Paris, 1803), pp. 1ff.
19 Experiments on tanning are reported in Davy's notebook 13c at the Royal Institution, with electrical experiments done at the same period, and with drafts of lectures and of his epic on Moses.
20 Paris, *Humphry Davy*, p. 104–5.
21 C. Hatchett, 'On an artificial substance which possesses the principal characteristic properties of tannin', *Philosophical Transactions*, 95 (1805), pp. 211–24, 285–315; 96 (1806), pp. 109–46.
22 *Encyclopedia Metropolitana* (29 vols., London, 1807–45), Vol. 8, pp. 542–56; this is part of a book-length contribution by Peter Barlow on 'manufactures', including copious statistics on the leather trade.
23 D. Valenze, 'Women's Work and the Dairy Industry, c1740–1840s', *Past and Present*, 130 (1991), pp. 142–69.
24 These are listed in my *Natural Science Books in English, 1600–1900*, 2nd edn, (London, 1989).
25 G. Sinclair, *Hortus Gramineus Woburnensis*, 3rd edn (London, 1826); the first edition contained actual pressed grasses.
26 H. Davy, *Agricultural Chemistry*.
27 J. B. Boussingault, *Rural Economy*, tr. G. Law, 2nd edn, (London, 1845); F. W. J. McCosh, *Boussingault*, (Dordrecht, N. Holland, 1984).
28 J. Liebig, *Chemistry in its Applications to Agriculture and Physiology*, tr. L. Playfair, 3rd edn, (London, 1843); W. H. Brock is writing a biography of Liebig.
29 Royal Institution, Davy MSS, Box 6, Laboratory Notebook, 1805–9.
30 Royal Institution, Davy notebook 13c, p. 101.
31 Paris, *Humphry Davy*, pp. 545–7 reprints Davy's will.
32 H. Davy, *Agricultural Chemistry*, pp. 264ff.
33 ibid.
34 ibid.
35 H. Davy, *Works*, Vol. 7, pp. 175ff.
36 J. Davy, *Memoirs*, Vol. 1, p. 263.
37 ibid., p. 136.
38 ibid., p. 262; John lived with Humphry and acted as his assistant from 1808 until he went to Edinburgh University in 1811. The future Cardinal Newman was reported similarly to have preached in a dirty surplice: S. Gilley, *Newman and His Age*, (London, 1990), p. 173.
39 J. Davy, *Memoirs*, Vol. 1, p. 262.
40 ibid.
41 Paris, *Humphry Davy*, p. 119.
42 ibid., pp. 128ff.
43 R. Holmes, *Coleridge*, p. 360.
44 Paris, *Humphry Davy*, p. 131.
45 Royal Institution, Davy MSS, Box 2, A/1.
46 D. A. Stansfield, *Thomas Beddoes*, p. 249.
47 J. Davy, *Memoirs*, Vol. 2, p. 393.

48 N. G. Coley, 'The Animal Chemistry Club', *Notes and Records of the Royal Society*, 22 (1967), pp. 173–85.
49 G. Averley, 'English Scientific Societies of the Eighteenth and Early Nineteenth Centuries', PhD thesis, CNAA (Teesside Polytechnic), 1989; and 'The Social Chemists', *Ambix*, 33 (1986), pp. 99–128.
50 J. Davy, *Memoirs*, Vol. 1, pp. 301.
51 R. Siegfried, 'Davy's "Intellectual Delight" and his lectures at the Royal Institution', in S. Forgan (ed.) *Sons of Genius*, pp. 182–95 describes these lectures; and see R. Siegfried and R. H. Dott (eds.), *Humphry Davy on Geology*.
52 J. Davy, *Memoirs*, Vol. 1, p. 267.
53 ibid., p. 301.

CHAPTER 5 *Electric Affinity*

1 T. S. Kuhn, *The Structure of Scientific Revolutions*, 2nd edn, (Chicago, University of Chicago Press, 1970).
2 F. W. J. Schelling, *Ideas for a Philosophy of Nature*, tr. E. E. Harris and P. Heath, intr. R. Stern (Cambridge, CUP, 1988). See also A. Cunningham and N. Jardine (eds.), *Romanticism and the Sciences*; H. Schnädelbach, *Philosophy in Germany 1831–1933*, tr. E. Matthews (Cambridge, CUP, 1984).
3 See Barry Gower's important paper 'Speculation in Physics', *Studies in the History and Philosophy of Science*, 3 (1973), pp. 301–56; and J. W. Ritter, 'Die Begrundung der Elektrochemie', in A. Hermann (ed.), *Ostwalds Klassiker der Exacten Wissenschaften*, new series 2 (Frankfurt am Main, 1968).
4 J. Davy, *Memoirs*, Vol. 1, pp. 62–3; A. Pritchard (ed.), *Poetry*, p. 17.
5 J. Priestley, *Electricity*.
6 H. Cavendish, *Electrical Researches*, ed. J. C. Maxwell, (Cambridge, CUP, 1879), pp. 194–215; this first appeared in *Philosophical Transactions* in 1776.
7 W. H. Wollaston, 'Experiments on the Chemical Production and Agency of Electricity', *Philosophical Transactions*, 91 (1801), pp. 427–34.
8 For a more detailed discussion of such ideas, see my *Transcendental Part of Chemistry*, ch. 2.
9 The classic account is by C. A. Russell, 'The Electrochemical Theory of Sir Humphry Davy', *Annals of Science*, 15 (1959), pp. 1–25; 19 (1963), pp. 255–71; it is extremely valuable.
10 M. Faraday, *Experimental Researches in Electricity*, (London, 1839), Vol. 1, p. 136.
11 C. A. Russell, *Annals of Science*, 19 (1963), p. 270.
12 Royal Institution, Davy MSS, Box 6, Laboratory Notebook 1805–1809.
13 ibid., expt 43.
14 H. Lyons, *The Royal Society 1660–1940* (Cambridge, CUP, 1944), p. 192.
15 J. J. Berzelius, *Traite de Chimie*, tr. A. L. Jourdain & Esslinger, (Paris, 1829–33), Vol. 1, p. 164.

16 T. Thomson, *The History of Chemistry*, 2nd edn, (London, 1830–1), Vol. 2, p. 260.
17 H. Davy, 'The Bakerian Lecture, on some chemical Agencies of Electricity', *Philosophical Transactions*, 97 (1807), pp. 1–56; the plate that follows shows the agate tubes, gold cones and amianthus bridges.
18 ibid., p. 12.
19 ibid., pp. 38ff.
20 ibid.
21 ibid., pp. 55–6.
22 Paris, *Humphry Davy*, pp. 168–9.
23 Laboratory Notebook, 19, October 1807.
24 ibid., fo. 85.
25 ibid., fo. 101.
26 ibid., fo. 103.
27 H. Davy, 'The Bakerian Lecture, on some new Phenomena of Chemical Changes produced by Electricity', *Philosophical Transactions*, 98 (1808), pp. 1–44.
28 ibid.
29 ibid.
30 ibid.
31 ibid.
32 ibid.
33 ibid.
34 See my *Ordering the World: a History of Classifying Man*, (London, Burnet, 1981).
35 H. Davy, 'New Phenomena of Chemical Changes'.
36 E. Home, 'A Description of the Anatomy of *Ornithorhyncus paradoxus*', *Philosophical Transactions*, 92 (1802), pp. 67–84; J. W. Gruber, 'Does the platypus lay eggs? The History of an Event in Science', *Archives of Natural History*, 18 (1991), pp. 51–123.
37 H. Davy, 'New Phenomena of Chemical Changes'.
38 ibid.
39 ibid.
40 ibid.
41 H. Davy, 'Electro-Chemical Researches', *Philosophical Transactions*, 98 (1808), pp. 333–70.
42 M. Ignatieff, *A Just Measure of Pain: the Penitentiary in the Industrial Revolution* (Harmondsworth, Penguin, 1989).
43 J. Davy, *Memoirs*, Vol. 1, pp. 390–2; Pritchard, *Poetry*, pp. 21ff.
44 See the essay on him by Wilhelm Oldenberg in T. Frangsmyr (ed.), *Science in Sweden* (Canton, Mass., 1989), pp. 124–47.
45 J. J. Berzelius, *Essai sur la Theorie des Proportions Chimiques et sur l'influence chimique de l'electricite*, [Paris 1819], reprint usefully intr. C. A. Russell, (New York, 1972).
46 J. Davy, *Memoirs*, Vol. 2, p. 216.

CHAPTER 6 *Forces, Powers and Chemistry*

1 J. Priestley, *Disquisitions relating to Matter and Spirit*, 2nd edn, (Birmingham, 1782); see Royal Society of Chemistry, Special Publication 48, *Oxygen and the Conversion of Future Feedstocks*, (1983), esp. the paper by R. E. Schofield, pp. 410–31.
2 E. Mendelsohn, *Heat and Life* (Cambridge, Mass, Harvard University Press, 1964); my paper, 'The Vital Flame', *Ambix*, 23 (1976), pp. 5–15.
3 N. G. Coley, 'Medical Chemistry at Guy's Hospital, 1770–1850', *Ambix*, 35 (1988), pp. 155–68.
4 See my paper, 'Accomplishment or Dogma: Chemistry in the introductory works of Jane Marcet and Samuel Parkes', *Ambix*, 33 (1986), pp. 94–8.
5 There are various essays on physiology in A. Cunningham and N. Jardine, *Romanticism and the Sciences*; and see my *Ideas in Chemistry*, ch. 6.
6 N. G. Coley, 'The Animal Chemistry Club', *Notes and Records of the Royal Society*, 22 (1967), pp. 173–85.
7 B. C. Brodie, *Psychological Inquiries*.
8 Coley, 'The Animal Chemistry Club', pp. 177ff. H. Hartley, *Studies on the History of Chemistry* (Oxford, OUP, 1971); ch. 5 is on the place of Berzelius in the history of chemistry.
9 F. Greenaway, in D. S. L. Cardwell (ed.), *John Dalton and the Progress of Science*, (Manchester, Manchester University Press, 1968), pp. 206ff. A. Thackray, *John Dalton: Critical Assessments of His Life and Work* (Cambridge, Mass, Harvard University Press, 1972), pp. 56, 78, 103ff, 106.
10 Dalton's writings and those of others, including Davy, Thomson and Wollaston, are reprinted in facsimile in my *Classical Scientific Papers: Chemistry* (London, Mills & Boon, 1968).
11 H. Davy, 'The Bakerian Lecture, on some new Phenomena of Chemical Changes produced by Electricity', *Philosophical Transactions*, 98 (1808), p. 44.
12 T. Thomson, 'On Oxalic Acid', *Philosophical Transactions*, 98 (1808), pp. 63–95.
13 W. H. Wollaston, 'On Super-acid and Sub-acid Salts', *Philosophical Transactions*, 98 (1808), pp. 96–102.
14 On this distinction, see my *Atoms and Elements*, 2nd edn, (London, Hutchinson, 1970), and A. J. Rocke, *Chemical Atomism*.
15 M. J. Nye, *The Question of the Atom* (Los Angeles, Tomash, 1984).
16 W. H. Brock (ed.), *The Atomic Debates* (Leicester, Leicester University Press, 1967).
17 H. Davy, *Works*, Vol. 4.
18 W. H. Wollaston, 'A Synoptic Scale of Chemical Equivalents', *Philosophical Transactions*, 104 (1814), pp. 1–22.
19 H. Davy, *Works*, Vol. 4, pp. 364–5, 374–5.
20 ibid.
21 ibid.

22 W. H. Brock, *From Protyle to Proton: William Prout and the Nature of Matter, 1785–1985* (Bristol, 1985).
23 The writings of Prout, Thomson, Berzelius and others in the nineteenth century on this question are reprinted in facsimile in my *Classical Scientific Papers: Chemistry, series 2* (London, Mills & Boon, 1970).
24 M. Faraday, 'A Speculation touching Electric Conduction and the Nature of Matter', *Experimental Researches in Electricity* (London, 1844), Vol. 2, pp. 284–93; L. P. Williams, *Michael Faraday*.
25 H. Davy, *Works*, Vol. 7, pp. 93–9.
26 ibid.
27 ibid.
28 ibid.
29 See my *Transcendental Part of Chemistry*.
30 See my 'Tyrannies of Distance in British Science', in R. W. Home and S. G. Kohlstedt (eds.), *International Science*, pp. 39–53.
31 C. W. Scheele, *Chemical Essays*, tr. F. X. Schwediaur (London, 1786; facsimile reprint London, Dawson, 1966), pp. 90ff. See also *Annals of Science*, 14 (1968), pp. 259–73 for Linder and Smeaton's account of the translation process.
32 Scheele, *Chemical Essays*.
33 M. P. Crosland, *The Society of Arcueil*.
34 'The Early History of Chlorine', *Alembic Club Reprints*, 13 (1897), pp. 11–31; also contains translations of papers by Scheele, Berthollet, Guyton, and Gay-Lussac and Thenard.
35 Guyton de Morveau, *Alembic Club Reprints*, 13, pp. 32–3.
36 M. P. Crosland, *Gay-Lussac*.
37 M. P. Crosland, 'Davy and Gay-Lussac: Competition and Contrast', in S. Forgan (ed.), *Sons of Genius*, pp. 95–120.
38 J. Coyac, M. Fetizon, M. Sadoun-Goupil and R. Taton (eds.), *Actes du colloque Gay-Lussac* (Palaiseau, École Polytechnique, 1980).
39 This and many other classic papers in the history of chemistry are published in English in H. M. Leicester and H. S. Klickstein, *A Source Book in Chemistry* (Cambridge, Mass., 1952); see pp. 292ff.
40 *Alembic Club Reprints*, 13, pp. 34–49.
41 The detailed account is by J. H. Brooke, 'Davy's Chemical Outlook: the Acid Test', in Forgan (ed.), *Sons of Genius*, pp. 121–76.
42 J. A. Paris, *Humphry Davy*, pp. 202–3.
43 ibid.
44 ibid., pp. 200ff.
45 ibid.
46 His writings are usefully collected in H. Davy, 'The Elementary Nature of Chlorine', *Alembic Club Reprints*, 9 (1894).
47 H. Davy, 'Researches on the Oxymuriatic Acid', *Philosophical Transactions*, 100 (1810), pp. 231–57; Royal Institution, Davy notebook 13i, p. 37.
48 Berzelius introduced the letter symbols of atoms, but before the 1860s

there was little agreement about formulae. I have used them here and elsewhere as convenient anachronisms.

49 H. Davy, 'Oxymuriatic Acid'.
50 ibid.
51 H. Davy, 'On some of the Combinations of Oxymuriatic Gas and Oxygene, and on the chemical Relations of these Principles, to inflammable Bodies', *Philosophical Transactions*, 101 (1811), pp. 1–35, esp. pp. 28ff.
52 ibid.
53 H. Davy, *Works*, Vol. 8, p. 313.
54 J. H. Brooke, 'Davy's Chemical Outlook: the Acid Test', in Forgan, *Sons of Genius*, pp. 134f.
55 H. Davy, *Chemical Philosophy*, p. 485.
56 H. Davy, 'On the fluoric Compounds', *Philosophical Transactions*, 104 (1814), p. 71.
57 S. T. Coleridge, *The Friend*, ed. B. Rooke, (London, Routledge, 1969), Vol. 1, p. 471; this reprints the version of 1818.

CHAPTER 7 *A Chemical Honeymoon, in France*

1 J. Davy, *Fragmentary Remains*, p. 155.
2 ibid., p. 156.
3 ibid., pp. 141ff.
4 See 'Davy, Jane' in *DNB*.
5 Corinne was the eponymous heroine of a novel about an Italian poetess: written by Mme de Staël, in 1807.
6 *DNB*.
7 J. A. Paris, *Humphry Davy*, p. 220.
8 J. Z. Fullmer, *Davy's Published Works*, pp. 67, 70.
9 H. Hartley, *Humphry Davy*, pp. 94ff.
10 J. Davy, *Fragmentary Remains*, p. 142.
11 ibid.
12 ibid.
13 ibid., p. 71.
14 Paris, *Humphry Davy*, pp. 220ff.
15 ibid.
16 ibid.
17 ibid., p. 225.
18 J. Davy, *Memoirs*, Vol. 1, p. 457.
19 ibid., p. 459.
20 ibid., p. 143.
21 J. Davy, *Fragmentary Remains*, pp. 163–4; notebook entries from this period.
22 ibid.
23 ibid., p. 169.
24 Paris, *Humphry Davy*, p. 267.

25 J. Davy, *Memoirs*, Vol. 1, pp. 460–1.
26 H. B. Jones, *Faraday*, Vol. 1, p. 75.
27 J. Davy, *Fragmentary Remains*, p. 322.
28 W. Falconer, *An Universal Dictionary of the Marine* (London, 1780).
29 J. Davy, *Memoirs*, Vol. 1, p. 463.
30 Jones, *Faraday*, Vol. 1, pp. 77–8.
31 Ibid.
32 E. G. Holt, *The Triumph of Art for the Public*, (Princeton, Princeton University Press, 1983), pp. 75ff., 90ff. Anon, *Notice des Tableaux*.
33 Paris, *Humphry Davy*, p. 268.
34 ibid.
35 ibid.
36 Jones, *Faraday*, Vol. 1, p. 84.
37 Paris, *Humphry Davy*, p. 270.
38 Jones, *Faraday*, Vol. 1, p. 87. See also Paris, *Humphry Davy*, pp. 272ff; J. Davy, *Memoirs*, pp. 463ff. Davy kept no journal in Paris.
39 Jones, *Faraday*, Vol. 1, p. 90.
40 H. Davy, 'On fluoric Compounds', *Philosophical Transactions*, 104 (1814), pp. 62–73.
41 H. Davy, 'Some Experiments and Observations on a new Substance which becomes a violet coloured Gas by Heat', *Philosophical Transactions*, 104 (1814), pp. 74–93, 487–507.
42 J. Davy, *Memoirs*, Vol. 1, pp. 468ff.
43 See Davy's 'explanation' in *Quarterly Journal of the Royal Institution*, 3 (1817), pp. 378f.
44 Jones, *Faraday*, Vol. 1, p. 92.
45 J. Davy, *Memoirs*, Vol. 1, p. 472.
46 H. Davy, 'Some Experiments on the Combustion of the Diamond', *Philosophical Transactions*, 104 (1814), pp. 557–70.
47 H. Davy, 'Some experiments on a solid compound of iodine and oxygene', *Philosophical Transactions*, 105 (1815), pp. 203–13, 214–19. See J. Davy, *Memoirs*, Vol. 1, pp. 497ff on further controversy with Gay-Lussac.
48 H. Davy, 'Experiments and Observations on the Colours used in Painting by the Ancients', *Philosophical Transactions*, 105 (1815), pp. 99–100.
49 On Canova, see E. G. Holt, *Triumph of Art*, pp. 16, 30ff.
50 H. Davy, 'Experiments and Observations'.
51 ibid.
52 ibid.
53 J. Davy, *Memoirs*, Vol. 1, p. 488.
54 ibid, pp. 489–90; A. Pritchard, *Poetry*, pp. 29–30.
55 J. Davy, *Memoirs*, Vol. 1, p. 507.

CHAPTER 8 *The Safety Lamp*

1 J. Davy, *Memoirs*, Vol. 1, p. 481.
2 ibid., p. 501.

3 Paris, *Humphry Davy*, p. 308.
4 S. Carnot, *Reflexions sur la puissance motrice du feu*, ed. R. Fox, (Paris, 1978), p. 62.
5 An account of the dreadful accident which happened at Felling Colliery', *Annals of Philosophy*, 1 (1813), pp. 355–65, 438–47.
6 W. R. Clanny, 'On the means of procuring a steady Light in Coal Mines without the danger of Explosion', *Philosophical Transactions*, 103 (1813), pp. 200–5 and plate XV.
7 ibid.
8 ibid.
9 G. Basalla, *The Evolution of Technology*, (Cambridge, CUP, 1988).
10 Paris, *Humphry Davy*, p. 309.
11 H. T. de la Beche (ed. & tr.), *A Selection of the Geological Memoirs contained in the Annales des Mines* (London, 1828); my copy was his.
12 Paris, *Humphry Davy*, p. 310.
13 ibid. pp. 311ff.
14 J. Davy, *Fragmentary Remains*, p. 208; W. R. Dawson (ed.), *The Banks Letters* (London, 1958), p. 253.
15 See his later letter to Gray, in Paris, *Humphry Davy*, p. 314.
16 H. Davy, 'On the fire-damp of coal mines, and on methods of lighting the mines so as to prevent its explosion', *Philosophical Transactions*, 116 (1816), pp. 1–22 and plate I.
17 ibid., p. 3.
18 ibid., p. 9.
19 ibid.
20 ibid., p. 12.
21 H. Davy, 'Further experiments on the combustion of explosive mixtures confined by wire-gauze', *Philosophical Transactions*, 106 (1816), pp. 115–19.
22 ibid.
23 H. Davy, 'On the Wire-gauze Safe-lamps', *Quarterly Journal of Science and the Arts*, 1 (1816), pp. 1–5.
24 J. Buddle, 'Letter to Sir H. Davy', *Quarterly Journal of Science and the Arts*, 1 (1816), pp. 302–7.
25 ibid.
26 ibid.
27 J. Davy, *Fragmentary Remains*, p. 210.
28 C. Smith and M. N. Wise, *Energy and Empire: a biographical study of Lord Kelvin*, (Cambridge, CUP, 1988), ch. 23.
29 Paris, *Humphry Davy*, pp. 330ff; J. Davy, *Fragmentary Remains*, p. 211.
30 Paris, *Humphry Davy*, p. 337.
31 ibid., p. 338.
32 On Tennant, see the obituary in *Annals of Philosophy*, 6 (1815), pp. 1–11, 81–100.
33 H. Davy, 'Further experiments', p. 8 note.
34 H. Hartley, *Humphry Davy*, p. 119.
35 Obituary, *Annals of Philosophy*, 6 (1815), p. 98.

36 J. B. Longmire, 'Remarks on the Wire-Gauze lamp recently constructed by Sir H. Davy', *Annals of Philosophy*, 8 (1816), pp. 31–4, and the rejoinder and continuations, pp. 265–71, 420–9, 429–33.
37 J. Murray, 'Observations on the Fire-Damp in Coal-Mines, with a Plan for Lighting Mines so as to guard against its Explosion', *Annals of Philosophy*, 8 (1816), pp. 406–20 and plate facing p. 352.
38 H. Davy, 'Notice of some Experiments and New Views respecting Flame', *Quarterly Journal of Science and the Arts*, 2 (1817), pp. 124–7; and compare his 'Some Researches on Flame' in *Philosophical Transactions*, 107 (1817), pp. 45–76, 77–85, and plate of wire-gauze lamp.
39 H. Davy, 'Notice of some Experiments'.
40 F. Accum, *Description of the Process of manufacturing coal-gas*, (London, 1820).
41 H. Davy, 'Notice of some Experiments'.
42 M. Faraday, *A Course of Six Lectures . . . on the Chemical History of a Candle* (London, 1861). Faraday reported on Davy's experiments in the *Quarterly Journal of Science and the Arts*, 2 (1817), pp. 463–5.
43 H. Davy, 'Notice of some Experiments'.
44 ibid.
45 H. Davy, 'Some Researches on Flame', pp. 77ff.
46 Paris, *Humphry Davy*, pp. 347ff.
47 ibid., p. 348.
48 ibid.
49 J. Davy, *Fragmentary Remains*, pp. 220f.
50 J. Davy, *Memoirs*, Vol. 2, p. 101.
51 H. Davy, 'Some observations and Experiments on the Papyri found in the ruins of Herculaneum', *Philosophical Transactions*, 111 (1821), pp. 191–208.
52 J. Davy, *Fragmentary Remains*, p. 221.
53 ibid., p. 222.
54 ibid.
55 J. Davy, *Memoirs*, Vol. 2, p. 126.
56 Paris, *Humphry Davy*, p. 369.

CHAPTER 9 *A Son in Science: Davy and Faraday*

1 This chapter is closely based upon my article 'Davy and Faraday: Fathers and Sons' in the very useful symposium, David Gooding and Frank James (eds.), *Faraday Rediscovered*, pp. 33–49.
2 M. Shelley, *Frankenstein*, pp. 24ff, 31, 95ff.
3 H. Davy, *Works*, Vol. 2, pp. 318, 320, 321.
4 ibid.
5 J. C. Beaglehole (ed.), *The Journals of Captain James Cook*, (Cambridge, CUP, 1969), Vol. 2, p. 322.
6 H. B. Jones, *Faraday*, Vol. 1, p. 210.
7 This is explored in G. Cantor, *Michael Faraday*, a fascinating biographical study.

8 H. B. Jones, *The Royal Institution*, (London, 1871), p. vii.
9 *Fraser's Magazine*, 13 (1836), p. 224.
10 Cantor, *Michael Faraday*, pp. 115ff.
11 J. Tyndall, 'Faraday as a Discoverer' [1868], reprinted in facsimile in *Royal Institution Library of Science, Physical Sciences* (London, 1970), Vol. 2, pp. 50–123, esp. pp. 65f.
12 J. Davy, *Fragmentary Remains*, pp. 106f.
13 There is an excellent brief sketch of Faraday's early life in F. A. J. L. James (ed.), *Correspondence*, Vol. 1, pp. xxvii–xxxviii; a superb edition.
14 Jones, *Faraday*, p. v.
15 M. Berman, *Social Change and Scientific Organization* (London, 1978), ch. 5, is critical of Faraday's social attitudes; but see Cantor, *Faraday*, p. 104.
16 J. Morrell and A. Thackray, *Gentlemen of Science*.
17 F. W. J. McCosh, *Boussingault* (Dordrecht, 1984). Thenard himself came from a peasant family and had been patronized by L. N. Vauquelin.
18 Tyndall, 'Faraday', p. 51.
19 J. Davy, *Fragmentary Remains*, p. 69.
20 F. Greenaway (ed.), *Archives*, Vol. 4, pp. 223, 225, 273, 410; the introductions by M. Berman in these volumes are very useful indeed.
21 H. Hartley, *Humphry Davy*, p. 73.
22 Greenaway (ed.), *Archives*, Vol. 5, pp. 332–3.
23 See DNB, DSB. D. Hudson (ed.), *The Diary of Henry Crabb Robinson*, (London, 1967) has anecdotes of John and his wife, who became intimate with him and the Wordsworths.
24 Draft in Royal Institution, Davy notebook 13c, p. 107.
25 Greenaway (ed.), *Archives*, Vol. 4, p. 387.
26 James (ed.), *Correspondence*, Vol. 1, pp. 178ff.
27 J. Davy, *Physiological Researches* (London, 1863).
28 *Proceedings of the Royal Society*, 16 (1867), pp. lxxix–lxxxi.
29 ibid.
30 R. Dawkins, *The Selfish Gene*, (Oxford, OUP, 1976).
31 James (ed.), *Correspondence*, Vol. 1, p. 88.
32 Jones, *Faraday*, Vol. 1, p. 420.
33 ibid., Vol. 2, p. 321.
34 ibid., p. 447.
35 ibid., p. 445.
36 ibid., Vol. 1, p. 46.
37 Tyndall, *'Faraday'*, p. 51.
38 James (ed.), *Correspondence*, Vol. 1, p. 68.
39 Greenaway (ed.), *Archives* Vol. 5, pp. 353, 355.
40 Royal Institution MS F8.
41 These are the Laboratory Notebooks, Davy MSS Box 6.
42 Royal Institution MS F8.
43 James (ed.), *Correspondence*, Vol. 1, p. 128.
44 J. B. Dumas, *Eloge of Faraday* (1885), quoted in S. P. Thompson, *Michael Faraday: His Life and Work*, (London, Cassell, 1898), p. 20.

45 A. Pritchard, *Poetry*, p. 31.
46 James (ed.), *Correspondence*, p. xxxii.
47 Jones, *Faraday*, Vol. 1, pp. 240–1.
48 M. Faraday, *Experimental Researches in Chemistry and Physics*, (London, 1859), p. 1 note.
49 James (ed.), *Correspondence*, Vol. 1, p. 159; and cf. 179.
50 ibid., pp. 162ff.
51 J. Davy, *Memoirs*, Vol. 1. p. 483 (16/3/14); Vol. 2, p. 157 (11/8/21).
52 James (ed.), *Correspondence*, Vol. 1, p. 165. Englishmen did not generally use first names outside the family, in letters or conversation, until the mid-twentieth century.
53 ibid., p. 172.
54 ibid., pp. 178, 181.
55 ibid., p. 181.
56 ibid., p. 186.
57 J. Davy, *Memoirs*, Vol. 2, p. 157; Pritchard, *Poetry*, p. 37.
58 Royal Institution, Laboratory Notebook, 30 August, 1810.
59 H. Davy, *Consolations*, p. 252.
60 M. Faraday, 'The Liquefaction of Gases", *Alembic Club Reprints*, 12 (1896); James (ed.), *Correspondence*, Vol. 1, pp. 296f, 433.
61 James (ed.), *Correspondence*, Vol. 1, pp. 231–2, 228–9, 235, 322ff.
62 ibid., pp. xxxiv, 315.
63 Jones, *Faraday*, Vol. 1, p. 340.
64 J. Tyndall, *Fragments of Science* (London, 1895), Vol. 1, pp. 403ff, esp. 408–9.
65 ibid.
66 ibid.
67 Cantor, *Michael Faraday*, pp. 102, 129.
68 W. White, *The Journals*, (London, 1898), p. 256, but cf. 69, 120–1. John Stenhouse and Thomas Graham were eminent chemists, founder members of the Chemical Society; Graham was its first president.
69 L. P. Williams (ed.), *Selected Correspondence of Michael Faraday*, (Cambridge, CUP, 1971), Vol. 2, p. 975–6.
70 S. Thompson, *Faraday*, (London, 1898), p. 59.
71 J. Davy, *Memoirs*, Vol. 2, pp. 95f; I have preferred this version to that in Pritchard, *Poetry*, pp. 31–2.

CHAPTER 10 *President*

1 R. McCormmach, 'Henry Cavendish on the Proper Method of Rectifying Abuses', in E. Garber (ed.), *Beyond History of Science*, (Bethlehem and London, 1990), pp. 35ff.
2 G. de Beer, Preface in W. R. Dawson (ed.), *The Banks Letters* (London, British Museum (Natural History), 1958), p. vii.

3 Dawson, *The Banks Letters*, pp. xxff, xxviii.
4 M. B. Hall, *All Scientists Now*, H. Lyons, *The Royal Society, 1660–1940*, (Cambridge, CUP, 1944) ch. 7.
5 R. Briggs, 'The Académie Royale des Sciences and the Pursuit of Utility, *Past and Present*, 131 (1991), pp. 38–88.
6 J. Coyac, M. Fetizon, M. Sadoun-Goupil et R. Taton (eds.), *Actes du colloque Gay-Lussac* (Palaiseau, École Polytechnique, 1980).
7 On the French model for the police see C. Emsley, *Crime and Society in England, 1750–1900* (London, Longman, 1987), ch. 8.
8 C. A. Russell, *Science and Social Change, 1700–1900*, (London, Macmillan, 1983).
9 See S. F. Cannon, *Science in Culture* (New York, Science History, 1978), ch. 2, on the 'Cambridge Network'.
10 See my 'Tyrannies of Distance in British Science', in R. W. Home and S. G. Kohlstedt, *International Science*, pp. 39–53.
11 D. P. Miller, 'Between Hostile Camps: Sir Humphry Davy's Presidency of the Royal Society of London, 1820–1827', *British Journal for the History of Science*, 16 (1983), 1–47, is a very important study on which I have drawn heavily; its viewpoint is, on the whole, that of the Cambridge Network.
12 C. Babbage, *Reflections on the Decline of Science in England* (London, 1830), wrote indignantly on this and other topics.
13 H. Davy, *Works*, Vol. 7.
14 ibid., pp. 6–7.
15 ibid., p. 7.
16 ibid., pp. 14–15.
17 A. J. Paris, *Humphry Davy*, p. 414.
18 ibid., p. 418.
19 ibid., p. 437.
20 ibid.
21 See my 'William Swainson: naturalist, author and illustrator', *Archives of Natural History*, 13 (1986), pp. 275–90.
22 J. Davy, *Memoirs*, Vol. 2, p. 138.
23 ibid., p. 215.
24 ibid., pp. 149, 156–7.
25 ibid., p. 146.
26 J. Davy, *Memoirs*, Vol. 2, p. 218; A. Pritchard, *Poetry*, p. 40. On attitudes to death, see M. Wheeler, *Death and the Future Life in Victorian Literature and Theology* (Cambridge, CUP, 1990).
27 Wordsworth was alarmed that 'his constitution was clearly giving way': J. Davy, *Fragmentary Remains*, p. 256.
28 J. Davy, *Memoirs*, Vol. 2, p. 220.
29 ibid.
30 ibid.
31 J. Davy, *Fragmentary Remains*, p. 274.
32 J. Davy, *Memoirs*, Vol. 2, pp. 221ff.
33 ibid., p. 221.

34 ibid., p. 222.
35 ibid., p. 227.
36 ibid., p. 228.
37 ibid., pp. 229ff.
38 J. Davy, *Fragmentary Remains*, pp. 287–8.
39 J. Davy, *Memoirs*, Vol. 2, pp. 263ff.
40 ibid., p. 276.
41 ibid.
42 J. Davy, *Fragmentary Remains*, p. 284.
43 ibid., p. 291.
44 ibid., p. 296.

CHAPTER 11 *Salmonia*

1 H. Davy, *Consolations*, pp. 167ff.
2 P. Medawar, *Memoir of a Thinking Radish* (Oxford, 1986), p. 179.
3 There is some discussion of Davy in connection with Naturphilosophie and Neoplatonism in my *Transcendental Part of Chemistry*, ch. 3.
4 *Blackwood's Edinburgh Magazine*, 24 (1828), pp. 248–72. Reviewers are identified in W. E. Houghton *et al.*, *Wellesley Index to Victorian Periodicals* (2 vols., Toronto, 1966–72).
5 He had a predecessor: *The Gentleman Angler ... by a gentleman who has made Angling his diversion for upwards of 28 years*, 3rd edn (London, 1736).
6 J. Lowerson, 'Angling', in T. Mason (ed.), *Sport in Britain: a Social History* (Cambridge, CUP, 1989), pp. 12–43.
7 A. Geikie, *Life of Sir Roderick I. Murchison* (London, Murray, 1875), Vol. 1, p. 94.
8 G. Cuvier, *Eloges historiques* (Paris, n.d.), p. 355.
9 W. Yarrell, *British Fishes* (London, 1836), Vol. 2, pp. 80–2.
10 *Parliamentary Papers, 8, Reports from Committees*, 4, (London, 1824) pp. 144–5.
11 E. Jesse, *Gleanings of Natural History*, 2nd series (London, 1834), p. 65. On Jesse, see *DNB*.
12 Jesse, *Natural History*, p. 85.
13 ibid., p. 320.
14 ibid.,
15 W. Scrope, *Days and Nights of Salmon Fishing on the Tweed*, (London, Murray 1843; rep., Edinburgh, 1975). On Scrope, see *DNB*.
16 *Blackwood's Edinburgh Magazine*, 24 (1828), p. 243.
17 Scott, *Quarterly Review*, 38 (1828), pp. 503–35.
18 *Blackwood's Edinburgh Magazine*, 24 (1828), p. 243.
19 ibid.
20 J. Morley, *Nineteenth Century Essays*, ed. P. Stansky (Chicago, University of Chicago Press, 1970), p. 92.
21 S. T. Coleridge, *Biographia Literaria*, ed. J. Shawcross (Oxford, OUP, 1907), Vol. 2, p. 115.

22 J. Davy, *Memoirs*, Vol. 2, p. 287.
23 *Quarterly Review*, 38 (1828), pp. 503–35.
24 ibid.
25 ibid.
26 H. Davy, *Consolations*, pp. 170–1.
27 J. A. Paris, *Humphry Davy*, p. 547.
28 H. Davy, *Consolations*, pp. 172–3.
29 J. Davy, *Fragmentary Remains*, pp. 303, 314.
30 J. Davy, *Memoirs*, Vol. 2, p. 332.
31 H. Davy, *Salmonia*, pp. 98–9; W. Yarrell, *A History of British Birds*, 2nd edn, (London, 1845), pp. 20ff; A. Rutgers in *Gould's Birds of Europe* (London, 1966), p. 17; J. Davy, *Memoirs*, Vol. 2, p. 157, and compare Vol. 1, p. 377.
32 H. Davy, *Salmonia*, p. vii.
33 ibid.
34 ibid., dedication.
35 J. Davy, preface to 4th edn of *Salmonia*, (London, Murray, 1851), and *Memoirs*, Vol. 2, p. 303. Babington had interests outside medicine; see W. Campbell Smith, 'Early Mineralogy in Great Britain and Ireland', *Bulletin of the British Museum of Natural History* (*Hist. Ser.*), 6 (1978), 49–74, pp. 59–60. On pp. 75–108 of the same issue, see the paper by A. E. Gunther on J. G. Children.
36 The wine pint in Britain was 16 fl. oz, that used for other liquids being 20 fl. oz: for units see *Annals of Philosophy*, 1 (1813), 452ff.
37 H. Davy, *Salmonia*, pp. 125–6; Wilson, *Blackwood*'s, pp. 267–8.
38 H. Davy, *Salmonia*, p. 126.
39 ibid., p. 127.
40 J. Davy, *Memoirs*, Vol. 2, p. 336.
41 J. Davy, *Fragmentary Remains*, p. 307.
42 ibid., p. 310.
43 H. Davy, *Salmonia*, 1828 edn, p. 271; 1832 edn, p. 325; refs to Home are on the previous page.
44 J. Davy, *Fragmentary Remains*, p. 242.
45 The 2nd and 3rd editions have engravings by Edward Finden of the scenes; the eagle facing p. 95 seems to be after Bewick (C. E. Jackson, *Wood engravings of Birds* (London, 1978), p. 41), reversed; the grayling is on p. 175 of the 1st edition, p. 212 of the 3rd; in *Works* Vol. 9 there are only the entomological plates.
46 Paris, *Humphry Davy*, p. 457; J. Davy, *Fragmentary Remains*, pp. 15, 302.
47 G. S. Myers, intro. p. 14, in R. Playfair and A. C. Gunther, *The Fishes of Zanzibar* (rep. Kentfield, Calif., 1971).
48 J. Davy, *Memoirs*, Vol. 2, p. 340; Yarrell, *British Fishes*, Vol. 2, p. 20; Scott, *Quarterly Review*, 38 (1828), 503–35.
49 J. Davy, *Memoirs*, Vol. 2, pp. 286–7.
50 ibid., Vol. 1, p. 263.
51 W. Kirby and W. Spence, *An Introduction to Entomology* (4 vols., London, 1815–26).

52 W. Carter, *The Sheep and Wool Correspondence of Sir Joseph Banks*, (Sydney, Library Council of New South Wales, 1979).
53 H. Davy, *Salmonia*, pp. 91, 290–1. On trade and science, see D. F. S. Scott, *Luke Howard, 1772–1864*, (York, 1976), p. 4.
54 J. Davy, *Fragmentary Remains*, pp. 280, 284f, 292, 311–12.
55 Paris, *Humphry Davy*, p. 547.
56 H. Davy, *Salmonia*, pp. 114–15; J. Davy, *Fragmentary Remains*, pp. 237–8; On Mackenzie, see *DNB*; on Berthollet, see M. Sadoun-Goupil, *Le chimiste C. L. Berthollet*, 1748–1822 (Paris, 1977), pp. 68–71.
57 H. Davy, *Salmonia*, pp. 160–1; W. Kirby, *Bridgewater Treatise*, ed. T. R. Jones (London, 1853), Vol. 2, pp. 162–3; J. Davy, *Memoirs*, Vol. 2, pp. 89–90; W. H. Thorpe, *Purpose in a World of Chance* (Oxford, OUP, 1978), ch. 2.
58 H. Davy, *Salmonia*, p. 174.
59 R. Dawkins, *The Blind Watchmaker* (London, 1986); J. H. Brooke, *Science and Religion* (Cambridge, CUP, 1991).
60 H. Davy, *Salmonia*, p. 173.
61 ibid., p. 237; T. A. Knight, *Physiological and Horticultural Papers*, pp. 46ff.
62 J. Davy, *Fragmentary Remains*, p. 287 for Davy on young Knight.
63 H. Davy, *Salmonia*, p. 197; J. Davy, *Memoirs*, Vol. 1, p. 149.
64 H. Davy, *Salmonia*, p. 197.
65 J. Davy, *Memoirs*, Vol. 2, p. 88.
66 H. Davy, *Salmonia*, p. 313.

CHAPTER 12 *Consolations*

1 J. Davy, *Fragmentary Remains*, pp. 314f. For an obituary, see *Annual Register*, 71 (1829) pp. 504–21.
2 W. H. Brock is writing a biography of Liebig.
3 H. Davy, *Consolations*, pp. 69ff.
4 J. Tobin, *Journal of a Tour*, p. v.
5 J. Davy, *Fragmentary Remains*, p. 305.
6 ibid., p. 308.
7 I owe this explanation to Professor J. R. Watson.
8 J. Davy, *Memoirs*, Vol. 2, pp. 343ff.
9 That and the fifth had chiefly been written down earlier in notebooks.
10 J. Davy, *Memoirs*, Vol. 2, p. 384.
11 A. Treneer, *The Mercurial Chemist*, p. 253.
12 J. Davy, *Fragmentary Remains*, p. 313.
13 Royal Institution, Davy notebook 13d, pp. 14f.
14 J. Tobin, *Journal*, p. 120.
15 J. Davy, *Fragmentary Remains*, p. 307.
16 H. Davy, *Consolations*, pp. 20–1, 30, 34–5, 57, 228.
17 ibid., p. 51. This seems to owe something to speculative writings of Swedenborg and perhaps of Kant.
18 ibid., p. 93.

19 ibid., p. 94.
20 W. Paley, *Natural Theology*, (London, 1802, and many later editions); see J. H. Brooke, *Science and Religion* (Cambridge, CUP, 1991).
21 H. Davy, *Consolations*, p. 277ff.
22 ibid.
23 K. Coburn (ed.), *Inquiring Spirit*, pp. 381f. For an interesting comment on science and religion by Coleridge to Davy, see J. Davy, *Fragmentary Remains*, pp. 110–11.
24 J. Davy, *Memoirs*, pp. 375f.
25 G. Cantor, *Michael Faraday, Sandemanian and Scientist*, 1991.
26 H. Davy, *Consolations*, p. 221.
27 J. Hutton, *System of the Earth etc.*, ed. V. A. Eyles and G. White, (Darien, Conn., 1970), p. 128.
28 J. Playfair, *Illustrations of the Huttonian Theory of the Earth*, [1802], (rep. New York, Dover, 1964).
29 C. Lyell, *Principles of Geology* [1830–3], rep. intr. M. Rudwick, [Lehre, Cramer, 1970], Vol. 1, title page and pp. 144–5.
30 ibid.
31 H. Davy, *Consolations*, pp. 141ff.
32 ibid.
33 ibid.
34 Tobin, *Journal* , pp. 154ff.
35 J. Davy, *Memoirs*, Vol. 2, pp. 260–1; A. Pritchard, *Poetry*, p. 43.
36 H. Davy, *Consolations*, pp. 219f. On medical materialism in the next generation, see A. Desmond and J. Moore, *Charles Darwin* (London, 1991).
37 J. Davy, *Fragmentary Remains*, p. 293. I am grateful to Professor Dr Janez Batis of the Sloveman Academy of Sciences for information about Davy's Podkoren and Ljubljana.
38 ibid.
39 J. Davy, *Fragmentary Remains*, pp. 318ff. There is a glowing review of *Consolations* in the *Athenaeum*, 156 (1830), pp. 82–3.
40 ibid.
41 ibid.
42 ibid.
43 Tobin, *Journal*, pp. 210f.
44 ibid., pp. 240ff.

Select Bibliography

Anon, *Notice des Tableaux exposés dans la Galerie du Musée*, Paris, Dubray, 1814.

Berman, M., *Social Change and Scientific Organization: the Royal Institution, 1799–1844*, London, Heinemann, 1978.

Brodie, B. C., *Psychological Inquiries: in a Series of Essays intended to illustrate the Mutual Relations of the Physical Organization and the Mental Faculties*, 2nd edn, London, Longman, 1855.

Brooke, J. H., *Science and Religion: Some Historical Perspectives*, Cambridge, CUP, 1991.

Cantor, G., *Michael Faraday, Sandemanian and Scientist: a Study of Science and Religion in the Nineteenth Century*, London, Macmillan, 1991.

Christie, J., and Shuttleworth, S., *Nature Transfigured: Science and Literature, 1700–1900*, Manchester, Manchester University Press, 1989.

Coburn, K., *Inquiring Spirit: a New Presentation of Coleridge*, London, Routledge, 1951.

Cottle, J., *Reminiscences of Samuel Taylor Coleridge and Robert Southey*, London, Houlston and Stoneman, 1847.

Crosland, M. P., *Gay-Lussac*, Cambridge, CUP, 1978.

Crosland, M. P., *The Society of Arcueil: a View of French Science at the Time of Napoleon I*, London, Heinemann, 1967.

Crosland, M. P., *Science under Control: the French Academy of Science, 1795–1914*, Cambridge, CUP, 1992.

Cunningham, A., and Jardine, N., *Romanticism, and the Sciences*, Cambridge, CUP, 1990.

Cuvier, G., *Eloges Historiques*, Paris, Ducrocq, n.d.

Davy, H., *The Collected Works of Sir Humphry Davy*, ed. J. Davy, 9 vols., London, Smith, Elder, 1839–40.

Davy, H., *Consolations in Travel, or the Last Days of a Philosopher*, London, Murray, 1830.

Davy, H., *Elements of Agricultural Chemistry*, London, Longman, 1813.
Davy, H., *Elements of Chemical Philosophy, Pt. 1, Vol. 1*, London, Johnson, 1812.
Davy, H., *Researches, Chemical and Philosophical, chiefly concerning Nitrous Oxide . . . and its Respiration*, London, Johnson, 1799.
Davy, H., *Salmonia: or Days of Fly Fishing*, 3rd edn, London, Murray, 1832.
Davy, J., *Memoirs of the Life of Sir Humphry Davy*, 2 vols., London, Longman, 1836.
Davy, J., *Fragmentary Remains, Literary and Scientific, of Sir Humphry Davy*, London, Churchill, 1858.
Faraday, M., *Chemical Manipulation: being Instructions to Students in Chemistry, on the Methods of performing Experiments of Demonstration or of Research, with Accuracy and Success*, London, Murray, 1827.
Faraday, M., *Chemical Notes, Hints, Suggestions and Objects of Pursuit*, eds. R. D. Tweney and D. Gooding, London, Peregrinus, 1992.
Forgan, S. (ed.), *Science and the Sons of Genius: Studies on Humphry Davy*, London, Science Reviews, 1980.
Fullmer, J. Z., *Sir Humphry Davy's Published Works*, Cambridge, Mass., Harvard University Press, 1969.
Golinski, J., *Science as Public Culture: Chemistry and Enlightenment in Britain, 1760–1820*, Cambridge, CUP, 1992.
Gooding, D., and James, F. (eds.), *Faraday Rediscovered*, London, Macmillan, 1985.
Gooding, D., Pinch, T., and Schaffer, S., *The Uses of Experiment*, Cambridge, CUP, 1989.
Greenaway, F. (ed.), *Archives of the Royal Institution*, 15 vols. in 7, Ilkley, Scolar, 1971–6.
Hall, M. B., *All Scientists Now: the Royal Society in the Nineteenth Century*, Cambridge, CUP, 1984.
Hartley, H., *Humphry Davy*, London, Nelson, 1966.
Holmes, R., *Coleridge: Early Visions*, London, Hodder and Stoughton, 1989.
Home, R. W. and Kohlstedt, S. G., *International Science and National Scientific Identity*, Dordrecht, Kluwer, 1991.
James, F. A. J. L. (ed.), *The Correspondence of Michael Faraday*, London, I. E. E., 1991, Vol. 1.
Jeffries, A. E., *Michael Faraday: a List of His Lectures and Published Writings*, London, Chapman and Hall, 1960.
Jones, H. B., *The Life and Letters of Faraday*, 2 vols., London, Longman, 1870.
Knight, D. M., *The Age of Science: the Scientific World-view in the Nineteenth Century*, 2nd edn, Oxford, Blackwell, 1988.
Knight, D. M., *Ideas in Chemistry: a History of the Science*, London, Athlone, 1992.
Knight, D. M., *The Nature of Science: the History of Science in Western Culture since 1600*, London, Deutsch, 1976.
Knight, D. M., *The Transcendental Part of Chemistry*, Folkestone, Dawson, 1978.
Knight, T. A., *A Selection from the Physiological and Horticultural Papers*, London, Longman, 1841.

Levere, T. H., *Affinity and Matter: Elements of Chemical Philosophy 1800–1865*, Oxford, OUP, 1971.

[Marcet, J.], *Conversations on Chemistry*, 11th edn, 2 vols. London, Longman, 1828.

Mayhew, H., *The Wonders of Science: or Young Humphry Davy*, 2nd edn, London, Bogue, 1856.

Morrell, J., and Thackray, A., *Gentlemen of Science*, Oxford, OUP, 1981.

Paris, J. A., *The Life of Sir Humphry Davy*, London, Colburn and Bentley, 1831.

Priestley, J., *Experiments and Observations on Different Kinds of Air*, 3 vols., Birmingham, Pearson, 1790.

Priestley, J., *The History and Present State of Electricity*, 3rd edn, (2 vols., London: Bathurst and Lowndes, 1775).

Priestley, J., *A Scientific Autobiography*, ed. R. E. Schofield, Cambridge, Mass., MIT, 1966.

Pritchard, A. (ed.), *The Poetry of Humphry Davy*, Penzance, Penwith D.C., 1978.

Robinson, E., and McKie, D. (eds.), *Partners in Science*, London, Constable, 1970.

Rocke, A. J., *Chemical Atomism in the Nineteenth Century*, Columbus, Ohio University Press, 1984.

Russell, C. A., Roberts, G. K., and Coley, N., *Chemists by Profession*, Milton Keynes, Open University Press, 1977.

Schofield, R. E., *The Lunar Society of Birmingham*, Oxford, OUP, 1963.

Shea, W. (ed.), *Revolutions in Science: Their Meaning and Relevance*, Canton, Mass., Science History, 1988.

Shelley, M., *Frankenstein*, intr. M. Hindle, London, Penguin, 1985.

Siegfried, R., and Dott, R. H. (eds.), *Humphry Davy on Geology: the 1805 Lectures*, Madison, Wisc., Wisconsin University Press, 1980.

Stansfield, D. A., *Thomas Beddoes*, MD, Dordrecht, Reidel, 1984.

Thackray, A., *Atoms and Powers*, Cambridge, Mass., Harvard University Press, 1970.

Thompson, B. (Count Rumford), *The Collected Works of Count Rumford*, ed. S. C. Brown, Cambridge, Mass., Harvard University Press, 1968–70.

Tobin, J. J., *Journal of a Tour . . . whilst accompanying the late Sir Humphry* Davy, London, Orr, 1832.

Treneer, A., *The Mercurial Chemist: A Life of Sir Humphry Davy*, London, Methuen, 1963.

Williams, L. P., *Michael Faraday*, London, Chapman and Hall, 1965.

Index

ROYAL

The Arnold and Caroline Rose Monograph Series of the American Sociological Association

The cooperative workplace

The Rose Monograph Series was established in 1968 in honor of the distinguished sociologists Arnold and Caroline Rose, whose bequest makes the series possible. The sole criterion for publication in the series is that a manuscript contribute to knowledge in the discipline of sociology in a systematic and substantial manner. All areas of the discipline and all established and promising modes of inquiry are equally eligible for consideration. The Rose Monograph Series is an official publication of the American Sociological Association.

The editor and board of editors gratefully acknowledge the contributions of Carl Milofsky, Yale University, and Howard Aldrich, University of North Carolina at Chapel Hill, as expert reviewers of this manuscript.

For other titles in this series, turn to p. 222.

The cooperative workplace

Potentials and dilemmas of organizational democracy and participation

Joyce Rothschild
University of Toledo

J. Allen Whitt
University of Louisville

Cambridge University Press

Cambridge
London New York New Rochelle
Melbourne Sydney

Published by the Press Syndicate of the University of Cambridge
The Pitt Building, Trumpington Street, Cambridge CB2 1RP
32 East 57th Street, New York, NY 10022, USA
296 Beaconsfield Parade, Middle Park, Melbourne 3206, Australia

First published 1986

Printed in the United States of America

Library of Congress Cataloging in Publication Data

Rothschild, Joyce, 1948–
The cooperative workplace.
(The Arnold and Caroline Rose monograph series of the American Sociological Association)
Bibliography: p.
Includes index.
1. Management—Employee participation. 2. Democracy.
3. Political participation. 4. Cooperation.
I. Whitt, J. Allen, 1940– . II. Title.
III. Series.
HD5650.R736 1986 658.3′152 86–8310

British Library Cataloguing in Publication Data

Rothschild, Joyce
The cooperative workplace: potentials and dilemmas of organizational democracy and participation.—(American Sociological Association Rose Monograph series)
1. Organization
I. Title II. Whitt, J. Allen III. Series
302.3′5 HM131

ISBN 0 521 32967 1

Contents

Acknowledgments

This book was born out of curiosity about how workplaces might be organized without a hierarchy of bosses and workers. But as anyone who has written a book knows, it is a long journey from creative inspiration to final book manuscript. This book represents many years of research, of interpreting and making sense of research observations, of distilling the work of others, and of rethinking and revising conclusions. Many people have contributed to its development along the way.

Long before this project was undertaken, numerous people contributed to our thinking about it. The seeds that later led to a professional interest in democracy were probably planted by Sam and Evelyn Rothschild, who understood the value of a democratic household and self-initiated action. Fellow graduate students at the University of California, Santa Barbara, now sociologists all around the country, provided an abundance of moral support and an ever-present sounding board for ideas not yet polished. Especially to be thanked are Laura Nathan, Tom Koenig, Brad Smith, Rosemary Taylor, and Robert Wolf. This analysis of democratic organizations began as dissertation research and we are grateful to the Department of Sociology at Santa Barbara for providing the kind of intellectually stimulating milieu in which students may learn how to formulate their own questions. Debts are owed especially to Professors Richard Flacks, Bettina Huber, Richard Appelbaum, and Harvey Molotch. Together they combined enthusiasm with considered criticism, and they coupled their high expectations of scholarship with personal autonomy.

In the course of working on this book we have benefited from the ideas of many other people who have given serious attention to the prospects for nonhierarchical forms of organization: Howard Aldrich, Paul Bernstein, Joseph Blasi, Henry Etzkowitz, Art Hochner, Rosabeth Kanter, Frank Lindenfeld, Ray Loveridge, Jane Mansbridge, Patricia Martin, Carole Pateman, Charles Perrow, Corey Rosen, Raymond Russell, Ann Swidler, and William Foote Whyte come especially to mind. Each of these people read one part or another of our manuscript, offered scholarly criticism and innumerable references,

and brought a valued sense of professional camaraderie.

A number of institutions provided the time and financial support to pursue different aspects of the work reported in this book: The New Systems of Work and Participation Program at Cornell University's School of Industrial and Labor Relations, supported by funds from the National Institute of Mental Health, and the Program on Non-Profit Organizations at Yale University's Institution for Social and Policy Studies.

This book benefited as well from the rigorous review of Ernest Q. Campbell and the anonymous reviewers he enlisted to review it for the Arnold and Caroline Rose Monograph Series, sponsored by the American Sociological Association.

Finally, this research could literally not have been done were it not for the openness of the participants in the alternative organizations studied. Because the names used in this study are fictitious, they cannot be mentioned by name here, but we sincerely wish to thank them for generously giving of themselves and their time. We hope that this work justifies their many confidences.

Remaining errors are quite obviously our own doing.

J.R., J.A.W.

Introduction

At least since the 1911 publication of Robert Michels's *Political Parties,* the process by which organizational democracy yields to oligarchy has been accepted – however regretfully – as inevitable. Organizations with no bosses and no followers, organizations in which all members have an equal say in running things, have largely escaped the notice of organizational analysts. Though democratic organizations have long existed, detailed study of them has been displaced by the assumption that they are fragile, short-lived structures or that they will eventually come under the control of one or a few leaders, thus losing their defining characteristic. This expectation has become a cornerstone of twentieth-century social science.

Today in the United States we are witnessing the birth and life of scores of grass-roots organizations – organizations calling themselves "collectives," "cooperatives," and "alternative institutions" – that aspire to be radically democratic in purpose and in practice. These organizations provide us with a unique opportunity to take a fresh empirical look at the supposed inevitability of oligarchy and bureaucracy. This book examines the nature, possibilities, and limits of direct democracy in such organizations. We develop a theory of democratic organizations and show how this theory is applicable to a broad range of directly democratic and related organizations.

This subject is relevant to anyone who would hope to live and work in a democratic society; it is relevant also to anyone who has written off the possibility of organizational democracy as utopian. These anomalous organizations reject bureaucracy and attempt to fashion an alternative, providing a natural laboratory for evaluating long-held assumptions about the universality of hierarchy and bureaucracy. To the extent that these organizations succeed, they promise to broaden our theory of organizations and to provide concrete models of alternative organizational practices.

Grounded in empirical observation of collectivist organizations in many different domains, this book has two major theoretical aims. The first is to try to construct a systematic, definitive model of the organizational properties of collectivist or cooperative organizations. *We define a collective or a cooper-*

ative as any enterprise in which control rests ultimately and overwhelmingly with the member–employees–owners, regardless of the particular legal framework through which this is achieved.[1] It is the priority given to democratic methods of control that is the essential characteristic of the contemporary cooperative. Since the right to govern rests ultimately with the collectivity of members and delegated authority is accountable to the group as a whole, members also call their enterprises "collectives." In the nineteenth century lexicon, these enterprises would have been called "producers' cooperatives." The term remains technically correct but the participants themselves seldom use this designation.[2] Although the terms *collectives, cooperatives,* and, more recently, *alternative institutions* have been used historically to denote a range of organizational types, we are interested in the central characteristic they all have in common – direct, democratic control by the members. For this reason, we often use these designations interchangeably.

The second theoretical goal of this work is to discover those conditions that undermine or support the most essential characteristic of cooperatives: decision-making procedures based on participatory democracy.

Cooperatives are important organizations. Throughout their long history in the United States and Europe, they have often formed the cutting edge of movements for social change and organizational innovation. They also carry forward and attempt to put into practice long-held dreams of people, dreams with deep roots in the social theory and philosophy of Western society. They are thus organizations that look to both the past and the future. Part I of the book examines the origins of alternative organizations and shows how they are one strand of a broad social movement currently producing several related types of democratically oriented workplaces.

In spite of the historic legacy of cooperatives, they are not organizations of a bygone era. The United States is currently experiencing the largest and most vital burgeoning of cooperatives in its history. Yet, ironically, we know next to nothing about cooperatives, particularly the specifics of their internal structures, processes, and conditions of operation. The case studies that exist are often descriptive and idiosyncratic. Mainstream organizational theory and research have almost entirely ignored these organizational forms. Ecological studies of the distribution and duration of cooperatives reveal general demographic patterns but little of their internal functioning. We are sorely in need of a theoretical model for understanding collectives and cooperatives *as organizations*. In Part II, we attempt to construct a general theory of democratic organizations. Chapter 3 begins the section by detailing the structural features that define the democratic organization.

Nothing is more central in the values of Western society than the ideals of

democracy and equality. Visions of direct democracy can be traced back to ancient Greece, resurfacing in many subsequent eras. Yet organizational democracy has been an elusive goal, rarely achieved in practice and chronicled mostly in the breach. The most interesting thing about cooperatives is that they are attempting to achieve something most social science tells us is impossible: viable participatory democracy. Our research on cooperatives convinces us that although they do not always succeed, neither do they always fail. We argue that the creation of organizational democracy is *conditional*. In Chapters 4 and 5 of Part II, we identify specific conditions, some internal to the organization and others external, appearing to favor, or in their absence to undermine, organizational democracy.

In discovering these conditions we hope to advance organizational theory and help to clarify matters for cooperative members. Cooperatives, like all organizations, embrace multiple and often competing goals. The desire for internal democracy, though central, is usually coupled with other legitimate goals. The nexus of these goals places the cooperative in numerous binds. For every condition we identify as supporting democracy, we show the structural dilemma this condition raises for the organization.

We hope this book will help both organizational theorists and practitioners in cooperatives to identify the organizational features essential to the collectivist form, the conditions that promote direct democracy within the organization, and the inherent trade-offs that go with the pursuit of democracy. Identification, however, will not make the necessary choices any easier for members.

Part III draws out the general significance of organizational democracy for the individual and for society. In Chapter 6 we examine not only our own cases, but the existing research literature on worker satisfaction in cooperative-type organizations. Here we reach some unexpected conclusions concerning the effects of democracy on the individual member. Chapter 7 looks at the future of democratic organizations and specifically asks how they perform economically vis-à-vis more conventional forms. We consider what role the government may play in their development, and what kind of evolution and life span we may expect of them in view of historical precedents. The final chapter provides a more philosophical overview of how autonomy and democracy in the workplace may help to transform the relationship between work and play.

As the first step in the analysis, it is necessary to understand the origins and nature of contemporary cooperatives. They are radically different from conventional organizations not only in their form and aims, but also in their unique intellectual and historical roots. For this reason, Chapter 1 begins by

pointing out how certain classical ideas have shaped contemporary cooperatives. Included here are the writings of Rousseau, Bakunin, and Marx. Consciously or unconsciously, members of cooperatives try to put into practice these venerable ideas. However, according to the work of Weber and Michels and the considerable body of research and theory following from their tradition, the prospects for organizational democracy are extremely remote. This tradition forces us to take seriously the forces of bureaucratization, specialization, and oligarchization confronting would-be democratic organizations. Chapter 1 therefore also examines the Weber/Michels challenge, and then looks at the cultural and economic forces that have favored the rise of the recent wave of cooperatives in the United States. Chapter 2 describes the specific organizations that we have studied and the analytical methods we have used.

Part I

Origins and types of alternative organizations

1. Cooperatives in the late twentieth century: the democratic impulse and the challenge of oligarchy

By dinner time that summer afternoon the weekly edition of the *Community News* had been delivered to the racks and newstands and most of the newspaper's staff had gone home for the day. John and Ann remained in the office, sharing the last of the coffee and casually examining the latest issue. For the past several months, John had worked as a reporter covering political affairs in the area. Ann, a photographer, had been in her present job for almost two years. Anyone at the paper, if asked, would have been quick to say both of them had been doing fine work and were well liked. The unusual amount of attention they were giving at that moment to a just-printed edition may have been due to the realization this might be the last time – at least for a good while – they would be doing their present jobs at the paper.

Although he would probably not have admitted it, John was proud to see that the story on the incorporation controversy, into which he had poured so much time and effort, was the week's lead. He didn't want it to be too obvious to Ann, but he was privately admiring his story's prominent position. Ann was more open with her feelings. She had two photos on the front page and another on page 4. She rested her index finger on the photo of the mayor, who was shouting in anger at a contrary City Council member. A slight smile came to her lips.

"This photo really is effective," she said, half to herself. Then her eyes fastened more critically on the page. "But maybe next time a little less back-lighting. . . ."

John said he thought the photo was great as it was.

"Yeah," she brightened, "I love doing photography."

He knew how she felt.

Had anyone been there to ask them, Ann and John would have also agreed they liked working at the *Community News*. It was not like other newspapers. Indeed it was unlike any other organization for which they had ever worked. Here, Ann and John and the others had a real say in how the paper would be run. Instead of having bosses at the top and workers at the bottom, the *News* was run democratically, as a collective. Whatever functional role they might

be playing at the time – reporter, photographer, proofreader, ad salesperson – everyone had an opportunity to attend weekly meetings at which all important matters of internal policy, finances, news coverage, and the like were jointly decided by a process of consensus. The meetings were sometimes long and sometimes tedious, but most people left afterward feeling their voices had been heard and they were committed to the decisions that had been reached. It was *their* newspaper in a profound way, and all were proud of what they had accomplished together. Most had never had another job they liked nearly as well, nor a job that seemed so important to the community. Although conflicts and unhappiness occasionally erupted, as in any work group, morale was generally high.

A few weeks before, John and Ann had been at the meeting in which all had agreed jobs at the newspaper should be rotated periodically. This would give everyone a chance to develop new skills, would introduce some variety, would equitably distribute the dull and exciting aspects of specific jobs, and would help everyone to develop a healthy sense of all that is required to operate a newspaper. Just as important, it would prevent the growth of a bureaucratic hierarchy. No one wanted that to happen. So they, along with the others, had heard all the arguments for and against the idea of job rotation, and had agreed it should be done.

There had been considerable talk around the paper about the upcoming rotation, and everyone was curious about what was involved in the various jobs. Most people appeared to be excited about breaking old routines and learning some new skills. John was among these. Even though he liked his reporting job and was good at it, he felt he would like the more technical side of the process as well. Maybe layout or pasteup, things he now knew little about. And that would probably work out fine for the paper since he was aware of three other people who wanted to switch out of production. John did not have any trouble listing his first and second job choices. His sheet of paper was in the box with all the rest, and he had been comparing notes with others and joining in the speculation about who might be moving into what job.

John said he had to be going in order to get home on his bike before the sun went down.

Ann folded the paper.

They noted they would see each other early Saturday morning at the meeting. A yellow piece of paper on the bulletin board above the desk announced that job rotations would be worked out at the meeting.

As he started toward the door, John light-heartedly commented: "Well, I guess we'll soon know what we'll be doing next week!"

He stopped, noticing what he took to be a troubled look cross Ann's face.

"I haven't put down my choices yet," she said softly.

He was surprised.

She said she'd thought about it a long time, but just couldn't come up with anything she thought she would enjoy nearly as much as being a photographer. She took out a cigarette. He had rarely seen her smoke before.

Lowering his backpack and claiming the corner of the desk opposite her, he reminded her of the reasons why it was the best thing to do, and of the careful consensus the group had reached.

As he talked, her cigarette made increasingly frequent trips to the ashtray.

"I'm not saying I don't agree with the principle," she interrupted, "I really do. It's just that I'm not sure what I want to do next. . . ." Her voice trailed off and she was quiet for a while.

Then, as if exploring a new idea, she said, "I guess I've been worried. I'm good at what I do. It's kind of scary to think about giving that up. Photography is a challenge and I'd like to learn to do it even better, not have to start all over again at something else."

Another silence.

She wondered if other people didn't share some of her feelings.

He said he'd not heard any real doubts expressed, but there might be some.

"Well, I know we're all committed to this and everyone wants to see it work, but once in a while I wonder," she added, suddenly rising from the desk and declaring in mock-dramatic humor, "if this is any way to run a newspaper!"

They both laughed.

This vignette, based on actual events at one of the organizations we studied, illustrates several of the general themes this book addresses. We are concerned with understanding what collectives are, how they work, what principles they follow, what possibilities and options they seem to present, what problems and limits they face, and what their impact may be on individuals and the larger society.

What is most unusual about the above story is not the existence of individual doubts and ambivalences. These exist in all organizations. What is most striking is the fact that this particular organization, this collective, was trying to do things that almost no other type of organization even attempts: to conduct itself as a pure democracy, deliberately rotating organizational roles among its members. These are bold steps.

The information contained in the story is suggestive, but incomplete. Where did such ideas and practices originate? Not simply in this particular

organization, for it is virtually identical in form with other cooperatives and collectives around the United States. More precisely, then, we need to know the source from which this whole class of organizations draws its inspiration. To find the answer, we must go far beyond the origins of the *Community News*. We must seek the beginnings of the ideas that gave birth to these organizations – and ultimately to dilemmas such as Ann's – in history.

The historical legacy

The idea of workers cooperatively producing goods or services is not new. In the United States, for example, workers' cooperatives have a long history, dating back to the revolutionary period. The historical record from 1790 to 1940 reveals more than 700 producers' cooperatives. These co-ops have not been randomly distributed in time, but have appeared in distinct waves – the 1840s, the 1860s, the 1880s and the 1920s–30s. What is intriguing is that these four historical periods immediately followed major movements for social change; the pattern of these events suggests a connection between social movements and cooperative formation (Aldrich and Stern, 1978).

In part, producers' cooperatives in the nineteenth century represented efforts by workers to retain highly skilled craft production in the face of increasing mechanization and standardization in industry. In addition, they attempted to put into practice the age-old ideals of democracy, equality, and community, turning direct control over the means and the product of production to the producers.

After a 50-year hiatus, a new, fifth wave of cooperatives has arisen in the United States. The thousands of cooperatives created in local communities since 1970 make the current wave larger than any previous one in American history. In fact, the past decade has seen the emergence of more cooperatives than the rest of American history combined. This latest wave, too, comes on the heels of some of the largest and most vigorous social movements in U.S. history: the civil rights, antiwar, environmental, women's, and student movements of the 1960s and 1970s.

Most contemporary cooperatives are involved in high-quality craft production, retail sales, or the provision of human services. They are made up largely of college-educated young people who were active in, or who were influenced by, the social movements of the 1960s and 1970s. Unlike the nineteenth century artisans, they are not trying to maintain a former state of autonomy, but hope to create entirely new (for them) opportunities and conditions of work and community life. The development of cooperatives in the 1970s appears to be the natural outgrowth of counter-cultural values and sentiments developed and expressed in the social movements of the 1960s. It was

a short step from insisting that blacks, women, and the Vietnamese be given control over the conditions of their own lives to insisting on the same for oneself. In the 1970s, as the people who had taken part in these social movements graduated from college, they came to focus on the institutions that most touched their day-to-day lives – the community and the workplace. They created community-oriented workplaces such as alternative newspapers, arts and handicrafts shops, food co-ops, publishing houses, restaurants, health clinics, legal collectives, natural foods bakeries, auto repair cooperatives, and retail stores.

These organizations continue to spread at a remarkable rate. A national directory (Gardner, 1976) lists some 5,000 alternative organizations and estimates that about 1,000 new ones are created yearly. A recent estimate (Jackall and Crain, 1984) places the number of producer co-ops at over 1,000, with a median size of 6.5 employees/firm. Other researchers estimate at least 1,300 alternative schools (Moberg, 1979, p. 293), between 5,000 and 10,000 food co-ops (Zwerdling, 1979, p. 90), and several thousand communes (Moberg, 1979, p. 285).

With a few notable exceptions, most of the worker co-ops contain no more than 10 or 20 members, but taken together they account for the employment of thousands of people. Since they often choose to remain small so they can retain their democratic structure, members may create additional co-ops in an area if the market is too large for one. And as each co-op has a demonstration effect, it often spawns another. Most have been organized at the grass-roots level without the aid of government agencies, banks, or other established institutions.

The apparently sudden growth of collectivist organizations in the 1970s is understandable only if we recognize that one such organization spawns another, and that they are manifestations of a social movement. Five thousand alternative organizations do not represent 5,000 isolated, independent social inventions. They derive from, and for the most part they continue to identify with, larger movements seeking societal change.

Since collectivist organizations are oriented toward goals of social or personal change, they can be considered social movement organizations (Zald and Ash, 1966). Their development, however, has not been directed by some centralized leadership with clearly defined means, ends, and dogma. There has been a great deal of spontaneity and experimentation in local cooperative enterprises. And yet, because they are part of a movement, the basic organizational forms that have developed from place to place are virtually identical.

Contemporary co-ops represent attempts to build organizations that are parallel to, but outside of, established institutions and that fulfill social needs (for food, health care, education, etc.) without using bureaucratic authority.

That is, they often produce goods or services currently provided by conventional organizations, but in a manner members consider more responsive and socially progressive. For this reason, participants may refer to their organizations as "alternative institutions." In this sense alternative institutions represent rejection of mainstream organizations and an attempt by members to live out other values. For members, to use the feminist slogan, "the personal is political." They believe their most important political message lies in the very act of doing cooperative work: Insofar as members accomplish the job at hand without resorting to hierarchical patterns of authority, they demonstrate that democratic management can work.

From our observations of alternative organizations of many different types, we attempt to draw out the characteristic features that define and unify these organizations. We hope to identify the specific conditions that facilitate, or conversely, that undermine participatory democracy in individual organizations, and finally, to locate the structural dilemmas in which these cooperatives find themselves.

By any standard, cooperatives are radical organizationally. They struggle to resist the hierarchical and bureaucratic practices that we all take for granted and to establish in their place participatory-democratic practices. This makes them unusual in a society based at the organizational level on bureaucracy and subordinate/superior relations, and at the political level on representative democracy. In order to understand how cooperatives came to exist, we must examine the intellectual roots from which they draw their inspiration.

Intellectual roots

The theories, philosophies, and ideas of one era are the common currency of the next. The intellectual dialogues of earlier ages not only become the conscious heritage of future scholars but also give a diffuse coloration to the larger culture through which they percolate. Members of social institutions are not always aware of the origins of the ideas that gave birth to those institutions. So it is with cooperatives. Members are often unaware of the specific historic roots of their ideas and practices.

Democracy is a much-used term in our society. Members of cooperatives, however, use it to mean something quite different from what is traditionally meant. To understand their meaning, one must grasp the distinction between *participatory* democracy and *representative* democracy.

Participatory democracy. Participatory democracy has a firm foundation in the intellectual history of Western society. Indeed, some of the major archi-

tects of Western political thought, such as Rousseau, J. S. Mill, and twentieth-century theorist G. D. H. Cole argued that direct participation by citizens in government and other institutions is crucial for a democratic society. They were thinking of something more than simple discussion of issues and voting, however. A far more direct and active process is essential to the creation and maintenance of a democratic polity, possessed of the skills and attitudes needed for such a system to work. Participation had an important educational function.

As Carole Pateman puts it, summarizing the views of J. J. Rousseau, J. S. Mill, and Cole: "The existence of representative institutions at [the] national level is not sufficient for democracy; for maximum participation by all the people at that level, socialization, or 'social training,' for democracy must take place in other spheres in order that the necessary individual attitudes and psychological qualities can be developed. This development takes place through the process of participation itself" (Pateman, 1970, p. 42). In short, one learns how to effectively participate only through participation. And, people will not develop such skills and attitudes if opportunities are found at only the national political level. This means that a truly participatory polity can exist only if other social institutions are also participatory. Of particular importance for these writers is industry, where most adults spend the greater part of their working lives. If individuals are to learn to expect and to exercise self-determination and democratic responsibilities in the political arena, then work and other local institutions must be organized in such a way that people can participate directly in decision making. The argument is that if people are given a good taste of self-governance, they will want more, and will come to know how to use it effectively.

Rousseau, J. S. Mill, and Cole, then, developed the classical notion of participatory democracy: self-governance in the workplace as well as in political institutions. These ideas form the intellectual background – conscious or not – of modern cooperatives. Co-ops are motivated both by the desire to build direct participatory democracy at the organizational level in economic enterprises and by the vision of someday achieving a more fully democratic society.

However, in reality most workplaces and other social institutions in the surrounding society are organized hierarchically, and the political system is built on representative democracy. How is it that the heritage of participatory democracy has been lost in society as a whole? Why do scholars and laypeople tend to think of democracy only in its representative rather than its direct form?

Pateman (1970) argues this one-sided thinking is due in part to the long-

held myth that there is but one "classical" theory of democracy – that one being representative democracy. On the contrary, her analysis of the work of Rousseau, Mill, and Cole reveals, not one, but two very different theories of democracy. Pateman traces this myth of a single theory primarily to the influential work of Joseph Schumpeter (1942). Schumpeter asserts that the classical theory of democracy rests on the idea of an institutional arrangement for making political decisions through the election of representatives. In order for this system to function, Schumpeter maintains, an unrealistically high level of public rationality, independence, and knowledge would be required of ordinary people. He goes on to charge that classical theory compounds the problem by virtually ignoring the (for him, essential) issue of leadership. Given these – for Schumpeter – unrealistic tenets of classical democracy, he rejects the classical model and constructs his own, presumably more practical, model of democracy. Pateman, however, maintains that Schumpeter misrepresents what the classical theorists had to say and also fails to distinguish the two very different theories about democracy found in those writings (Pateman, 1970, p. 18).

The influence of Schumpeter and others who followed his lead remains strong among contemporary political theorists who base their theories of democracy on the work of such writers as Bentham, Locke, and James Mill. These writers stress representative democracy, in which "participation" of the populace is limited to voting and discussion. For them, as for Schumpeter and most contemporary theorists, participation has a narrow aim: It is designed to protect private interests of citizens and, in that sense, to ensure good government. The educative and self-generating nature of participation is lost; here, participation is merely defensive. Democracy, according to representative democratic theory, rests on the competition of leaders (potential representatives) for votes and office. Thus, as Pateman (1970, p. 20) notes, "theorists who hold this view of the role of participation are, first and foremost, theorists of representative government."

The ascendancy of the ideas of representative democracy in our society obscures the equally rich heritage of participatory democratic theory. Whereas our society and theorists have largely forgotten the legacy of direct democracy, co-ops have not.

Anarchism. A second intellectual foundation of cooperatives, as strong as that of participatory democracy and perhaps even less known by members, is the philosophy of anarchism. From the anarchistic tradition co-ops have inherited crucial principles of organization, operating strategies, and goals.

The Greek word *anarchos,* the root of anarchism, means "without a ruler."

It signifies the existence of organization without external authority and thus implies social order can be achieved solely by internal discipline.[1] The terms *anarchism* and, more frequently, *anarchy* are unfortunately associated with the idea of chaos and disorder, and with nihilism and terrorism. This widespread misunderstanding, one sympathetic historian suggests, arises in part from the nature of anarchism: "few (doctrines) have presented in their own variety of approach and action so much excuse for confusion" (Woodcock, 1962, p. 9). Because of this tendency toward confusion, and because anarchist principles play a large role in co-ops, we wish to clarify these ideas and resolve the seeming paradox of an anarchist organization.

One caveat should be borne in mind: anarchism does not lend itself to a systematic and determinate social theory.[2] Its libertarian attitude admits a variety of viewpoints and actions. But, in spite of the range of anarchist thinking and its resistance to codification, all forms of anarchism do share common features.

The first premise of anarchist organization is what Colin Ward (1966, p. 389) calls the "theory of spontaneous order": "[G]iven a common need, a collection of people will, by trial and error, by improvisation and experiment, evolve order out of chaos – this order being more durable and more closely related to their needs than any kind of externally imposed order." This belief in a naturally evolving cooperative order can be traced to Kropotkin's (1903) proposal for an extensive network of mutual-aid institutions. With external authority abolished, Kropotkin believed, people's "natural" tendency toward "mutual-aid" would express itself in the evolution of countless local associations granting mutual support to each other.

Mutualism, collectivism, anarchist communism, and anarcho-syndicalism are all closely linked in the history of anarchist thought. Proudhon in the 1840s was the first writer to willingly adopt the term *anarchist,* knowing it to be perjorative, believing it to be otherwise. He used this term in the development of the idea of mutualism. By mutualism he meant a future society of federated communes and workers' cooperatives, with individuals and small groups controlling (not owning) their means of production, and bound by contracts of mutual credit and exchange to ensure each individual the product of his or her own labor. Mutualists established the first French sections of the International in 1865.

Collectivists, following Bakunin in the 1860s, continued to stress the need for federalism and workers' associations, but tried to adapt these ideas to industrialization by arguing that larger voluntary organizations of workers should control the means of production. Bakunin and his collectivist followers challenged Marx's leadership at the congress of the International at the

Hague in 1872. A fierce debate and struggle for power ensued, during which Bakunin's followers charged Marx with being "an authoritarian and centralizing communist." The anarchists lost the fight. So began a split in the history of socialist movements between those who support a central-management model of socialism and those who support a decentralized popular-control model. The outcome of the battle at the International ensured the ascendancy of the former.

In the main, the Marxists and anarchists were divided over the use of state authority. Engels (1959, pp. 481–485) argued bluntly that it was impossible to eliminate authority relations in work and industry. Marxists were unwilling to relinquish the state bureaucracy as an instrument of power and wished to use it to implement socialist authority, whereas anarchists were unwilling to accept hierarchical authority, proletarian or otherwise. Authority relationships were anathema to anarchists because they violated principles of individual liberty. Further, anarchists did not believe Marxian claims of the eventual "withering away of the state." A revolution for the anarchists would consist of dissolving the structure of authority, not merely switching who controls that structure.

Following Bakunin came Kropotkin and his fellow anarchist-communists in the 1870s, and, based in the French trade unions, the anarcho-syndicalists in the 1880s. The latter stressed the trade union as an organ of struggle and as a basis for future social organization.

Although differing in emphasis, all of these strains of anarchist thought have much in common. All would have two basic types of social institutions provide the organizational structure for a future free society: the "commune" or "soviet" and the "syndicate" or "workers' council." Both are seen as relatively small, collectively controlled, local units that would federate with each other to deal with large social and economic affairs and thereby benefit from mutual aid and exchange. Each local unit would retain its own autonomy. The commune or soviet would federate territorially; the syndicate or workers' council would federate industrially. Today, clear examples of a commune structure are found in the kibbutz community assemblies in Israel (Rosner, 1981) and in the syndicates or workers' councils existing throughout Yugoslavia and Western Europe.

Collectives and cooperatives in the United States have several characteristics of anarchistic forms of organization. One is their commitment to decentralized, small, voluntary associations. Another is their support of the anarchist federative principle, inasmuch as many of these groups are committed to the development of full-blown "alternative communities" within their own

locales, and form support networks regionally with similar alternative organizations. Furthermore, their goal, which is to create functional organizations without hierarchical authority, is anarchistic, as is their dual emphasis on community control over community functions and workers' control over workplace decisions.

The features mentioned so far relate to anarchist goals. Alternative organizations also utilize many of the methods and tactics of anarchism. A key feature of anarchist strategy is the insistence upon a unity of means and ends:

> Anarchism doesn't want different people on top; it wants to destroy the pyramid. In its place it advocates an extended network of individuals and groups, making their own decisions, controlling their own destiny. (Ward, 1972, p. 289)

Anarchist strategies stress the congruence of means and ends, and thus, for example, would not propose mandatory organizations to reeducate people for a free society. They would not advocate violent means to achieve a peaceful society; nor would they choose centralized means to attain a decentralized society. From the congruence of means and ends flows the conception of "direct action." Direct actions are directly relevant to the ends sought and are based on individual decisions as to whether or not to participate in the proposed action. Examples of direct action include the general strike, resistance to the draft, and the creation of food cooperatives, credit unions, and worker-run workplaces.

Members of collectives see themselves as providing working models for a future society. If they succeed, they may constitute what Buber (1960, pp. 44–45) calls "pre-revolutionary structure-making." That this is the political meaning members ascribe to their collectivist organizations is implicit in the words of a person at the alternative newspaper in this study:

> What I see that we are doing is trying to create soviets. We're creating organizations of people's power. . . . Capitalism is getting more and more into crisis everyday and in defense of themselves people are organizing . . . creating institutions that fulfill people's needs. . . . The more crisis, the more organization will occur.
>
> By the time the final crisis arrives [alternatives] will be everything. The new society will kind of just grow up through the roots and the old society will just be brushed aside.

Our point is that the decision of cooperators to build and work in organizations in the present embodying what they want to create on a societal level in the future is a political strategy in the anarchist mode. It is politics of function, a unity of means and ends.

Although the political strategy of cooperative enterprises fits well with anarchist principles, members may be unaware of the connections. As in the

case of participatory-democratic theory, their anarchist predispositions are largely unschooled and unconscious. It is much more common for people in alternative institutions to pay intellectual tribute to Karl Marx than to Bakunin or Kropotkin.

Marxism. Marxian ideas are a third foundation of co-ops. Generally, co-ops subscribe to: the critique of private property; the analysis of capitalism as riddled with contradictions; the concepts of alienation and exploitation; the materialist conception of history; and the vision of a future society that is classless and just, and one in which producers control the means of production.

Marx himself was ambivalent toward workers' cooperatives. Avineri (1968, pp. 179–180) points out that Marx saw co-ops as representing revolutionary new forms of property – social property – that would prefigure the coming of socialism. But he also believed cooperatives would have to be developed on a national scale, under public sponsorship, if they were to significantly improve working conditions for the masses or arrest the spread of monopoly in the economy. Most members of contemporary co-ops are probably not aware of Marx's specific ideas about cooperatives, and if they were they would likely prefer grass-roots development rather than centralized government sponsorship of co-ops.

The point at which Bakunin departed from Marx is the point at which the contemporary cooperatives depart from Marxian principles: the seizure of state power versus the dismantling of it. This is really a dispute over political means, not ends, but it has serious repercussions for the ends attained. Marx believed that without the state apparatus at its disposal, the proletariat would not be able to assert its will nor to suppress its capitalist adversaries who might attempt a counterrevolution. Thus, state power had to be invoked, if only temporarily. In opposition to this idea, Bakunin spoke of the dangers of "red bureaucracy" that would prove to be "the most vile and terrible lie that our century has created." He warned that state socialism would produce an overwhelming centralization of property and power, and ultimately a bureaucratic despotism. Instead, he urged the use of direct means of appropriating capital: workers' associations, federated in cooperative relationships with each other.

Although the contemporary collectives tend to attribute more of their intellectual debt to the ideas of Marx than to anarchism or classical participatory-democratic theory, in fact, Marxism offers them little direction in their day-to-day affairs. Their practices, we argue, owe far more to the ideas of anarchism and participatory democracy.

Alternatives to bureaucracy: why now?

Max Weber argued bureaucracy would increasingly come to dominate all aspects of social life, replacing all previous forms of organization in history and perpetuating itself indefinitely. More than almost any prediction in social science, his has stood the test of time. For Weber, "once . . . fully established, bureaucracy is among those social structures which are the hardest to destroy" (1946, p. 228), one reason being that it has a powerful set of beliefs supporting it. Bureaucracy, like the forms of organization preceding it, requires certain "legitimating principles" to sustain it, and it finds this legitimation in the modern belief that, for the sake of efficiency, a hierarchy of expertise and standardized procedures should be followed.

In a Weberian framework, then, it is puzzling that contemporary collectivists, or anyone else, could ever develop antibureaucratic sentiments and practices. If social beliefs support institutional arrangements, and vice versa, then how does change in either begin?

Despite Weber's belief that bureaucracy would prove to be revolution-proof (1946, p. 230), it is possible to ferret out of his work three sources of tension that have the potential to undermine the legitimacy of bureaucratic authority. First, Weber recognized that public officials who blatantly disobey the law would undermine the rule of (formal) law itself. In the post-Watergate era, examples are plentiful. Second, he noted that special laws passed for particular interests would constitute another assault on the universalistic basis of formal law. Most important, he realized that demands for substantive justice and special treatment would likely arise from the underprivileged in a desire to equalize life opportunities (1946, pp. 220–221).

Although a Weberian analysis suggests several general reasons why Americans might begin to question the justice and legitimacy of established institutions, it does not tell us why a particular group of people, American young adults of the 1970s and 1980s, have taken their questioning of the system to greater lengths, attempting to construct organizational embodiments of countervalues. Young American adults in alternative institutions are critical not only of bureaucracy: They also frequently oppose the fundamental principles of our capitalist economic system.

Joseph Schumpeter provided in 1942 perhaps the most cogent analysis and prediction of a coming change in attitudes toward capitalism. In brief, he argued capitalism is the most productive and efficient economic system in history (so far, both Marx and Weber would agree). However, "its very success undermines the social institutions which protect it and 'inevitably' creates conditions in which it will not be able to live" (1942, p. 61). For Schum-

peter, the capitalist process creates a critical frame of mind that ultimately turns against private property and bourgeois values. The very logic of capitalist society generates a hostile atmosphere, and creates a special social class of intellectuals who help to articulate and organize this general hostility. The defenses of capitalism are broken down further with the disintegration of the bourgeois family. As rationalization spreads to all domains of society, including private life, it renders children and home irrational in cost accounting terms, and thus removes an important force that had motivated people to save and invest. Without children, people become more present oriented and less willing to sacrifice for the future.

In one institution after the next, Schumpeter argues, the process of capitalism destroys its own institutional framework – its protective strata. The modern bureaucratic corporation narrows the scope of capitalist motivation until it eventually kills its roots. In time, the entrepreneurial model of thrift, hard work, risk taking, and self-denial is replaced by the values and mentality of the salaried bureaucratic worker. A consumption orientation overtakes a production orientation. Ultimately, "the bourgeois order no longer makes any sense to the bourgeoisie itself" (1942, p. 161).

The material well-being generated under a capitalist, bureaucratic system undermines the entire cluster of values that had supported it – what Weber called the Protestant Ethic. That is, from the individual's standpoint it no longer appears rational to be frugal, hard working, or dedicated to the intense pursuit of material wealth. As the Protestant Ethic is weakened, the compelling nature of the capitalist mode of production wanes. Not only does the evolution of capitalism tend to wear away its own socio-cultural bases, but at the same time "it shapes things and souls for socialism" (Schumpeter, 1942, p. 220), thereby clearing the way for (centralized) socialism as the heir apparent. Thus, although Schumpeter reaches Marxian conclusions, his mode of analysis is distinctly non-Marxian. His is an analysis of the contradictions inherent in the culture of capitalism, not in the economics.

If it is the rationalization, the technology, and the affluence generated by a capitalist mode of production that permit freedom from the necessity of human toil and that render the values of the Protestant Ethic obsolete, then we should expect the work ethic to lose its coherence first for people who have grown up under conditions of affluence. This expectation is supported by a variety of analyses of social change in advanced capitalist society (Flacks, 1971; Marcuse, 1962). Following Flacks (1971), we would expect young people who find conventional cultural values so incoherent that they create alternative values, motivation patterns, and life meanings (viz., they become participants in alternative institutions) to be disproportionately from affluent,

well-educated, intellectually oriented families. As we will show in the next chapter, the data of this study do fit such a model. Participants in the organizations we studied tend to come from relatively affluent, well-educated families. They have backgrounds and social characteristics quite similar to the social activists of the 1960s and 1970s, and have often taken part in those earlier protest movements. They have high expectations for their jobs and careers, and commonly have been frustrated by perceived and actual possibilities in mainstream organizations, with the result that they have created their own workplaces as an alternative.

A summary to this point

Current economic and cultural conditions have nourished the most recent wave of cooperatives in the United States. The new cooperatives embody the historic themes of anarchism, participatory democracy, and, in part, Marxian thought. Cooperatives thus attempt to be true to ideas with deep intellectual and historical roots, ideas that have lost out in the mainstream of both capitalist and centralized socialist societies.

As we have seen, two distinct models of democracy co-existed in the eighteenth century, but as the Western capitalist societies emerged from their revolutions, it was the ideal of representative democracy that was institutionalized, displacing the notion of participatory democracy. In socialist societies such as the Soviet Union, it was the centralized version of Marxian socialism that won out over the more decentralized, democratic vision of the anarchists. With the ascendancy of representative systems of democracy in the capitalist societies and of state socialism in the socialist societies, there was no system built on decentralized local control of community institutions and of workplaces by the people in them. Cooperatives represent this third road. They are based on principles of economic and political organization that are an alternative to the concentrated corporate power of capitalism and to the centralized state power of socialism. The world has yet to see an entire society based on a system of local, participatory democracy, yet that is the vision of many in cooperatives.

The challenge posed by Weber and Michels

In a modern bureaucratic society, cooperatives are sometimes seen as curious anomalies, inconsequential organizations without serious possibilities for growth or influence. Far from seeing co-ops as models for entire societies, many outside of co-ops are skeptical about the possibility of participatory

democracy even at the level of the single organization. In order to appreciate this point of view it is essential to return to the important work of Max Weber and to consider another enormously influential early theorist of organizations, Robert Michels.

For Weber, Bakunin's anarchist ideas were "naive" (Weber, 1946, p. 229). From a Weberian point of view, the anarchist ideal of organization without authority would be revolutionary because it would represent a structural transformation, not a mere change in who controls the state bureaucracy. But to Weber, it is a hopelessly utopian idea. As earlier noted, the requirements of modern society, Weber believed, would make bureaucratic authority a permanent and indispensable feature of the social landscape.

Cooperatives do not fit into the well-known Weberian typology of legitimate authority relations. For in their collectivist mode of organization, they do not grant authority on the grounds of formal legal-rational justifications, nor on the basis of tradition or the charisma of leaders. Instead they conform to a fourth basis of legitimate authority, a type mentioned but not elaborated in Weber's work, that of value-rationality. They are committed first and foremost to substantive goals, to an ethic, even where this overrides commitment to a particular organizational setting.

Weber defined "authority" as the power of command and the duty to obey. He used the term synonymously with "domination" (*Herrschaft*) to refer to situations in which both the ruler(s) and the ruled subjectively accept the legitimacy of commands from the former.[3] Domination, in Weber's view, requires an administrative apparatus to execute commands, and conversely, all administration requires domination.

However, in the alternative organizations studied here no one (ideally) has the right to command, and no one the duty to obey another. There can be no subordinates where there are no superordinates. They strive for the absence of domination. Legal-rational justifications for hierarchical authority have been actively challenged. In collectives, ultimate authority resides not in the individual (by virtue of positional incumbency or technical expertise) but in the collectivity of worker–owner–members as a whole.

The prime goal of collectivist organizations is escape from the Weberian imperatives of domination and hierarchical administration. Central to our analysis, therefore, is an examination of the means they use and an evaluation of how well they succeed.

Weber did acknowledge that under select conditions (e.g., small size, functional simplicity) organizations can avoid structures of domination and can maintain directly democratic forms (Weber, 1968, pp. 289–290). However, since he was primarily interested in the principles that legitimate domination

and the corresponding administrative machinery that implements it, he unfortunately did not examine the conditions supportive of genuinely democratic forms of organization, free from *Herrschaft*. It is here we hope to make a contribution.

Weber's insistence on the inevitability of bureaucracy has profoundly shaped later thinking about organizations. The work of his contemporary, Robert Michels, speaks even more directly to the problem of maintaining democracy within organizations. Boldly stated and well remembered as the "Iron Law of Oligarchy," it has had great influence. On the basis of his study of the German socialist party, Michels concluded: "Who says organization, says oligarchy" (Michels, 1962, p. 365). The main subject of Michels's study, the German Social Democratic Party, tried to be internally democratic in a manner consistent with its socialist principles. However, Michels argued, the party evolved a relatively closed and oligarchic structure, controlled by only a few of its central members, and abandoned the interests of its rank and file. He perceived similar situations in other socialist parties and trade unions. In the case of political parties, for example, he attempted to delineate the process of oligarchization: The primary goal of a political party is to attain political power. As the power of a party is largely a function of the size of its membership, growth is of the essence. In search of growth, members are seduced into abandoning their principles. Internal democracy gives way to rule by a small group of elite members who are able, through the control of information and patronage, to make themselves indispensable.

Michels wanted to generalize from his case studies of political parties and trade unions to organizations of other kinds. Indeed, the subtitle of his 1911 book, *Political Parties,* is *A Sociological Study of the Oligarchical Tendencies of Modern Democracy,* which suggests he thought his ideas had broad applicability. Even today, scholars treat Michels's work in just this way: The Iron Law is assumed to apply to all types of organization. Mainstream organizational and management theories typically ignore or dismiss the possibility of internal organizational democracy and, following Weber and Michels, take for granted the presence of hierarchical, superior/subordinate relationships, and oligarchic control.

Although Michels's work has had large and continuing influence in promoting the idea of the inevitablity of oligarchy, it has not stood without challenge. Some scholars of Michels's work, for example, have charged the history of the German Socialist Democratic Party does not fully support Michels's thesis. There is some evidence, too, that Max Weber, mentor to Michels, did not entirely agree with Michels's conclusions. A long correspondence between the two scholars indicates Weber wanted Michels to tone

down or qualify his blanket assertions about the inevitability of organizational oligarchy. As Scaff (1981, p. 1282) puts it:

> For Weber, the mere appearance of power, leadership or professional specialization in an organizational setting could not in itself eliminate all possibility of democratic rule. . . . Thus, unlike his colleague [Michels] Weber was prepared to see democratization as a typical development in modern societies, just as he was later able to characterize the "often-described organization" of the SPD [Social Democratic Party of Germany] as "strictly disciplined and centralized" but "within democratic forms."

Thus, Weber was somewhat more open to the possibility of organizational democracy than was Michels, but since Weber never really spelled out his own ideas regarding the topic, Michels's more absolute point of view captured the stage. It is thus the work of Michels that poses the most acute theoretical challenge to the direct democracy that cooperatives wish to create and maintain. In the analysis to follow, we assess to what extent cooperatives are able in practice to successfully evade the Iron Law.

2. The organizations studied, the methods used

Sociologists have long recognized and bemoaned "the gap between the character of current theories and the character of much current research" (Merton, 1964, p. 242). Sociologist Robert Merton (1957) suggests that "middle-range" theories, by being more amenable to verification than grand theories, might help to narrow this gap. Responding to this call, Glaser and Strauss (1967) have proposed a specific research approach to close the gap between ungrounded theories and empirical studies unguided by any theory. They urge that theory should be generated from the data, and that theory arrived at in this way, namely "grounded theory," will have more power to predict and to explain the subject at hand than will theory arrived at through only speculation or logical deduction. For Glaser and Strauss, it is the responsibility of sociologists to ground their theoretical categories and propositions firmly in empirical research. Further, they maintain that a comparative method of analysis based on systematic procedures for collecting, codifying, and analyzing data (particularly qualitative data) best lends itself to the discovery of grounded theory.

This study attempts to develop an original and grounded theory of organizational democracy using the qualitative, comparative method of analysis developed by Glaser and Strauss. In so doing, we develop middle-range theoretical propositions about the form, conditions, and dilemmas of democratic organization in modern society.

We base our analysis of collectivist organizations on two sources of data: intensive case studies of five cooperative workplaces we observed, and an extensive and systematic review of the work done by other researchers on similar or related organizations. As we develop our theory in Part II, which is based primarily on our own case studies, we also bring in the empirical observations of other researchers to further support our points. In Part III, however, we turn chiefly to the research of others to indicate the broader applicability of our theory.

The small-scale collectives and cooperatives that are the focus of our study are but one part – albeit an important part – of a much larger movement

toward the democratization of work organizations. As we noted in Chapter 1, contemporary cooperatives are motivated by a special set of values and by the desire of people to integrate their values with the need to earn a livelihood. However, for reasons that stretch beyond any new values that may be arising in our society, democratization is proceeding in more conventional sectors too.

The growing international competition for markets and the resulting pressure to improve productivity have led to Quality of Work Life (QWL) programs in a substantial number of *Fortune* 500 corporations (Simmons and Mares, 1983). Such programs typically try to enhance the degree of worker influence in decision making, even if only in an attempt to tap workers' ideas for solving problems. Although the actual level of worker influence or control varies from case to case and is hotly debated, the point is that this very large and vigorous corporate movement *has* appreciated the link between increased worker influence and improved productivity.

Another type of democratization that is proceeding apace can be seen in the thousands of Employee Stock Ownership Plans (ESOPs) that have been created in the last five years. Most of these give only a minority ownership position to the workers, but in an estimated 250 cases workers have acquired a majority of the voting stock. Clearly this new development carries with it the potential for worker control, but researchers have found that the vast majority of these cases do not extend substantial participation rights to workers in the actual process of production (Rothschild-Whitt, 1983, 1984). In general, ESOPs arise in part because the ability of capital to move internationally produces numerous shutdowns of what once were – and could again be – viable plants (Bluestone and Harrison, 1980). The foreseeable devastation to workers and communities produces cooperative efforts to "save" the plant under the terms of worker ownership. Still other ESOPs arise because of the tax advantages allowed in the law, and other material reasons. Again, our point is that this reflects a realization of the connection between worker motivation and productivity.

Other countries, too, have made substantial forays – indeed, they have often led the way – into experiments with various forms of worker control, including workers' councils and codetermination in Europe, self-management in Yugoslavia, cooperatives in Israel, and so on.

Still another kind of organization in which sustaining member interest and participation are often a central problem is the nonprofit, voluntary organization such as a charity or public interest research group. Our hypotheses concerning the conditions and dilemmas of democratic organizations should prove relevant here too.

If one takes the perspective of Karl Marx, one might expect the ownership structure to be the most determinative aspect of any production organization. Following the lead of Max Weber, on the other hand, one might take the bureaucratic control structure to be primary. Looking at the worldwide movement toward the democratization of work, we note that some models provide examples of democratizing ownership, but fail to give actual control to workers. Other models, like QWL, focus on the control structure and neglect the underlying structure of ownership. The "purest" forms of democratization, of course, represented by the cooperatives in this study, attempt to democratize *both* ownership and control.

To emphasize, Part II of the book builds a theory of organizational democracy on the basis of the purest examples of direct democracy we have available, the five cooperatives we observed intensively. Part III draws out the relevance of the theory and the cooperatives' experiences for other, related democratizing organizations.

As theoretical points are made in the chapters to come, we frequently refer to empirical examples from each of the organizations in our study. We therefore need to begin with an overview of these five organizations and of the research methods used to study them.

Research methods

Since the aim of this study is to create a new theory about the factors that nurture organizational democracy, we have selected for study organizations that, on close scrutiny, appear to share a sincere desire for democracy, but that are as varied as possible along several other dimensions: Some are relatively large, others are small; some perform complicated tasks and use a sophisticated technology to do so, others perform more simple tasks requiring little technological support; some produce goods, others provide services; some are self-supporting financially, others receive outside grants from foundations and government agencies. Because we also want to see how well participatory democracy can function in a variety of specific organizational settings, we picked five different kinds of collectivist organizations: a cooperatively run community medical clinic, an alternative high school, a food cooperative, a collective newspaper, and a legal collective. By making systematic comparisons among these organizations, we have tried to isolate for analysis the generic features of cooperative organizations.

Because of the limitations of space, we are able to report only a fraction of the extensive qualitative and quantitative data we have collected. In making a theoretical point, we have tried to select those examples that seem most char-

acteristic of *collectives in general,* not those that occur only rarely or that appear typical of only one or two kinds of organizations. We would argue therefore that the theoretical formulations in this work should prove applicable in a broad range of participatory democratic settings.

A number of research methods were employed, with data from one serving as a check on the other. Participant observation was undertaken in four of the five research settings. Field observations ranged in duration from six months to two years. Meetings were a focus of attention, and the researcher observed an average of 16 meetings per organization. Meetings ranged from a full day in some cases to just an hour or two. Detailed field notes were taken both during and immediately following observed events. The fifth setting, the legal collective, would not permit field observation of routine activities, and so documents (including a diary written by a participant) and interviews with organizational members were used. Although participant observation forms the core of this study, it is amplified in important ways by three other sources of data. First, intensive, semistructured interviews were conducted with a number of members from each of the five groups. Interviews were recorded on tape, and took from 1 to 5 hours, with the mean interview time being 2¼ hours. Second, written documents from each of the groups were examined; these included constitutions and by-laws, funding proposals, agenda and minutes of meetings, personal diaries, internal memoranda, budgets and financial statements, statistics on clients, and newsletters. Finally, observational material was supplemented by questionnaires returned from the members of the free medical clinic, the food cooperative, and the community newspaper. Several sets of questionnaires were used, with individual items worded similarly where appropriate.

The comprehensive questionnaires dealt with the socioeconomic-economic backgrounds of people in alternative organizations, their political attitudes, and their religious backgrounds. In addition, the questionnaires contained items pertaining to members' reasons for joining the organization, their perceptions of their own decision-making influence in the organization, their estimate of the distribution of power and influence in the organization, their sense of an optimal size, the degree of satisfaction and participation in the organization, the responsiveness of the staff and board of directors to members' interests and needs, and a variety of other organizational issues. Further, in each case, the researcher solicited questionnaire items from the members themselves, and added their suggestions – usually items of practical utility to the particular organization – to the final questionnaire.

In sum, we used data from many different sources – direct observation, interviews, organizational documents, and questionnaire surveys – to trian-

gulate on the research problems at hand. This strategy holds the advantage of allowing the researcher to obtain independent confirmation or disconfirmation of any generalization arrived at inductively (see Webb, 1970; Whyte, 1979).

The research settings

The five alternative organizations selected for this study were located in California in a medium-sized metropolitan area. To ensure that some permanence was involved, no organization was included in the study unless it was at least two years old. All organizations have been given fictitious names. The brief descriptions that follow are intended to provide contextual and historical information, so that examples introduced later for analytic purposes will be more meaningful.

Freedom High. The first alternative organization to be examined in this study is a "free" high school, hereafter referred to as Freedom High. This school was not an accredited school and was not part of any public school system. It was a legal alternative day school in the state of California (its students were not considered truant from public school, but they received no official accreditation). The school financed its meager budget from tuition payments, which were based on a sliding scale.

The initial conception of Freedom High emerged in March of 1970. At that time two young adults who were interested in educational reform began to hold weekly coffehouse discussions with high school students who were disaffected with the public schools; about 20 students attended. Discussions were organized by the student council president of a local high school. Meanwhile, some adults in the community were helping to set up a free school at the elementary level, and a number of contacts existed between these two groups.

On April 6, 1970, the free high school opened with 12 full-time students. Their enthusiasm spread by word of mouth, and by the end of May they had lined up about 50 students who wanted to attend Freedom High in the fall. Since the two nonstudents who were associated with the formation of the school would soon be leaving the area, they convinced two out-of-town friends to direct the high school. Upon arriving, the latter found a nascent organization and 50 eager students.

The high school was able to get on the nonprofit corporation charter of the elementary school and thereby avoid certain legalities. The charter allowed the school to be a legal alternative to the public system. Sharing a charter

meant the two schools shared the same board of directors. Although the board had certain policy-making rights and legal responsibilities, it opted from the beginning not to exercise its powers with regard to the high school. The board was made up of a number of wealthy, liberal adults who were instrumental in starting the elementary school. Since they were mainly interested in the elementary school, they did not raise money or make decisions for Freedom High. This left control of the school to the participants.

Freedom High was located in a storefront in a downtown area. It was staffed by a large number of part-time volunteers, most of whom were college students enrolled in a special course for which they received university credit. There were also a handful of full-time but poorly paid staff members. Classes at the school were generally shaped around the interests and the talents of available staff, and so a comprehensive program was not initially possible. As time went on, the school was able to stabilize its staff somewhat and to expand its program. By the second year of operation, it was asking volunteer teachers for a full year's commitment to the school (rather than just 2½ months) and was able to recruit volunteer staff to fill specific gaps in the program in response to students' interests (e.g., in oceanography, women's history, nutrition). In addition, resources in the community began to be utilized extensively for their pedagogic value, and job apprenticeships were arranged for students desiring them, for example, in carpentry and legal aid.

Decisions at Freedom High were made in three types of meetings: plenary "all school" meetings, staff meetings, and coordination meetings. All meetings were open to any staff member or student who wanted to attend, and decisions at all were taken by a process of consensus, not by vote. Meetings tended to be long and frequent, with participation regarded as a right and a responsibility of members. Full-time staff, part-time staff volunteers, and students were all considered "members" with equal rights whose responsiblity it was to participate in shaping the school. Even the usual dichotomy in schools between staff and students was purposefully blurred at the Free School.

The school contained from 27 to 41 enrolled students during the course of this study, which coincided with the school's first three academic years. A slight majority were from upper middle-class families and all were white. The large numbers of part-time volunteers from the university permitted an enviable student-teacher ratio of about 3:1, if part-time teachers are weighted .5. Classes ranged in size from individual tutorials to seven.

Tensions sometimes developed at the Free School between two main sets of goals: those oriented toward personal change and those oriented toward social change. Like many free schools, this one tried to combine these goals by stressing: learning outside the schoolhouse; affective development and its

integration with the cognitive; individualized instruction; no tracking or grading, though there were some written evaluations; a critical understanding of institutions in American society; and an activist orientation toward changing society.

Participant observation at Freedom High took place over a period of two years and three months, ending at the close of the school's third full year of operation. At this point, the school was losing its first generation of students, who had helped to found the school, and its second full generation of staff members. The need for a local free school no longer seemed so pressing, since the public high school had liberalized its program. Thus, the school was considering dissolving itself when a graduate student in education, heretofore not associated with the school, asked if he could take it over for the next year. Having no definite plans for the future of the school, the people at the school agreed. Thus, during its fourth year, Freedom High was the same organization in name only. It experienced a complete turnover in staff, it was able to recruit only 12 students, and what had been an enthusiastic, self-initiating student body was replaced by a largely passive student body, eager to escape the punitive aspects of public school, but uninvolved in creating a positive alternative. Field observations were not made during this last year, but interview reports indicate it provided a fruitless environment for the students. After an unsuccessful year, Freedom High closed its doors in June of 1974.

Research on the four other organizations, described below, began shortly after Freedom High closed and extended over an 11-month period. The same research methods and sources of data were used (participant observation, interviews, surveys, and documentary analysis), except that only interview and documentary material could be obtained in the case of the legal collective.

The Free Clinic. The Free Clinic began as a two-month pilot project in May 1971 under the auspices of a county health department. Originally,the project was initiated by professionals at the county health department, funded by a state grant, and targeted to reach three separate communities of youth: blacks, Chicanos, and the counterculture. Personnel and outreach representatives from each of these groups were recruited to run the pilot program. Although it was moderately successful, officials at the county health department decided not to continue a free medical clinic there because, in the words of one health department professional, "They felt there was too much conflict between the establishment way of operating clinics and this group's way."

The demonstration project did arouse the enthusiasm of the people who had worked on it, and they decided to launch an autonomous free clinic. Within two months they had obtained a nonprofit corporation charter for the new

clinic and they opened a few months later. Although this was among the earliest free clinics, its founders were well aware of similar efforts around the country. From the start, they defined their medical goals very broadly, as captured in their motto: "health care for the whole person, mind and body." Unlike many other clinics at the time, they were not devoted solely to providing venereal disease or drug treatment.

The social base of the Free Clinic turned out to be rather narrow. Some of the early organizers learned from their experience in the pilot program that "these three groups [blacks, Chicanos, counterculture] do not interrelate too well," and decided to tailor their services toward the counterculture. That the Free Clinic would have had to design itself differently to appeal to each of these three groups was explained by one founder of the clinic:

> The counterculture group is very casual . . . barefoot, casual, wild colors . . . and the Chicano and the Black have grown up with an image of medical care as being – if it's too casual it's second class, and they want the best. . . . They want the doctor in the white coat, everything spic and span, just right. Otherwise we're saying to them we're giving you second-class medicine. That's how they interpreted it to us.

Thus, the early organizers of the Free Clinic instead settled on the strategy of providing a very broad range of medical services to a restricted target population of counterculture youth.

By becoming incorporated, the clinic was forced by law to assemble a board of directors. Like Freedom High, they sought out the most prestigious members of the community they could find to be board members, hoping in this way to lend an air of legitimacy to their alternative institution. But, unlike Freedom High, they endowed this board with organizational power and responsibilities that the board of Freedom High did not have or desire. This set the Free Clinic up for bitter staff-board conflicts that Freedom High never had.

As Sarason (1972) suggests, all the decisions made in the "before the beginning" stage of the Free Clinic – its early relationship to the county health department, its independence from them, its countercultural orientation, the breadth of its conception of health, and its powerful board – significantly influenced later events at the clinic.

When the Free Clinic opened in late 1971, one of the organizers of the pilot project, and later of the clinic itself, became its director. A number of assistants worked with him. Although the clinic was quite participatory in its early days, it was not collectivist. It resembled any human relations organization with a rather horizontal structure of authority. The director, however, was not pleased with this mode of organization, and tried to develop in the clinic a number of "component coordinators" to replace the present director-assistant relationship. After instituting these changes, he realized he would always be

regarded as the de facto "director," a common problem in the transition to collectivist organization. He soon left to get a Ph.D. in public health, leaving the clinic with a health education coordinator, a counseling coordinator, and two medical coordinators, arranged in an egalitarian manner. Though it continued to grow, the clinic retained that structure.

At the time of observation, the Free Clinic had six full-time staff coordinators: two on medical, two on administration, and one each on health education and counseling. Each of the four components had a trained staff of volunteers to implement the components' programs: patient advocates, lab technicians, counselors, receptionists, and health education speakers and writers.

Decisions were made at component meetings, at weekly staff meetings, and at monthly board of directors meetings. After the original director left, staff insisted upon a collectivist mode of operation (e.g., equal pay, no "director") over the strong objections of the same prestigious board members the clinic had eagerly sought in the beginning. While staff struggled to retain "collective" control over the clinic (i.e., staff control), the board tried to reassert hierarchy among staff, and board control, by appointing a director accountable to the board. For two years dozens of staff-board conflicts raged over day-to-day concerns involving this issue. Finally, the staff succeeded in gaining the resignations of those members of the board who had wanted to reestablish hierarchy, and filled those vacancies with friends and clinic volunteers who supported collectivism.

The philosophy of the Free Clinic remained progressive. To quote from a position paper written by Free Clinic staff:

> We are committed to providing an alternative health care system showing the way health care should be. . . . We believe illness is a social-economic problem, not just biological. . . . We are dedicated to total health care, of mind and body. . . . We believe that prevention is as important as treatment of illness.

The clinic tried to affirm these principles in its everyday practices. In addition, a later philosophical statement reads:

> No one should gain excess profit from the sickness, death, or misery of others. . . . Health care workers, patients and community members should determine the priorities of health institutions. . . . The roles and interrelationships of health workers should be redefined to be nonsexist and nonelitist.

At first, the Free Clinic was entirely supported by city and county revenue-sharing grants, private foundation grants, and donations from patients and others. This allowed it to be, as the name implies, entirely free to patients. Even such services as analysis of laboratory tests, dispensing of prescribed drugs, physician examinations, and private counseling sessions were free.

The average patient donation was $1.00 to $2.00 per visit. However, toward the close of its fourth year of operation, the clinic faced the serious probability of not receiving further revenue-sharing funds for the 1975-76 fiscal year. Although the clinic was well regarded, local government officials did have a general rule of not supporting any community organization for more than three years. The threat of losing this base of financial support and the prospect of having to close the clinic jolted the staff into making plans to become more financially self-supporting. They decided to move gradually toward a fee-for-service system. This entailed instituting a liberal sliding scale of nominal fees, and where possible, billing insurance companies and Medi-Cal for services rendered.

The Free Clinic provided an extensive range of services. The medical component had several afternoon and evening medical clinics each week, a weekly dental clinic, and a weekly children's clinic. Clinic hours were staffed by one or two physicians, a medical coordinator (who was also a nurse), and 6 to 8 volunteer patient advocates, lab technicians, and receptionists. The medical component generally had about 50 volunteer "patient advocates" from which to draw at any one time. These people had gone through a two- to three-month training program, and attended component meetings at which they decided issues of relevance to the medical component and were kept abreast of new medical information useful in their jobs. Patient advocates, an innovation of free clinics, took medical histories, tried to ensure that the physician understood the patients' problems and that the patient understood the physicians's diagnosis and treatment, and in general tried to personally give patients health care information that would help them to care for themselves.

The counseling component provided individual psychological counseling with its staff of trained volunteer counselors. In addition, it held evening group sessions on such topics as transactional analysis, verbal self-defense for women, and psychodrama.

The main purpose of the health education component was to "demystify medicine." Following the clinic's belief that "educating people about their health problems is an important and basic part of any treatment process," the clinic's health educators tried to raise public awareness of health care treatment and prevention. Toward this end, they wrote a series of free pamphlets on such topics as Medi-Cal, dental care, herpes, hepatitis, venereal disease, and drugs. These pamphlets were widely respected in the medical community; they were translated into Spanish and were distributed at several community clinics, at general hospitals, and at the county health department. Health education also had a speakers' program in which trained volunteers spoke by invitation to a variety of organizations such as schools, colleges,

juvenile hall, and a Lion's Club on such topics as sexuality, venereal disease, birth control, sex roles, drugs, and nutrition. The health education component also put on a regular radio program. Subjects covered in this show included self-healing, nutrition, pre-orgasmic counseling, male sexuality, and child abuse. At any given time, the Free Clinic had 60–100 active volunteers, and each component was to hold meetings with these people.

Over the years the Free Clinic had grown from an average clientele of 130 patients per month by the end of its first year of operation to a mean of 479 patients per month by the close of its third year. Most of the patients (79 percent) were between 15 and 30 years of age and were white (86 percent). Typically, they came to the clinic for venereal disease testing and treatment (23 percent), followed in frequency by treatment for the skin (18 percent), and gynecology (16 percent). In addition, the counseling program reached an average of 69 people per month, and the health education speakers' program reached an estimated 675 people per month. As free clinics go, this one was large in the number of patients it saw and comprehensive in the services it provided.

The Community News. The alternative newspaper in this study was just that – an alternate – it was neither "underground" nor "establishment." Referring to similar alternative newspapers in Boston, Kopkind (1974) describes them as "sea-level."

The genesis of the *Community News* can be traced to the summer of 1971. At that time a number of student activists who later were to found the paper were working on an underground paper. One member described the group as, "very political, almost to a Weatherman position . . . we were studying self-defense and guerrilla warfare." However, by mid-1971 the group began to reanalyze the political situation:

> We came to the realization that the revolution will be a very gradual thing. . . . The collective had thought about the problems of ideological hegemony and how it was maintained by the media. We came to see all media as propaganda . . . toyed with the idea of having an alternative media to present leftist propaganda instead of the usual capitalist propaganda.

What transformed this from a pipe dream into a feasible plan was the coincidental discovery of a fellow West Coaster who had reached the same conclusion and who had just received $5,000 from a wealthy liberal patron to set up an alternative paper in California. All through the summer and fall the group held discussions about the paper they were to launch. In the words of one of the founders:

> It was, like all media, to be a political organ, not an impartial voice. We wanted to activate liberals and open them up to a left perspective . . . try to reach a broad base, not just the students and the underground . . . concentrate on local issues. The Left had been associated with national issues, with Vietnam. We would resist being labeled and ignored as "radicals" if we emphasized community issues. Also, we felt it was important to learn to apply a left analysis to local isues and that these issues, being more concrete, would be better able to arouse people . . . if we wanted to reach nonradicals. We really had a commitment to speaking in a language and style that anyone could understand and accept.

These were the consensual goals of the *Community News* when it began publishing in December 1971, and they remained its essential goals.

However, specific points of disagreement existed among the organizers of the *Community News*. For instance, all agreed the paper should appeal to a "broad base," not only students, but some favored a labor orientation, whereas others wanted to cultivate their most "natural audiences" – liberals, professionals, and the middle class.

Evidence seems to indicate the paper did best at attracting the latter. According to a readership survey conducted by the researcher, 52 percent of the readers of the paper identified themselves as professional or as middle class. But the paper had also self-consciously tried to attract the interest of the working class. A content analysis of articles in the paper over a 10-month period reveals that stories about labor were among those most frequently published. Of the respondents to the readership survey, 18.5 percent identified themselves as working class. Since its earliest days, people at the *Community News* sought to develop a broad community-based constituency rather than a student-based one. Consequently, the content analysis shows stories about the university and the student community were among the type least frequently covered. Only 13 percent of the respondents of the readership survey classified themselves as students and, only 21 percent of the readers of the paper were under 23 years of age, but 76 percent were between 18 and 40 years old. The median educational attainment of the readers of the paper was a B.A., but the median income was reported at only $7,500 to $10,000. Thus, the *Community News* did attract an audience wider than the student community.

There is also evidence that the *Community News,* again in line with its objectives, appealed to liberals and perhaps had some impact on their perspectives. Twenty-five percent of the readers of the paper viewed themselves as radicals, 53 percent identified themselves as liberal, and only 4 percent classified themselves as conservative. Twenty-one percent of the readers of the paper thought the paper was radical, while 71 percent believed it was liberal on most issues. Of more importance was the impact the paper had on

its readers. Two-thirds of the readers reported that the paper's coverage had changed their opinion about specific local issues. Of the remaining one-third who reported no opinion change, this was usually (in 71 percent of the cases) because the individual already agreed with the paper's opinion, and seldom (in 15 percent of the cases) because the reader was unconvinced by the paper. Fifty-four percent of the readers reported they always or usually followed the endorsements of the *Community News* in voting. The *News* was proud of a variety of liberalizing political changes that had taken place locally since the creation of the newspaper (for example, the election of progressives to city council and county Board of Supervisors positions), although it is difficult to tie such changes directly to the effect of the paper.

When the paper first began publishing it was a biweekly of about 12 pages. As a newly created newspaper, it had to build its circulation and its advertising from a base of zero. Two years later, it was putting out a weekly paper of 28–40 pages, and by its third year it boasted a circulation of 22,000. As a proportion of the local population, the readership of the *Community News* was higher than the 1974 combined readership (150,000) of the *Real Paper* and the *Boston Phoenix* (Kopkind, 1974).

The growth and success of the alternative press came at a time when "straight" community newspapers were folding or shrinking. The Los Angeles metropolitan area alone had more than 300 straight community newspapers. Many were suffering severe financial problems owing to the rapidly rising cost of newsprint, recessionary advertising cutbacks, and a decline in readership of local newspapers (*Los Angeles Times*, pt. I, pp. 1, 25–18, June 22, 1975). However, in marked contrast to the *Community News* and the alternative press in general, most of these papers saw their purpose as covering such traditional community items as the PTA, high school football, and civic clubs. The *News* tried from its inception to emphasize community issues, but this did not mean it covered high school football scores. At the alternative paper, a community orientation meant applying a left perspective to community issues, local investigative reporting, and "advocacy journalism." A member of the paper explained:

> People who started [the paper] kept using the words "advocacy journalism" . . . [to suggest] that no media is objective. The mere choice of a headline, what's included and what's excluded, are political judgments. We hoped to be upfront about that and to provide an alternative voice to the [city newspaper]. . . . We didn't like the idea of a one-paper town and wanted to provide the other side of their stories, as well as covering a lot of stories the [city newspaper] wouldn't touch.

The *News* was run by a full-time staff collective of 12 to 18 people at any one time. In addition, the paper had a number of part-time people on the

periphery. When the staff collective was unable to find one of its own to do particular tasks (e.g., typesetting), it was forced to hire a few people for wages:

> "We've always had marginal people who weren't necessarily in the collective, but they weren't outside either. A distinction was always made though between the hired staff and the collective, who could be expected to work for nothing if need be."

Staff members at the *News* became co-owners of the paper after six months.

At the paper, participatory democracy was taken very seriously. Decisions were made via consensus at weekly meetings of the full staff. All manner of decisions were made at these meetings and they covered both editorial and organizational issues. The job of chairing the meetings was rotated among staff members. As noted in the story about Ann and John in Chapter 1, jobs were also rotated at the paper.

The Food Co-op. The food cooperative began in 1970 as a buying club for about 100 people, mostly students, operating out of members' living rooms. On Wednesdays members would get together to place their orders, on Saturdays they would get their food, and in between they would rotate the tasks of tabulating orders, buying the food, distributing it, bookkeeping, and so forth. Altogether, this came to 10–15 hours per week for each household. As one early member put it, "There weren't enough people with the energy to keep it going. It just took too much time to get your food."

By mid-1973 about 12 core members began to work on plans for a direct-charge food co-op. This meant the co-op would take the form of a grocery store (for members only) that would charge each member a small monthly service fee to cover overhead expenses but would sell the food at cost. Members would each donate at least one hour's work per month toward running the store. In September 1973 a membership drive was launched, and in January 1974 the Food Co-op opened its store. In this new form, the co-op was able to attract 1,102 members after one year of operation. The sales were equally impressive: about $35,000 per month during the academic year. This figure would decline substantially during summers when most student members were out of town.

Food purchases took place in a relatively small storefront located in a student community adjacent to a university. The 1,400-square-foot store was packed with food from top to bottom, had bulletin boards for members' suggestions and messages, and was decorated with contributed plants and artwork with such titles as "exploding orange on desert highway" or "exploding corn in field with Indian family looking on."

Almost all of the Food Co-op's members were young (90 percent between

18 and 23), and 69 percent were college students. Members of the Food Co-op had very little money, at least for the time being: Only 6 percent made $7,000 per year or more, and the vast majority (85 percent) received $3,000 per year or less. Politically, 79 percent of the general membership (and 100 percent of the staff and board members) considered themselves to be between "liberal" and "radical" on the political spectrum. Eighty percent voted in elections, but even more (89 percent) had taken part in demonstrations, and 72 percent reported they had engaged in civil disobedience at some time. People at the Food Co-op tended not to believe solutions to American socioeconomic-economic problems would come through more progressive leadership (28 percent), but even fewer believed an armed struggle would be needed to change things (17 percent). Instead, they tended to think solutions to our country's problems would come through massive changes in people's basic values (87 percent), and by organizing self-governing communities and direct democracy (74 percent).

The Food Co-op was incorporated as a nonprofit corporation, meaning it had to have a board of directors. The board was elected by the general membership at quarterly meetings. It was their function to act as a "coordinating committee" to administer the co-op's operations, but major policy changes were reserved for the general membership meetings. In addition to the nine elected board members, any member who wished to attend the weekly board meetings was allowed a full vote there. A paid staff of three to four people ran the store and coordinated the volunteer labor of the 1,100 members.

Any mutual-benefit association must constantly struggle to maintain the interest and participation of its membership (Blau and Scott, 1962, pp. 45–49), and food cooperatives are no exception (see, e.g., Giese, 1974). Because the acquisition of food is not problematic to most members, it may be especially difficult to sustain interests in food cooperatives.

The Food Co-op, like any food cooperative, could not claim the participation of all of its 1,100 members. However, the majority of members did put their hour per month into the store, and most (81 percent) reported working longer. Ninety-eight percent said they read the co-op's newsletter, and 64 percent of the general members thought they had "quite a bit" of say in decisions at the co-op.

Financially, the Food Co-op was self-supporting. At its inception it did receive $6,500 in seed money from the student government of a nearby university, but none thereafter. The capital costs of establishing the store (equipment, initial inventory, remodeling the storefront) were largely covered by members' deposits ($10 per individual or $30 per household). These shares of the co-op were refundable when members left. The operating costs were

fully covered by a monthly service charge of $2.50 per member. The food was sold at cost. Individuals reported spending an average of $10 per week at the store. With this financial arrangement, the co-op was able to operate in the black, but several financial problems lingered. Since it was not a profit-making organization, it operated with almost no cushion, leaving sparse savings with which to finance one-time expenses such as the down payment on a mortgage for its store. Also, like many other co-ops (see Zwerdling, 1975), the Food Co-op found it had to pay much higher prices for wholesale food than the supermarkets did. Thus, after one year of operation, the co-op invested a considerable sum as a deposit so it could buy groceries through a major wholesaler, thereby ensuring lower prices for the co-op's members in the future.

The goals of the co-op were "economy, ecology, and community." Its motto stated: "Through economy we are trying to provide the commodity most basic to our survival, food, to co-op members at the lowest possible prices." In addition, the co-op tried to minimize environmental waste in its use of resources, for example, by using reusable containers and trying to sell only food of high nutritional value. Finally, it attempted to develop a sense of community both within the co-op and in the larger community of which it was a part. Its organizers saw the co-op as the first of a series of community development corporations. It supported efforts to create other community-run economic and political institutions in the town where it existed. In members' words:

> Part of the reason the [Food Co-op] was formed was so that we in [name of town] might begin to gain control over some of the economic institutions which affect our lives.

The Law Collective. The legal collective in this study began in 1971. Most of its members first met during a trial stemming from charges against participants in a protest demonstration in 1970. As a result of the trial, some of the defense attorneys and some of the defendants realized there was a need for a legal collective in town and began to make joint plans to form one. The Law Collective defined collective law practices as "experiments in the delivery of legal services to people whose lack of money and power severely limits their access to any form of legal assistance."

Over time, the Law Collective developed from defending oppressed individuals in need of legal aid to defending groups of people, particularly workers. By 1974–75 the latter were a priority, and they were providing legal counsel about labor contracts to a number of established unions. In addition, they had helped cab drivers, waitresses, and sea urchin fishermen to organize.

Before the Law Collective could take on a case for no fee, the case had to be, in its judgment, of political significance. However, most of their cases were more or less routine, and for these they received competitive or somewhat lower fees.

The Law Collective categorically refused to take some types of cases. They would not, for example, defend accused rapists, because the "prime defense tactic in rape cases is to attack the integrity of the prosecutrix" (document, 1974). They also refused to accept the cases of men in contested divorces, or cases for landlords or corporations.

Taken together, the cases the collective considered worthy of free legal attention and the cases they refused to take at any price reflected its political perspective. In fact, the collective spent many hours at meetings deliberately trying to bring its practice in line with its political ideals. In the words of one member of the collective: "We see ourselves as one front in a common struggle against bureaucracy and capitalism."

The Law Collective had 6 to 10 members in it at any one time. Usually 3 or 4 of these were attorneys, and the others were legal workers. The collective managed to pay each member a base salary of $250 per month in 1975 plus an additional amount for any dependents they might have, so a new legal worker who entered with no experience and with one child was paid $325 to start.

Members were very cautious about who was admitted into the collective. As one member put it, they had to be sure "each new person can generate the energy and the money to allow the collective to support one more person." Another criterion they used to decide on potential members was:

> They have to have a certain amount of past experience in political work . . . something really good and significant that checks out. We never take anyone in cold. . . . They have to share the same basic assumptions as far as politics goes and they have to be willing . . . to accept the collective way of doing things, which is on the basis of consensus.

This criterion meant the Law Collective was a much more homogeneous group than any of the other organizations in our study. It was also more exclusive. It required that new members undergo a longer initiation period, had definite organizational boundaries, and demanded rather exclusive commitment from the members. (As noted, it did not permit participant observation by researchers.) There was no room here for the part-timers, the hangers-on, and the volunteer labor we saw in abundance at all of the other alternative organizations we studied.

Tasks at the Law Collective were accomplished in a collectivist manner. There was no division of labor between, for instance, typists, receptionists,

and lawyers. Although legal workers entered the collective with little or no former training in law, each member was responsible for closing the gaps in knowledge, and was not to perpetuate them by dividing the labor on the basis of such differences. Thus, attorneys and legal workers alike shared the tasks of typing, reception work, cleanup, legal research, writing pleadings, appearing in court, and so forth. The Law Collective apparently had considerable success in developing in its legal workers the skills of effective jurists. A number of these workers later pursued admission to the bar through the alternative means of apprenticing under an experienced attorney (namely, through their work at the Law Collective). After three years, this work experience qualified them for the bar in the state of California.

Overview

Similarities and differences among the organizations. As a way of summarizing some of the differences among these five organizations, consider their board–staff–member relationships. At the Free Clinic, the board was self-perpetuating; it appointed its own new members to fill vacancies. The board at the clinic had formal powers, but when staff and board interests clashed, the staff tended to prevail. At Freedom High, the board was also originally appointed, but it never changed its composition, and it had no powers, board members being merely figureheads. At both the clinic and the school, the board was made up of prestigious, sometimes wealthy, patrons of the alternative institution. However, at the Food Co-op the board was elected by the membership, and was composed of fellow members of the co-op. It exercised some decision-making authority, but only within parameters carefully circumscribed by the general membership. The most committed participants – those from whom the greatest sacrifices could be expected in the interests of the organization – were the staff at the Free Clinic and the board at the Food Co-op. Staff at the clinic tried to fill many new board positions with more committed members, and the board at the co-op tried to fill new staff positions with more committed members. The *Community News* and the Law Collective did not have to work out a special staff-board relationship because they were completely self-governing staff collectives.

Clients or members of these alternatives also had varying relationships to the organization. Staff at the Free Clinic and the Law Collective tried to be attentive to clients' interests, and, as part of this responsiveness, they often tried to "demystify" medicine or the law in order to make the client less dependent on the organization's continued help. At the Free Clinic, some of their most interested patients underwent training and became volunteers for

the clinic. But, at the Law Collective, clients had no place to fit into the internal organization. The *Community News,* of course, had customers rather than clients, and although the paper sought community feedback, it was, by the very nature of a newspaper, more distant from the individuals it served. At Freedom High, the clients (the students) were so integrated into the decision-making structure of the school that they could be considered full-fledged members. As such, they exercised considerable influence over the selection of staff, curriculum, the issue of where the school should locate, and other important decisions. At the Food Co-op, the member-owners were, in a sense, its "clients," and the co-op was constantly aware it had to justify itself in terms of member satisfaction and participation.

Characteristics and backgrounds of participants. Questionnaire data on the social base of alternative institutions were gathered from three of the observed organizations (the Food Co-op, the Free Clinic, and the *Community News*). At the *News,* full-time staff members (of whom there are 14 to 18 at any one time) grew up in families whose mean annual income was about $29,000, a high figure for the times (early 1970s). A random sampling of the general membership at the Food Co-op (consisting of 1,100 people) revealed an average parental income of $19,500 (though these young members themselves are quite poor, if only temporarily, with 85% of them receiving annual incomes below $3,000). The more active members of the co-op, the staff and board, had a mean parental income of approximately $46,000. At the Free Clinic, the part-time volunteers (of whom there are 60 to 100 at any one time) reported an average parental income of about $25,000.

In addition to having rather financially privileged origins, people in grassroots collectives tend to come from well-educated families. In all three organizations, more than half the mothers of members had at least some college education; fathers on the average had acquired some graduate or professional training beyond the B.A., except at the Free Clinic, where fathers had a median education of some college. As these educational levels would suggest, only 14 to 18 percent of the members of each group report family occupations that could be categorized as working class (again, with the notable exception of staff at the Free Clinic).

The social origins of people in alternative institutions correspond strikingly with those of the New Left students. Moreover, it appears members of collectivist enterprises are not only the same types of people that Flacks (1967) and Keniston (1968) found in the social movements of the 1960s, but they are often the very same people, now a few years older.

Almost all of the founders of the collectives we studied had been active in

the student-based movements of the 1960s. At the *Community News*, for instance, members had taken part in an average of 14 demonstrations apiece, and all had engaged in acts of civil disobedience at some time. At the Food Co-op, more of the members had utilized demonstrations as a means of political expression (89 percent) than voting (80 percent). Fully 72 percent reported engaging in civil disobedience, such as resistance to arrest or dispersion at demonstrations. Even at the Free Clinic, the least explicitly political of the five organizations studied, 68 percent of its member-volunteers reported participating in some demonstrations. Between 79 and 100 percent of the members of each organization identified themselves as falling between "liberal" and "radical" on the political spectrum. Between 57 and 71 percent of the staff members of each group reported feeling a sense of solidarity with the New Left. Very similar levels of identification with the women's movement and the ecology movement were found in these groups. However, of the general (consumer) members of the Food Co-op, only 30 percent reported feeling a sense of solidarity with the New Left, probably because they are significantly younger than the "political generation" (Zeitlin, 1970) of the 1960s student activists.[1]

These statistics on social origins and political identifications reveal a marked continuity between people in collectivist organizations at present and former New Left students. Activists of the 1970s and 1980s are the same types of people, and often enough, are the very same individuals who participated in social movements in the 1960s, but they have changed their political strategy. They have moved from a strategy of student-based protest and political confrontation to one of building decentralized, community-based, collectivist organizations. This is an important cause of the recent rise of alternative organizations aiming to embody countercultural values, motivations, and work forms.[2]

These data on participants also suggest to us other social and economic causes of the rise of cooperatives in the 1970s. The work of Abraham Maslow (1954) tells us that once people satisfy their basic needs for subsistence and security, they become more concerned with fulfilling higher-order needs for self-esteem and self-actualization. Most of the people in the groups we observed were born and socialized under conditions of relative affluence and security. Not having to worry much about the basics, they began to see future jobs not just as sources of income or necessary drudgery to be endured, but as mainsprings of personal growth, satisfaction, and social meaning. They expected to put a lot into their work and to gain a lot from it. But to many growing up in the 1950s and 1960s, it appeared that the reality of work, as experienced by their parents and peers, would fall short of their own high aspirations.

Unemployment and underemployment hit young people particularly hard. In 1978, almost half of all unemployment was among 16- to 24-year-olds (Thurow, 1980, p. 187). Similarly, Sullivan (1978, p. 89) concludes, after a detailed analysis of marginality among workers, that the young and the old are more likely to be "underutilized" than other workers (i.e., either unemployed, enduring inadequate working hours, subject to low income, or mismatched in terms of occupation and education).

Adding to young adults' difficulties in finding satisfactory employment was the master trend of the past couple of decades toward a more concentrated economy. With the size of both private corporations and public bureaucracies growing, personal control over institutions was becoming increasingly remote. The inflation of educational credentials was pushing career entry requirements higher and higher. Among parents, relatives, and friends, children could see personal evidence of the pervasive "blue-collar blues" and "white-collar woes," as many jobs of both kinds became less autonomous, less skilled, and less satisfying (HEW, 1973; Braverman 1974). A 1977 survey found, for example, that 60 percent of American workers would prefer a job other than the one they presently hold, a rise of 16 percent over 1969 (Quinn and Staines, 1977, p. 210). Compared to workers in general, young workers – along with nonwhites and females – have significantly higher levels of job dissatisfaction (Sullivan, 1978, p. 202). Many young workers would have agreed with a young woman, working as a magazine editor, who said, "I think most of us are looking for a calling, not a job. Most of us . . . have jobs that are too small for our spirit. Jobs are not big enough for people." (Terkel, 1975, p. xxix).

Coupled with a desire for more meaningful and challenging work has been an increased willingness on the part of some young adults to forego traditional life-styles. Often they trade the consumption patterns of their parents for their collectively run workplaces with slimmer incomes, giving up split-level suburban ranch homes for communal living, station wagons for shared old VWs. In addition to being the common currency of much of the new generation, these modes of living give support to – and are sometimes necessitated by – the creation of alternative workplaces. People do not get rich in co-ops. Yet many people apparently do often find work there that they feel is psychically and socially enriching.

In a true participatory democracy, any new member (whether staff, volunteer, or client) with the inclination and the energy to participate in the affairs of the organization, would be able to exercise decision-making influence after only a few months or less. All of the organizations in this study tried to achieve this goal. The degree to which it was attained varied among organizations and within a single organization over time. As we will see in the initial

chapter of Part II, the exact nature of participatory democracy departs from bureaucratic forms of organization to such an extent that it must be considered an alternative model of organization.

In closing this chapter, we must reiterate that the above five case studies form only the point of departure, the grounding for our theory. Our analysis is in no way limited to them. The two parts of the book to follow examine many other organizations and organizational forms to broaden and corroborate our theoretical observations.

PART II

A theory of democratic organization

3. The collectivist organization: an alternative to bureaucratic models

For many decades the study of organizations has been, in effect, the study of bureaucracy and its many variations. This decade, however, has given rise to a wide array of organizations that self-consciously reject the norms of bureaucracy and identify themselves as "alternative institutions" or "collectives." In view of the history of cooperatives and the recent emergence of these counterbureaucratic organizations, we need a new model of organization that is able to encompass their alternative practices and aspirations. This chapter attempts to construct such a model. As a first step, it is necessary to distinguish between formal and substantive rationality, using the work of Max Weber.

Weber (1947; 1954) recognized a tension between formal and substantive rationality. For Weber, formal rationality – an emphasis on instrumental activity, formal laws, and procedural regularity – would have its main locus of expression in bureaucracy. But formal rationality would come into inevitable conflict with people's desire to realize their substantive goals and values such as peace or equality or good health, what Weber called substantive or value-rationality. Modern bureaucracy would be built on the procedural regularity of formal law, but in Weber's view it could never eliminate all moral, subjective concerns (Bendix, 1962, pp. 391–438). Nevertheless, in his classic statement on bureaucracy, Weber (1946, pp. 196–244) sets forth the characteristics of bureaucracy as if it could eliminate all substantive, moral considerations.

Taking a historical view, Weber believed virtually all organizations fit into one of three types. In the first and oldest type, authority was inherited and orders were accepted as legitimate and were obeyed because this was the way things had been done since time immemorial. Traditional societies gave rise to a patriarchal mode of organization. In a second type, charismatic leaders could depart from long-held traditions, since their authority rested on their own inspiring or magnetic qualities, their special gifts of grace. This situation gives rise to a communitarian form of organization, whose core consists of the charismatic leader and disciples. To Weber, a third type, legal-rational

authority, would eclipse the first two and come to dominate modern society. Here authority is followed because it represents consistently applied rules implemented by those with the appropriate position or expertise. Bureaucracy is the form of organization that implements legal-rational authority. Thus, for Weber, each legitimating principle develops a special mode of organization to implement it.

Weber notes in passing that a fourth type of legitimation and organization, that of substantive rationality, might be possible. Unfortunately, however, he does not elaborate. This chapter attempts to develop Weber's neglected type. The contemporary collectives in our study reject traditional, charismatic, and bureaucratic authority. Decisions are taken to be morally binding only if they reflect the will of the collectivity and are arrived at through a process of democratic consensus. Herein lies the importance of these organizations. They point the way to a fourth type of organization in history that, like the first three, is cohesive and internally consistent, but, unlike them, turns on principles of substantive rationality.

Just as the ideal of bureaucracy, in its monocratic pure type, is probably not attainable (Mouzelis, 1968), so the ideal of democracy, in its pure and complete form, is probably never achieved. In practice, organizations are hybrids.

We seek to develop an ideal-typical model of collectivist–democratic organization by which to delineate the form of authority and the corresponding mode of organization following from value-rational premises. The ideal-typical approach will allow us to understand collectivist democracies, not only in terms of bureaucratic standards they do not share, but in terms of the countervalues they do hold (see Kanter and Zurcher, 1973). Further, the use of an ideal type permits us to classify actual organizations along a continuum, rather than forcing us to place them in discrete categories. There are *degrees* of collectivism.

The collectivist organization: characteristics

Collectivist–democratic organizations can be distinguished from bureaucratic organizations along at least eight dimensions,[1] each of which will be taken up in turn. Table 3.1, at the end of this chapter, summarizes these eight characteristics.

Authority.

When we're talking about collectives, we're talking about an embryonic creation of a new society. . . . Collectives are growing at a phenomenal rate all over this country. The new structures have outgrown the science of analyzing them. Sociology has

to catch up with reality. . . . Collectivism is an attempt to supplant old structures of society with new and better structures. And what makes ours superior is that the basis of authority is radically different. (Staff member, *Community News*)

The words of this activist get right to the heart of the matter – authority. Perhaps more than anything else, it is the basis of authority that distinguishes the collectivist organization from any variant of bureaucracy. The collectivist–democratic organization rejects bureaucratic justifications for authority. Here authority rests not with the individual, whether on the basis of incumbency in office or expertise, but resides in the collectivity.

This notion stems from the ancient anarchist ideal of "no authority." It is premised on the belief that social order can be achieved without recourse to authority relations (Guerin, 1970). Thus it presupposes the capacity of individuals for self-disciplined, cooperative behavior. Like the anarchists, collectives seek not the transference of power from one official to another, but the abolition of the pyramid in toto: organization without hierarchy. In his empirical study of decision making in kibbutzim enterprises, Rosner (1981) finds hierarchy in the workplace limits the degree to which people can sit as equals in both community and workplace assemblies. In other words, hierarchy inhibits equality of influence in decision-making bodies.

An organization, of course, cannot be made up of a collection of autonomous wills, each pursuing its own personal ends. Some decisions must be binding on the group. Decisions become authoritative and binding in collectivist organizations to the extent they arise from a process in which all members have the right to full and equal participation. This democratic ideal, however, differs significantly from conceptions of "democratic bureaucracy" (Lipset et al., 1962), "representative bureaucracy" (Gouldner, 1954), or even, as noted earlier, representative democracy. In its directly democratic form, it does not subscribe to the established sets of rules of order and protocol with which we are all acquainted: It does not take formal motions and amendments, it does not usually take votes, majorities don't rule, and there is no two-party system. Instead there is a "consensus process" in which all members participate in the collective formulation of problems and negotiation of decisions.[2] All major policy issues, such as personnel hiring, firing, salaries, division of labor, the distribution of surplus, and the shape of the final product or service, are decided by the collective as a whole. Only decisions that appear to carry the consensus of the group behind them carry the weight of moral authority. Only these decisions, changing as they might with the ebb and flow of sentiments in the group, are taken as binding and legitimate.

Empirically, this consensus process seems to be well suited to the alternative organization. It reveals that the group is not as concerned with simple

efficiency as it is with exemplifying in its internal process the values of self-expression and group cohesion. For nothing is ruled out of order, no vote cuts short discussion because a majority are in agreement. Issues are discussed for as long as it may take for a consensus to be negotiated. Further, this process seems appropriate because it does not imply any authority structure; leadership of meetings can be rotated or meetings may not be chaired at all. Thirdly, it reflects, and perhaps serves to instill, cooperation and self-determination, attributes these groups usually want to foster.

In Weberian terms, then, we are concerned with organizations that aspire and claim to be free of *Herrschaft*.[3] They are organizations without domination in that ultimate authority is based in the group as a whole, not in the individual. Individuals, of course, may be delegated carefully circumscribed areas of authority, but authority is delegated and defined by the collectivity and subject to recall by the collectivity.

Rules. Collectivist organizations challenge too the bureaucratic conception that organizations should be bound by a formally established, written system of rules and regulations. Instead, they seek to use as few rules as possible. But just as the most bureaucratic of organizations cannot anticipate, and therefore cannot circumscribe, every potential behavior in the organization, so the alternative organization cannot reach the theoretical limit of zero rules. Collectivist organizations can, however, reduce drastically the number of spheres of organizational activity subject to explicit rule governance.

The most simple of the collectivist organizations we observed, Freedom High, formulated only one explicit organizational rule: no drugs in school. This rule was agreed upon by a plenary meeting of the school's students and staff primarily because its violation was perceived to threaten the continued existence of the school. Other possible rules were discussed as well, rules that might seem self-evident in ordinary schools, such as "each student should take *X* number of classes" or "students are required to attend the courses for which they are registered," but these did not receive the consensual backing of the school's members.

In place of the fixed and universalistic rule use that is the trademark of bureaucracy, operations and decisions in alternative organizations tend to be conducted in an ad hoc manner. Decisions generally are settled as the case arises, and are suited to the peculiarities of the individual case. No written manual of rules and procedures exists in most collectives, although norms of participation clearly obtain. Although there is little attempt to account for decisions in terms of literal rules, concerted efforts are made to account for decisions in terms of substantive ethics. This is like Weber's (1968, pp. 976–

978) *Kadi* justice, based on substantive ethics, and is a long distance from the formal justice that guides rational bureaucratic action.

A chief virtue of extensive rule use in bureaucracy is that it allows predictability and the appeal of decisions to higher authority. The lack of universalistic standards in prebureaucratic modes of organization did invite arbitrary and capricious rule. In bureaucracy at least, decisions can be calculated and appealed on the basis of their correspondence to the written law. In collectivist organizations, on the other hand, decisions are not arbitrary. They are based on substantive values (such as equality) applied consistently, if not universally. This permits some calculability on the basis of knowing the substantive ethic that will be invoked in a particular situation.

Despite the desire for minimal rules, some rules may be useful to collectivist organizations. In cooperatives that have a fast turnover of members, written rules may help to communicate to new members the agreements and expectations of former members. For example, in seven housing co-ops studied by Gillespie (1981), those that did incorporate rules were perceived by members to be as democratic, communal, and egalitarian as those that did not. In fact, the use of rules helped the co-ops to rapidly convey their norms to new members; this feature was important since the average length of residence of the members was only 12.7 months (Gillespie, 1981, p. 7). However, because people in most worker co-ops expect to be in the organization for a longer period of time and there is daily social interaction, it is easier to provide each other with an oral history of events and there may be less need for written rules. The fact that turnover is usually more rapid in consumer co-ops than in worker co-ops may increase their need to write down a history and rules with which to socialize new members. But there is still an important difference between collective and bureaucratic rule use. In a collective, rules are always subject to group negotiation and change. They are not carved in stone or ritualized, nor are they passed down from above.

Although collectives tend to be suspicious of rules, rules need not be antidemocratic. Certain rules may actually enhance democratic control. For example, the Food Co-op made a rule that if any member attended three consecutive meetings of the board, then they were allowed to vote as a board member. Clearly, this encouraged member participation.

In comparison with bureaucratic organizations such as a government agency or a corporation, collectives are characterized by the minimal use of formal, written rules. Of course, in no case that we observed or have heard about, have written rules been entirely eliminated. Indeed, collectives do have many informally shared expectations and agreements, as do all organizations. The degree to which collectives commit such understandings to writ-

ing may depend upon their turnover rates, needs for rapid socialization, legal requirements, and other practical as well as philosophic concerns. It is important to keep in mind that the extent of rule use is *not* the most fundamental characteristic of collectivist organizations – the locus of authority is.

Social Control. From a Weberian point of view, organizations are tools, instruments of power for those who head them. But what means does the bureaucracy have of ensuring that lower-level personnel, people who are quite distant from the centers of power, will effectively understand and implement the aims of those at the top? This issue of social control is critical in any bureaucracy. Perrow (1976) examines three types of social control mechanisms in bureaucracies. The first type of control is the most obvious – direct supervision. The second is less obtrusive, but no less effective: standardized rules, procedures, and sanctions. Gouldner (1954) shows that rules can substitute for direct supervision. This allows the organization considerable decentralization of everyday decision making and even the appearance of participation, for the premises of those decisions have been carefully controlled from the top. Decentralized decision making, when decisional premises are handed down from the top via standardized rules, may be functionally equivalent to hierarchical authority (Blau, 1970; Bates, 1970; Perrow, 1976). "Top-down" rules have the same consequence as centralized authority.

Collectivist organizations generally refuse to legitimate the use of centralized authority or standardized rules to achieve social control. Instead, they rely upon personalistic and moralistic appeals to provide the primary means of control. In addition, the search for a common purpose, a continuing part of the consensus process, is a basis for collective coordination and control. Examining free schools, Swidler (1979, p. 179) finds them "obsessed with the search for common goals" and shows this search to be an important source of social control. For Etzioni (1961), compliance here is chiefly normative. One person appeals to another, "do *X* for me," "do *X* in the interest of equality," and so forth. *The more homogeneous the group, the more such appeals can hold sway.* Thus, where personal and moral appeals are the chief means of social control, it is important, perhaps necessary, that the group select a membership sharing their basic values and worldview. An effort was made to do just that in all five of the alternative organizations under study. At the Law Collective, for instance, members were asked how they decide whether to take in a new member. One commented:

They have to have a certain amount of past experience in political work . . . something really good and significant that checks out. . . . same basic assumptions as far

as politics goes and they have to be willing to accept the collective way of doing things.

Such recruitment criteria are common in alternative work organizations.

Like Perrow (1976), we might conclude that alternative organizations avoid first- and second-level controls, but accept third-level controls. Third-level controls are the most subtle and indirect of all: selection of personnel for homogeneity. On this level, social control is achieved in traditional bureaucracies by selecting for top managerial positions only people who "fit in" – people who read the right magazines, go to the right clubs, and share the same style of life and worldview. Unless recruitment for homogeneity is considered a device for social control, it is difficult to understand why it so often takes precedence over criteria of competence in the corporate world (Kanter, 1977; Perrow, 1976). And so it is in collectivist organizations. Where people are expected to make major decisions – and this means everyone in a collective, and high-level managers in bureaucracy – consensus is crucial, and people who are likely to challenge basic assumptions are avoided. A person who eagerly reads the *Wall Street Journal* would be as suspect in applying for a position at the Law Collective as a person who avidly reads the *New Left Review* would be at General Motors. Both kinds of organizations utilize selection for homogeneity as a mechanism for social control.

Social relations. Impersonality is a key feature of the bureaucratic model. Personal emotions are to be prevented from distorting rational judgments. Relationships between people are to be role based, segmental, and instrumental. Collectivist organizations, on the other hand, strive toward the ideal of community. Relationships are to be wholistic, affective, and of value in themselves. In the extreme, the search for community may even become an instance of goal displacement, as when, for example, a free school comes to value community so highly that it loses its identity as a school and becomes a commune (see, e.g., Kaye, 1972).

Recruitment and advancement. Bureaucratic criteria for recruitment and advancement are resisted in the collectivist organization. Employment is not based on specialized training or certification, nor on any universalistic standard of competence. Instead, staff are generally recruited and selected on the basis of friendship and social-political values. Personality attributes seen as congruent in the collectivist mode of organization, such as self-direction and collaborative styles, may also be consciously sought in new staff (see, e.g., Torbert, 1973). Further, employment does not constitute the beginning of a career in collectivist organizations in the usual sense, for the organization

does not provide a li[illegible]ng ladder to ever-higher positions. Work may be volunteer or paid, and it may be part-time or full-time (or even 60 hours per week), but it is not thought of as a career. Whereas career advancement in bureaucracy is based on seniority and/or achievement, in collectives it is not a meaningful concept; for where there is no hierarchy of offices, there can be no individual advancement in positional rank (though there may be much change in positions).

How can a collectivist organization recruit competent and skilled personnel if selection criteria explicitly emphasize friendship networks, political values, and personality traits? To illustrate, during the year in which the Free Clinic was observed, four full-time staff positions were filled, and between 9 and 65 applications were received for each position. Yet each of the four positions went to a friend of staff members. The relevant attributes cited most frequently by the staff making these decisions were: articulation skills, ability to organize and mobilize people, political values, self-direction, ability to work under pressure, friendship, commitment to the organization's goals, cooperative style, and relevant experience. These selection criteria are typical of alternative organizations. In spite of their studied neglect of *formal* criteria of competence (e.g., certification), alternative organizations often attract highly qualified and able people.[4] In many ways, their selection criteria are well suited to their needs for multitalented and committed personnel who can serve a variety of administrative and task-oriented functions and who are capable of comanaging the organization.

Incentive structure. Organizations use different kinds of incentives to motivate participation. Most bureaucratic workplaces emphasize remunerative incentives, and few employees could be expected to donate their services if their paychecks were to stop. Collectivist organizations, on the other hand, rely primarily on a sense of shared purpose, secondarily on friendship ties, and only tertiarily on material incentives (Zald and Ash, 1966). As Etzioni has shown (1961), this kind of normative compliance system tends to generate a high level of moral commitment to the organization. Specific structural mechanisms that produce and sustain organizational commitment are identified by Kanter (1972a). Because collectivist work organizations require a high level of commitment, they tend to utilize some of these mechanisms as well as value-purposive incentives to generate it. Indeed, work in collectives is construed as a labor of love, and members may pay themselves very low salaries and may even expect each other to continue to work during months when the organization is too poor to afford their salaries.

Alternative organizations often appeal to symbolic values to motivate

people to join and to participate. The range of these values is considerable. At the Free Clinic, for instance, a member describes motivation:

> Our volunteers are do-gooders. . . . They get satisfaction from giving direct and immediate help to people in need. This is why they work here.

In contrast, at the *Community News,* the following was more illustrative:

> Our motives were almost entirely political. . . . We wanted to create a base for a mass left. To activate liberals and open them up to left positions. To tell you the truth, the paper was conceived as a political organ.

At the Food Co-op, the value of community was most stressed, and the co-op helped to create other community-owned and controlled institutions in its locale.

However, we should guard against an overly idealistic interpretation of participation in alternative organizations. In these organizations, as much as in any, there exists an important *coalescence of material and ideal interests.* Even volunteers in these organizations, whose motives on the face of it would appear to be wholly idealistic, also have material incentives.

For example, staff members at the Free Clinic suspected that some volunteers donated their time to the clinic "only to look good on their applications to medical school." Likewise, some of the college students who volunteered to teach at Freedom High believed that in a tight market, this would improve their chances of obtaining a paid teaching job. And, for all the talk of community at the Food Co-op, many members undoubtedly joined simply because the food was cheaper. Because material gain is not part of the acceptable vocabulary of motives in these organizations, public discussion of such motives is suppressed.

Nonetheless, for staff members, as well as for volunteers, both moral and material incentives operate. At the Law Collective, for instance, legal workers often used their experience there to pursue the bar, since California law allows eligibility for the bar through the alternative means of apprenticing under an attorney for three years. At the *Community News,* a few staff members confided they had entered the paper to gain journalistic experience.

Yet members of alternative institutions often deny the existence of material considerations and accept only idealistic motivations. In the opinion of one long-time staffer at the *Community News:*

> I don't think anyone came for purely journalistic purposes, unless they're masochists. I mean it doesn't pay, the hours are lousy, and the people are weird. If you want professional journalistic experience you go to a straight paper.

In many ways, she is right: Alternative institutions generally provide woefully inadequate levels of remuneration by the standards of our society. But,

it does not impugn the motives of participants to recognize that these organizations must provide some material base for their members if they are to be alternative places of employment at all.

At the Free Clinic, for illustration, full-time staff were all paid $500 per month during 1974–75; at the Law Collective they were paid a base of $250 per month plus a substantial supplement for dependents; and at the *Community News* they received between $150 and $300 per month, in accordance with individual "needs." These pay levels were negotiated in open discussion by all members, as were decisions regarding the entire labor process. Thus, if these wage levels appear exploitative, it is a case of self-exploitation. It is these subsistence wage levels that permit the young organization to accumulate capital and to reinvest this surplus in the organization rather than pay it out in wages. This facilitates the growth of the organization, and hastens the day when it may be able to pay high salaries.

Many collectives have found ways to help compensate for meager salaries. The Law Collective stocked a refrigerator with food so members could eat at least a free meal or two per day at the office. The collective also maintained a number of cars its members could share, thereby eliminating the need for private automobile ownership. Free Clinic staff decided to allow themselves certain fringe benefits to compensate for what they regarded as underpaid work: two weeks of paid vacation each year, plus two additional weeks of unpaid vacation, if desired; one day off every other week; and the expectation that staff would regularly work 28–30 rather than 40 hours a week. These are compensations or supplements for a generally poor income, and like income, they do not motivate people to work in alternative organizations; they only make work there possible.

The main reason people come to work in an alternative organization is because it offers them substantial control over their work. Collective control means members can structure both the product of their work and the work process itself. Hence, the work is purposeful to them. It is frequently contrasted with alienating jobs they have had, or imagine, in bureaucracies:

> A straight paper would have spent a third of a million dollars getting to where we are now and still wouldn't be breaking even. We've gotten where we are on the sweat of our workers. They've taken next to no money when they could have had $8,000 to $15,000 in straight papers doing this sort of job. . . . They do it so they can be their own boss. So they can own and control the organization they work in. So they can make the paper what *they* want it to be. (Member of *Community News*)

Social stratification. In the ideal-typical bureaucracy, the dimensions of social stratification are consistent with one another. Specifically, social prestige

and material privilege are to be commensurate with one's positional rank, and the latter is the basis of authority in the organization. Thus, a hierarchical arrangement of offices implies an isomorphic distribution of privilege and prestige. In this way, hierarchy institutionalizes and justifies inequality.

In contrast, egalitarianism is a central feature of the collectivist–democratic organization. Large differences in social prestige or privilege, even where they are commensurate with level of skill or authority in bureaucracy, would violate this sense of equity. At the Free Clinic, for instance, all full-time staff members were paid equally, no matter what skills or experience they brought to the clinic. At the Law Collective and the *Community News* pay levels were set "for each according to his or her need." Here salaries took account of dependents and other special circumstances contributing to need, but explicitly excluded considerations of the worth of the individual to the organization. In no case that we observed was the ratio between the highest pay and the lowest pay greater than two to one.

In larger, more complex democratic organizations wages are still set, and wage differentials strictly limited, by the collectivity. For example, in the 65 production cooperatives that constitute the Mondragon system in Spain, pay differentials are limited to a ratio of 3:1 in each firm (Johnson and Whyte, 1977). In the worker-owned and managed refuse collection firms in San Francisco, the differential is only 2:1, or less (Russell, 1982a). Schumacher (1973, p. 276) reports a 7:1 ratio between the highest and the lowest pay in the Scott Bader Co., a collectively owned firm in England. The cooperatively owned plywood mills in the Pacific Northwest pay their members an equal wage (Bernstein, 1976, pp. 20–21). By comparison, the wage differential tolerated today in Chinese work organizations is 4:1; in the United States it is about 100:1 (Eckstein, 1977).

Prestige, of course, is not easily equalized, as is pay. Nonetheless, collectivist organizations try in a variety of ways to indicate they are a fraternity of peers. Through dress, informal relations, task sharing, job rotation, the physical structure of the workplace, equal pay, and the collective decision-making process itself, collectives convey an equality of status. As Mansbridge observes (1977), reducing the sources of status inequality does not necessarily lead to the magnification of trivial differences. Likewise, decreasing the material differentials between individuals in a collectivist organization does not ordinarily produce a greater emphasis on status distinctions.

Differentiation. A complex network of specialized, segmental roles marks any bureaucracy. Where the rules of Taylorism hold sway, the division of labor is maximized: Jobs are subdivided as far as possible. Specialized jobs

require technical expertise. Thus, bureaucracy ushers in the ideal of the specialist-expert and defeats the cultivated, renaissance man of an earlier era (Weber, 1946, pp. 240–244).

In contrast, differentiation is minimized in the collectivist organization. Work roles are purposefully kept as general and wholistic as possible. The aim is to eliminate the bureaucratic division of labor that separates intellectual worker from manual worker, administrative tasks from performance tasks. Three means are commonly utilized toward this end: role rotation, task sharing, and the diffusion, or "demystification," of specialized knowledge.

Ideally, universal competence of the collective's members would be achieved in the tasks of the organization. It is the *amateur-factotum* who is ideally suited for the collectivist organization. In the completely democratized organization, everyone manages and everyone works. This may be the most fundamental way in which the collectivist mode of organization alters the social relations of production.[5]

This alteration in the division of labor is perhaps best illustrated by Freedom High, an organization in which administrative functions were quite simple and undifferentiated. Freedom High had no separate set of managers to administer the school. Whenever administrative tasks were recognized, "coordination meetings" were called to attend to them, but these were open to all interested teachers and students.

"Coordinators" were those who were willing to take responsibility for a particular administrative task, such as planning the curriculum, writing a press release, organizing a fund-raiser. A coordinator for one activity was not necessarily a coordinator for another project, though a few core members were involved in many. Further, since administrative tasks were assumed to be part-time, these would be done alongside of one's other responsibilities. Coordinators, then, were self-selected, rotated, and part-time. No one was allowed to perform administrative tasks exclusively. By explaining and simplifying administration and opening it up to the membership at large, the basis and pretense of special expertise were eliminated.

The school even attempted to break down the basic differentiation between students and staff, regarding students not as clients but as members, with decision-making rights and responsibilities. Few organizations can allow their clients so full a participatory role as Freedom High, but all of the alternative service organizations in this study did try to integrate their clients into the organization. The Free Clinic created spaces on its board of directors for consumers of medical care, and recruited many of its volunteers from the ranks of its patients.

Most alternative organizations are more complex than Freedom High. They

cannot assume that everyone in the organization knows how (or would want to know how) to do everything. Thus, they must develop explicit procedures to achieve universal competence. Such procedures, in effect, attack the conventional wisdom of specialized division of labor and seek to create more integrated, multifaceted work roles.

The *Community News,* for example, utilized task sharing (or team work), apprenticeships, and job rotations. Instead of assigning one full-time person to a task requiring one person, it would more likely assign a couple of people to the task part-time. Individuals' allocations of work often combined diverse tasks, such as 20 hours of selling advertisements with 20 hours of writing. In this way, the distribution of labor combined satisfying tasks with more tedious tasks and manual work with intellectual work. People did not enter the paper knowing how to do all of these jobs, but the emphasis on task sharing allowed the less experienced to learn from the more experienced. Likewise, if a task had few people who knew how to perform it well, a person might be allocated to apprentice with the incumbent. Internal education was further facilitated by occasional job rotations. Thus, although the *News* had to perform the same tasks as any newspaper, it attempted to do so without permitting the usual division of labor into specialties or its concomitant monopolization of expertise.

Minimizing differentiation is difficult and time-consuming. The *Community News* spent 1 hour and 40 minutes of regular meeting time deciding when and how the staff should do a complete role rotation, another 8 hours and 40 minutes at a special Sunday meeting called for that purpose, and not having finished, they spent an additional 5 hours and 20 minutes at the following regular staff meeting. Attendance at these meetings was 100 percent. This particular job rotation, then, absorbed 15 hours and 40 minutes of formal meeting time and countless hours of informal discussion. The time and priority typically devoted to internal education in collectivist organizations make sense only if they are understood to be part of a struggle against the division of labor. The creation of an equitable distribution of labor and wholistic work roles is an essential feature of the collectivist organization.

We have developed an ideal-type model of collectivist-democracy that is analogous to Weber's ideal-type characterization of bureaucracy.[6] Table 3.1 summarizes the ideal-type differences between the collectivist mode of organization and the bureaucratic. The eight dimensions discussed are not all equal. The basis of authority and the type of division of labor are plainly the most essential. Democratic control is the foremost characteristic of collectivist organization, just as hierarchical control is the defining characteristic of bureaucracy. For this reason, collectivist organizations transform the social

Table 3.1. *Comparisons of two ideal types of organization*

Dimensions	Bureaucratic organization	Collectivist–democratic organization
1. Authority	1. Authority resides in individuals by virtue of incumbency in office and/or expertise; hierarchal organization of offices. Compliance is to universal fixed rules as these are implemented by office incumbents.	1. Authority resides in the collectivity as a whole; delegated, if at all, only temporarily and subject to recall. Compliance is to the consensus of the collective, which is always fluid and open to negotiation.
2. Rules	2. Formalization of fixed and universalistic rules; calculability and appeal of decisions on the basis of correspondence to the formal, written law.	2. Minimal stipulated rules; primacy of ad hoc, individuated decisions; some calculability possible on the basis of knowing the substantive ethics involved in the situation.
3. Social control	3. Organizational behavior is subject to social control, primarily through direct supervision or standardized rules and sanctions, tertiarily through the selection of homogeneous personnel, especially at top levels.	3. Social controls are primarily based on personalistic or moralistic appeals and the selection of homogeneous personnel.
4. Social relations	4. Ideals of impersonality. Relations are to be role based, segmental, and instrumental.	4. Ideal of community. Relations are to be wholistic, personal, of value in themselves.
5. Recruitment and advancement	5a. Employment based on specialized training and formal certification.	5a. Employment based on friends, social-political values, personality attributes, and informally assessed knowledge and skills.
	5b. Employment constitutes a career; advancement based on seniority or achievement.	5b. Concept of career advancement not meaningful; no hierarchy of positions.
6. Incentive structure	6. Remunerative incentives are primary.	6. Normative and solidarity incentives are primary; material incentives are secondary.

7. Social stratification	7. Isomorphic distribution of prestige, privilege, and power (i.e., differential rewards by office); hierarchy justifies inequality.	7. Egalitarian; reward differentials, if any, are strictly limited by the collectivity.
8. Differentiation	8a. Maximal division of labor: dichotomy between intellectual work and manual work and between administrative tasks and performance tasks.	8a. Minimal division of labor: administration is combined with performance tasks; division between intellectual and manual work is reduced.
	8b. Maximal specialization of jobs and functions; segmental roles. Technical expertise is exclusively held: ideal of the specialist-expert.	8b. Generalization of jobs and functions; wholistic roles. Demystification of expertise: ideal of the amateur factotum.

relations of production. Bureaucracy maximizes formal rationality precisely by centralizing control at the top of the organization; collectivist organizations decentralize control in such a way that it may be organized around the alternative logic of substantive rationality.

Constraints on organizational democracy

Research and observation over the years have pointed out that many factors limit the actualization of the ideal features of bureaucracy as drawn by Weber. Similarly, we have found that various factors limit the actual attainment of organizational democracy. In practice, collectivist democracy, like bureaucracy, can be approximated, but not perfectly attained. This section outlines some of the more important of these constraints.

Judgments about the relative importance of the following constraints are intricately tied to cultural values. Alternative organizations may be assessed incorrectly when seen through the prism of the norms and values of the surrounding bureaucratic society. As in anthropology, one must seek to understand a different culture on its own terms.

Time. Democracy takes time. This is one of its major constraints. Two-way communication structures may encourage higher morale, innovative ideas, and more adaptive solutions to complex problems, but they are undeniably slow (Leavitt, 1964, pp. 141–150). Quite simply, a boss can hand down a bureaucratic order in a fraction of the time it would take a group to decide the issue democratically.

The time absorbed by meetings can be extreme in democratic groups. During the early stages of the *News,* three days out of a week were taken up with meetings. Between business meetings, political meetings, and "people meetings," very little time remained to do the tasks of the organization. Members quickly learn that this arrangement is unsatisfactory. Meetings are streamlined. Tasks are given a higher priority. Even so, constructing an arrangement that both saves time and ensures effective collective control may prove difficult. Exactly which meetings are dispensable? What sorts of decisions can be safely delegated? How can individuals be held accountable to the collectivity as a whole? These sorts of questions come with the realization that there are only 24 hours in a day.

Even under the best of circumstances, however, there is a limit to how streamlined collectivist meetings can be. It is true that with practice, planning, and self-discipline, groups can learn to accomplish more during their meeting time. Nevertheless, once experience is gained in conducting meet-

ings, the time given to them appears to be directly correlated with level of democratic control. The Free Clinic, for instance, could keep its weekly staff meetings down to an average of 1 hour and 15 minutes only by permitting individual decision making outside the meeting to a degree that would have been unacceptable to members of the *Community News,* where a mean of 4 hours was given over to the weekly staff meeting.

Time spent at meetings is not wasted, however, because participation in considering options and making decisions appears to breed commitment to the outcome. When people have had input into a decision, its implementation may be expected to go smoother. Unilateral decisions, albeit quicker, would not be seen as binding or legitimate in collectives, and there would be no enthusiasm for implementation. In bureaucracies, too, the nonparticipatory nature of decisions may contribute to the blocking or sabotage of organizational action. In his study of kibbutzim enterprises, Rosner (1981) has found that the frequency of meetings and the importance of issues raised are strongly correlated with a more equal distribution of influence. In turn, the greater a person's sense of influence, the more that individual will feel personally committed to the organization and the more trust he or she will have in its management. In this way, time spent on democratically deciding things at meetings may enhance commitment and implementation.

Emotional intensity. The familial, face-to-face relationships in collectivist organizations may be more satisfying than the impersonal relations of bureaucracy, but they are also more emotionally threatening. Intense emotions may constrain participatory organization.

Interpersonal tension is probably endemic in the directly democratic situation, and, for better or worse, members often perceive their workplaces to be emotionally intense. At the Law Collective, a member warned that "plants die here from the heavy vibes." And at the *Community News,* people reported headaches and dread before meetings in which divisive issues were to be raised. A study of the New England town meetings (Mansbridge, 1973a) found citizens reporting headaches, trembling, and even fear for one's heart as a result of the meetings. Altogether, a quarter of the people in a random sample of the town spontaneously suggested that the conflictive character of the meetings disturbed them.

To allay these fears of conflict, townspeople utilize a variety of protective devices: Criticism is concealed or at least softened with praise, differences of opinion are minimized in the formulation of a consensus, private jokes and intimate communications are used to give personal support during the meetings. Such avoidance patterns have the unintended consequence of excluding

the not fully integrated member, withholding information from the group, and violating the norms of open participation. Further, these same avoidance patterns and fears of conflict are in evidence even in groups that are highly sensitive to these issues and in which many members have been trained in group process (Mansbridge, 1982).

The existence of such feelings in all of the groups we observed in California, in the Boston area alternatives discussed by the Vocations for Social Change collective (1976), and in Mansbridge's democratic organizations suggests that these feelings are rooted in the very structure of collectivist decision making. Although participants generally attribute conflict to the stubborn, wrongheaded, or otherwise faulty character of others, we maintain that it is a structurally induced, inherent cost of participatory democracy.

First, the norm of consensual decision making makes the possibility of conflict all the more threatening because unanimity is required. In contrast, because majoritarian systems are based on voting, they can institutionalize and absorb conflicting opinions. In collectivist organizations, the existence of conflict means the discussion must continue. Second, the intimacy of face-to-face decision making personalizes the ideas that people espouse and thereby makes the rejection of those ideas harder to bear. A formal bureaucratic system, to the extent that it disassociates an idea from its proponent, makes the criticism of ideas less interpersonally risky.

Non-democratic habits and values. Because of the nature of their prior experience, many people are not very well prepared for participatory democracy. They have not learned the attitudes and behaviors that will be required in cooperative enterprises. This, too, is an important constraint on the development of such organizations.

It is a fundamental premise of sociology that people's behavior, attitudes, and personalities are to a great extent shaped by their environment. If work encourages, or sometimes requires, people to be competitive, narrowly specialized, obedient to authority from above, and willing to give orders below, then it should not be surprising that people accurately receive and act on these messages. Indeed, Kanter (1977) argues that it is the structural features of the modern corporation, much more than individual attributes, that determine the organizational behavior of men and women. In the educational sphere, Jules Henry (1965) poignantly shows how the norms of capitalist culture become the hidden curriculum of the school system. Even at the preschool level, the qualities of the bureaucratic personality are unconsciously, but relentlessly conveyed to children (Kanter, 1972b). Bowles and Gintis (1976) argue that the chief function of the entire educational apparatus is to reproduce the division of labor and hierarchical authority of capitalism.

In the face of these pervasive behavior-shaping institutions, it is difficult to sustain collectivist personalities. It is asking, in effect, that people in collectivist organizations constantly shift gears, that they learn to act one way inside their collectives and another way outside. The difficulty of creating and sustaining collectivist attributes and behavior patterns reflects a cultural disjunction, deriving from the fact that alternative organizations are as yet isolated examples of collectivism in an otherwise capitalist–bureaucratic context. Where collectives are not isolated, that is, where they are part of a network of cooperative organizations, such as the Mondragon system in Spain (Johnson and Whyte, 1977), this problem is mitigated.

One owner of a small graphics firm, himself a convert to democratic management, was dismayed to discover, when he first tried to implement participation, that "the group voted that management was going a good job and that they didn't want to be held accountable for any failures or take on other management responsibilities" (Hendricks, 1973). Slowly, progress has been made in this particular workplace, but as Hendricks says, "it would be easy just to gather up people who could comprehend simplicity and equality and eliminate any consciousness-raising problems, but that doesn't really lead to a new society." What the experience of the alternative institutions has shown is that even handpicking people with collectivist attitudes does not guarantee that these attitudes will be effectively translated into cooperative behavior (see, for example, Swidler, 1976; Taylor, 1976; Torbert, 1973).

Nevertheless, a number of recent case studies of democratic workplaces reveal that the experience of democratic participation can alter people's values, the quality of their work, and ultimately, their identities (Perry, 1978; Jackall, 1976; Bart, 1981). In a comparative examination of many cases of workers' participation, Bernstein (1976, pp. 91–107) finds democratic consciousness to be a necessary element for effective workers' control.

Fortunately for collectives, the solution to this problem of creating democratic consciousness and behavior may be found in the democratic method itself. As was noted in Chapter 1, Pateman has amassed a considerable body of evidence from research on political socialization in support of the classical arguments of Rousseau, Mill, and Cole. She concludes:

> We do learn to participate by participating and . . . feelings of political efficacy are more likely to be developed in a participatory environment. . . . The experience of a participatory authority structure might also be effective in diminishing tendencies toward non-democratic attitudes in the individual. (Pateman, 1970, p. 105)

Recent research provides further empirical support for Pateman's position that participation helps to develop feelings of political efficacy (Elden, 1976). Similarly, Gillespie (1981) finds that housing co-ops are able to rapidly socialize new members to accept participatory norms. It is evident that collec-

tivist organization requires that people be cooperative, self-directing, participatory, and open and responsive to the ideas of others. It is equally evident that the condition of relative powerlessness, the experience of most people in hierarchical organizations, thwarts the sense of efficacy needed for participation elsewhere (see Blumberg, 1973, pp. 70–138). Thus, if collectivist–democratic organizations are to expand beyond their currently limited social base, they must, in addition to getting a job done, serve an important educative function.

Indeed, for Pateman (1970) the theory of participatory democracy rises or falls on this educative function. But other social scientists (see especially Argyris, 1974) remain unconvinced that participation in collectivist–democratic processes of organization can produce the desired changes in people's behavior. For Argyris, unilateral, defensive, closed, mutually protective, non-risk-taking behavior, what he calls Model I behavior, is nearly universal: It permeates not only Western bureaucracies but also counterbureaucracies such as alternative schools as well as collectivist organizations in contemporary China and Yugoslavia. Change in organizational behavior, then, cannot be expected to follow from fundamental change in the mode of production, for Model I behavior is rooted in the pyramidal values of industrial culture and in the finiteness of the human mind as an information-processing machine in the face of environmental complexity.

Contrarily, we argue that where people do not have participatory habits, it is because they have not generally been allowed any substantive control over important decisions. Nondemocratic (pyramidal) habits are indeed a problem for democratic groups, but they are not a problem that a redistribution of power could not resolve. Admittedly, the evidence is not yet entirely in on this issue, but much of it does indicate that the practice of democracy itself develops the capacity for democratic behavior among its participants (Blumberg, 1973; Pateman, 1970).

Environmental constraints. Alternative organizations, like all organizations, are subject to external pressures. Because they often occupy an adversary position vis-à-vis mainstream institutions, such pressures may be more intense. Extra-organizational constraints on the development of collectivist organizations may be legal, economic, political, or cultural.

It is generally agreed among free-schoolers, for instance, that municipal building and fire codes are most strictly enforced for them (Kozol, 1972; Graubard, 1972). This is usually only a minor irritant, but in extreme cases it may involve a major disruption of the organization in that the organization may be forced to move or close down. One small, collectively run solar power

firm was forced to move its headquarters several times through this sort of legal harassment. At one site, the local authorities charged it with more than a hundred building "violations" (Etzkowitz and Schaflander, 1978). An even more far-reaching legal obstacle is the lack of a suitable legal statute for incorporating cooperatively owned and controlled firms. The *Community News,* for example, had to ask an attorney to put together corporate law in novel ways in order to ensure collective control over the paper.[7]

The law can be changed, but the more ubiquitous forces against collectivism are social, cultural, and economic. Alternative organizations often find that bureaucratic practices are thrust on them willy-nilly by established institutions. Freedom High, for example, began with an emphatic policy of no evaluative records of students. In time, however, it found that in order to help students transfer back into the public schools or gain entrance into college, it had to begin keeping some records. The preoccupation of other organizations with records and documents may thus force record keeping on a reluctant free school. In another free school, the presence of a steady stream of government communications and inspectors (health, building, etc.) pushed the organization into creating a special job to handle correspondence and personal visits of officials (Lindenfeld, 1982).

Economically, alternative organizations strive to be self-sustaining and autonomous, but, without a federated network of other cooperative organizations to support them, they often cannot. Usually they must rely on established organizations for financial support. This acts as a constraint on the achievement of their collectivist principles. In order to provide free services, the Free Clinic needed, and received, financial backing from private foundations, as well as from county revenue-sharing funds. This forced the staff to keep detailed records on expenditures and patient visits, and to justify their activities in terms of outsiders' criteria of cost-effectiveness.

In less fortunate cases, fledging democratic enterprises may not even get off the ground because they cannot raise sufficient capital. Attempts by employee groups to purchase and collectively manage their firms reveal the reluctance of banks to loan money to collectivist enterprises, even where these loans would be guaranteed by the government.[8] From the point of view of private investors, collective ownership and management may appear, at best, an unproven method of organizing production, and at worst, a high-risk method.

For a consistent source of capital, collectivist enterprises may need to develop cooperative credit unions as the Mondragon system has done (Johnson and Whyte, 1977) or an alternative investment fund. In March 1980 the National Consumer Cooperative Bank began providing loans, mostly to con-

sumer cooperatives. In many worker cooperatives, the poorly paid labor of the founders forms the initial "sweat equity" of the organization that makes possible some measure of financial autonomy. In any case, the larger issue of organization-environment relations remains problematic, particularly when we are considering collectivist organizations in a capitalist context.[9]

Individual differences. All organizations contain persons with very different talents, skills, knowledge, and personality attributes. Bureaucracies try to capitalize on these individual differences, so that ideally people with a particular expertise or personality type will be given a job, rewards, and authority commensurate with it. In collectives, such individual differences may constrain the organization's ability to realize its egalitarian ideals.

Inequalities in influence persist in the most egalitarian of organizations. In bureaucracies the existence of inequality is taken for granted. In collectivist organizations, however, this is less true. Because authority resides in the collectivity as a unit, the exercise of influence depends less on position and more on the personal attributes of the individual. Members who are more articulate, responsible, energetic, glamorous, fair-minded, or committed carry more weight in the group.[10] John Rice, a teacher and leader of Black Mountain, an educational succession that anticipated the free school movement, argued that Black Mountain came as close to democracy as possible: The economic status of the individual had nothing to do with community standing. Beyond rare cases, however, "the differences show up . . . the test is made all day and every day as to who is the person to listen to" (Duberman, 1972, p. 37).

Some individual differences are accepted in the collectivist organization, but not all, particularly not differences in knowledge. In bureaucracy, differences of skill and knowledge are honored. Specialized jobs accompany expertise, and people are expected to protect their expertise. Indeed, this is a sign of professionalism, and it is well known that the monopolization of knowledge is an effective instrument of power in organizations (Weber, 1968; Crozier, 1964). Collectivist organizations, being aware of this, make every attempt to eliminate differentials in knowledge. Expertise is considered not the property of the individual, but an organizational resource. Individually held knowledge is diffused and critical skills are redistributed through internal education, job rotation, task sharing, apprenticeships, or any plan seen as serving this end. Bart (1981) shows how the skills needed to provide abortion services were shared with all of the members of a feminist health collective, despite the fact that such medical procedures are ordinarily reserved for physicians.

The diffusion or "demystification" of knowledge, although essential to help

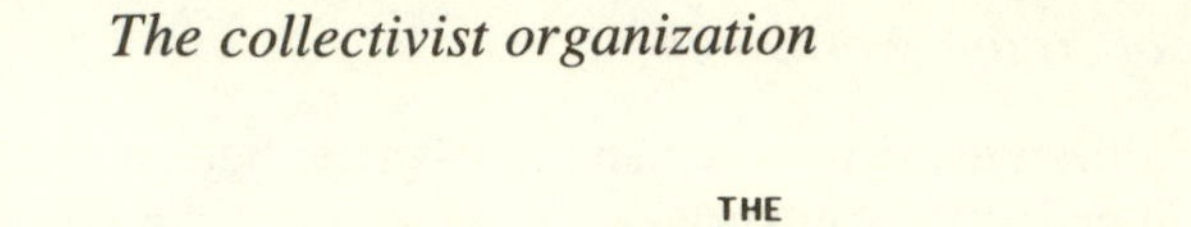
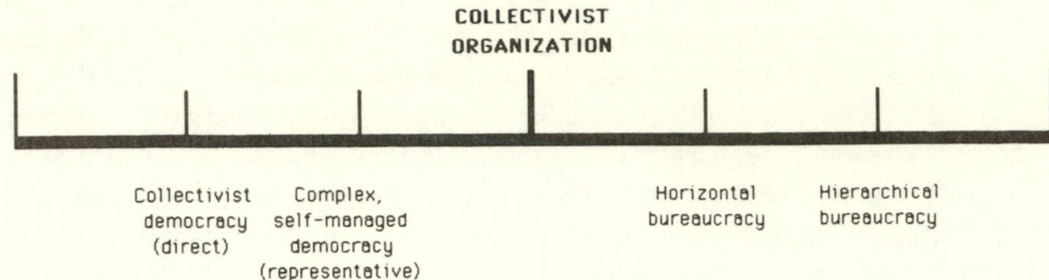

Figure 3.1. Range of organizational forms

equalize patterns of influence, involves certain trade-offs. Allowing new persons to learn to do task *X* by rotating them to that job may be good for their growth and development, but, as in the case of Ann at the *Community News*, it may displace an experienced person who had received a sense of satisfaction and accomplishment in job *X*. Further, encouraging novices to learn by doing may be an effective form of pedagogy, but it may detract from the quality of goods or services that the organization provides, at least until members achieve general competence in the tasks.

Even in the organization that might achieve universal competence, other sources of unequal influence would persist, such as commitment level, verbal fluency, and social skills.[11] The most a democratic organization can do is to remove the bureaucratic bases of authority: positional rank and expertise. The task of any collective workplace – and it is no easy task – is to eliminate all bases of individual power and authority, save those that individuals carry in their own persons.

Conclusion

The organizations in this study are uncommon ones. For this reason they are of great theoretical significance. By approaching the polar opposite of bureaucracy, they allow us to establish the parameters or limits of organizational reality. These parameters appear to be wider than students of organizations generally have imagined. Once the parameters of the organizational field have been defined, concrete cases can be put into broader perspective. Professional organizations, for example, although considerably more horizontal than the strictly hierarchical bureaucracy (Litwak, 1961), are still far more hierarchical than the collectivist–democratic organization. Thus, we may conceive of the range of organizational possibilities illustrated in Figure 3.1.

By contrasting collectivist democracy and monocratic bureaucracy along eight continuous dimensions, this chapter has emphasized the quantitative differences between the two. In many ways, this understates the difference. At some point differences of degree produce differences of kind. Fundamentally, bureaucracy and collectivism are oriented to qualitatively different principles. Whereas bureaucracy is organized around the calculus of formal rationality, collectivist democracy turns on the logic of substantive rationality.

If, in the Weberian tradition, we take the basis of authority as the central feature of any mode of organization, then organizations on the right half of Figure 1 empower the *individual* with authority (on the basis of office or expertise), whereas organizations on the left side grant ultimate authority only to the *collectivity* as a unit. Moreover, if, following Marx's lead, we take the division of labor as the key to the social relations of production, organizations on the right side of Figure 1 maintain a sharp division between managers and workers, whereas organizations on the left side are integrative: *Those who work also manage*. We take both – collectivist authority and the dedifferentiation of labor – to be essential defining characteristics of collectivist democracy. The other six features, although important, clearly are not as fundamental as these two.

As a dramatic departure from established modes of organization, the collectivist organization may be considered a major "social invention" (Coleman, 1970) deserving far more attention than it has yet received. Organization theory has for the most part considered only the right half of this spectrum, and, indeed, the majority of organizations in our society do fall on the right side of the continuum. Still, we gain perspective on all organizations by putting them into a broader frame of reference. With the proliferation of collectivist organizations both in this society and in others such as China, Spain, Yugoslavia, and Israel, we need an alternative model of organization – one toward which they themselves aspire – by which to assess their impact and success. Emphatically, collectivist organizations should be assessed not as failures to achieve bureaucratic standards, but as efforts to realize wholly different values. It is in the conceptualization of alternative forms of organization that organizational theory has been weakest, and it is here that the experimentation of collectives will broaden our understanding.

4. Internal conditions that facilitate collectivist–democratic organizations

It is not easy to maintain participatory–democratic organizations in a bureaucratic society. The surrounding sea of hierarchical organizations, market relationships, and traditions of representative democracy threatens to inundate brave islands of direct democracy. There are, however, defenses that collectives can erect to help them survive and even flourish as democracies. Members have a lot of choice in creating internal conditions, and some choice in creating external conditions, which assist them in maintaining direct democracy. These internal and external conditions are identified in this chapter and the next (see Table 5.1, at the end of Chapter 5 for a summary of all of these conditions).

To Weber, modernity meant that every aspect of life would become more impersonal, rule-bound, and bureaucratic. Today this idea arouses little controversy. It has become a coin of the sociological realm. For Weber, the inexorable process of bureaucratization is based on the technical superiority of bureaucracy vis-à-vis all other modes of organization in history (Weber, 1968, pp. 973–980) and on bureaucracy's indispensability as an instrument of power for those who head it. For this reason, bureaucracy, once firmly entrenched, renders revolution (i.e., a fundamental change in the structure of authority) impossible, and replaces it with mere changes in who controls the bureaucratic apparatus (1968, pp. 987–989). At the same time, the permanence and growth of bureaucracy hold potentially grave consequences. Scholars as diverse as Weber (1968), Jacques Ellul (1964), and C. Wright Mills (1959) have warned that as bureaucracy spreads, control over organizations will become more centralized and remote, thereby abridging individual freedom and control.

This belief in the inevitability of bureaucratic domination can be seen as a "metaphysical pathos," a pessimistic turn of mind not subject to proof or disproof (Gouldner, 1955). Instead, Gouldner urges sociologists to focus on the possibility of nonoppressive bureaucracy. Studying a gypsum mine, he finds that representative democracy, via labor unions, is possible (1954). His case is amplified by other students of organizations who have searched for the

social-historical conditions that encourage democratic processes, as in the International Typographical Union (Lipset, Trow, and Coleman, 1962) and in other unions (Edelstein, 1967).

But the quest for conditions that would permit democratic organizations to exist stops short here, for most sociologists of organizations have found patently non-democratic situations wherever they have looked. Firmly rooted in the work of Weber and Michels, the literature on social movement organizations is replete with case studies that indicate the fragility of participatory democratic systems and their tendency to develop oligarchies that displace original goals.

Various explanations for such a conservatizing process of goal displacement and the attendant process of oligarchization have been adduced: (1) Organizational goals may become increasingly accommodated to contrary values in the surrounding community, as in the Tennessee Valley Authority (Selznick, 1949); (2) organizations, such as the March of Dimes, may essentially accomplish their original goals and then shift to more diffuse ones in order to maintain the organization per se (Sills, 1957); (3) organizations, such as the Women's Christian Temperance Union, may find it impossible to realize their original goals and may then develop more diffuse ones (Gusfield, 1955); (4) procedural regulations and rules (means to attain goals) may become so rigid that they are converted into ends in themselves (Merton, 1957); and (5) organization maintenance and growth may be transformed into ends in themselves, as in the German Socialist Party, because it is in the interest of those at the top of the organization to preserve their positions of power and privilege within it (Michels, 1962). These processes of oligarchization and goal displacement – taken as they are to be near constants – represent substantial problems for the social movement organization, for they may destroy the raison d'être for which it was created.

The theoretical model introduced by Zald and Ash (1966), however, views these transformation processes as conditional. Zald and Ash suggest that transformation processes may run counter to those predicted by the Weber-Michels model. For example, increased rather than decreased radicalism may occur (Jenkins, 1977) or organizational coalition rather than factionalism may take place.

More recently, the question of whether democracy is possible within bureaucratic organizations has been approached from a different direction (Bennis and Slater, 1968; Toffler, 1970). Bennis argues that democratic organizations are not only possible but inevitable if organizations wish to survive in a society experiencing rapid technological change. Organizations will have the intellectual resources to adapt to changing and complex technological prob-

lems only if they direct the talents of specialists from many disciplines into project groups that are run democratically, groups that are dissolved upon completion of the project at hand.

To conclude, the literature on the prospects of democratic organization has come full circle. It begins with the tradition of Weber and Michels, which stresses that democratic control over bureaucracies is not possible, and ends with the Bennis forecast that democracy is inevitable in the bureaucracy of the future.

Our view is that democratic modes of organization are neither impossible, nor inevitable. They are conditional. Since grass-roots cooperatives aspire to be directly democratic, they are ideal vehicles in which to investigate the conditions under which democratic aspirations are realized or undermined. We seek to identify the structural conditions that allow at least some organizations to maintain democratic forms instead of yielding to oligarchy, and to adhere to their original goals instead of experiencing goal displacement.

The capacity of a cooperative, or any organization, to be directly democratic is conditioned by both internal and external factors. If our hypothesized conditions are correct, these factors should aid in the achievement of a non-authoritarian, collectivist structure. The absence of any of these conditions should constitute a source of tension or contradiction for the collectivist organization. These hypothetical conditions may be seen as antidotes to specific problems long thought to be endemic in organizations: conservatism of organizational purpose (through goal displacement, succession, or accommodation), rigidification of rules and ossification, oligarchization of power, and maintenance of the organization as an end in itself.

Practitioners in alternative institutions seldom if ever articulate their everyday problems in the sociologists' terms of "goal displacement," "oligarchization," "ossification," and the like. They do, however, constantly wrestle with those problems as conceived in a more specific way: "We're all supposed to be equal, right? So why does ——— have more say than the rest of us?" "How can we keep our commitment to total health care for mind and body, if we don't have the money to support our counseling component?" "With all this talk about promotional campaigns, we sometimes forget that we need a good newspaper to promote." "How do we stop ——— from usurping power for himself?" The hypothesized conditions speak directly to these concerns of participants, though they are necessarily posed on a more general level.

Although we hope practitioners will find this chapter and the next relevant to their concerns, this work is not intended to constitute a handbook in the field. Much as we have tried to identify the organizational conditions that support the participatory ideal, these conditions are not a recipe for organi-

zational success. Each carries with it important trade-offs and raises perplexing dilemmas for the would-be collectivist organization. In the abstract it may be easy to commit one's newly forming group to a democratic course, but in the concrete practice of everyday life, continued commitment to democracy is neither obvious nor easy. For each condition that we identify, we ascertain the corresponding dilemma that follows.

The *conditions* that we identify and examine are organizational features over which members have *some control*. The *constraints* discussed in Chapter 3 refer to received situations or facts of life about which the collective has *little choice*. Hence, the conditions are given a more extensive treatment.

Each of the following propositions has been generated and supported by the evidence of our comparative studies in a wide variety of collectivist organizations. In addition to its empirical grounding, each proposition is logically tied to the processes of oligarchization and goal displacement. This chapter, then, attempts to demonstrate, logically and empirically, that (1) the hypothesized organizational conditions facilitate participatory-democratic modes of organization, (2) that their absence undermines the desired collectivist form of organization, and (3) that the concrete choices participants must make about the conditions pose critical dilemmas for any attempted alternative to bureaucracy.

Provisional orientation

Just as fish are not aware of the water, students of organizations are sometimes unaware of their own most fundamental assumptions. One such assumption, as basic as it is unexamined, is that organizations desire to be permanent. Even research on alternative communal organizations assumes that longevity is desirable (Kanter, 1972a). Because this is assumed in nearly all of the literature on organizations, temporal orientation is rarely treated as an organizational variable (Palisi, 1970). Attitudes toward time can range from the expectation that the organization will be of limited duration to the expectation that it will last indefinitely. By treating such attitudes as an independent dimension of organizations, we can explore organizational consequences. The effect of temporal orientation on the organization can be profound.

One of the most notable features of collectivist organizations is that their members tend to reject the dominant cultural belief in the possibility, or even the desirability, of permanence. As a result, members tend to be oriented to the present. What distinguishes collectives from bureaucratic organizations is not the actual time they span – which may be short or long – but their members' *attitudes* toward time.

Members of collectives are not unique in their relatively short time horizons. Other researchers have pointed out that a rapidly changing environment requires, and brings into being, more adaptive organizations. Temporary organizations or subdivisions within organizations such as project teams are seen as best able to develop innovative solutions to complex problems, and in this way, to meet the challenges of a turbulent environment. For this reason, some scholars believe that such organizations will become increasingly prevalent in modern society (Bennis and Slater, 1968; Toffler, 1970).

Although this makes temporary systems sound like a very futuristic concept, temporary organizations have existed since antiquity. Yet they have received little study. Examples of organizations that are established with the intention of being short-lived include presidential commissions, task forces, theatrical productions, construction projects, negotiating committees, and campaign election organizations. The frequent adaptation required of participants in temporary systems may produce more personal stress and role strain (Bennis and Slater, 1968; Toffler, 1970; Keith, 1978). On the other hand, researchers of theatrical production companies have found that their temporary nature enhances professional growth and innovation (Goodman and Goodman, 1976). Our purpose here is not to address the meaning for the individual (a subject we take up in Chapter 6) but to draw out the meaning of a transitory orientation for organizational transformation.

Taking the long view, we can say all organizations are transitory and the idea of permanence is an illusion. Here we are concerned not with organizational longevity as such, but with members' subjective understandings of the appropriate lifetime of the organization. Many organizations in our society last a short time. For example, 50 percent of the small businesses in the United States go out of business within two years. Yet, insofar as the entrepreneurs desire growth and permanence, these closures are regarded as failures. In this section we focus on organizations that may come to associate the disbanding of the organization not with failure, but with achievement of goals.

Why has this provisional orientation come to be the rule, rather than the exception, in contemporary collectivist enterprises? There are at least three reasons. First, many members come to these collectives from social movements like the New Left. Their overriding commitments are to the broad goals of that movement. They are attracted to particular collectivist organizations as instruments for the achievement of movement subgoals. When the organization fails to live up to its promise, or conversely, when it attains all its goals, it may be considered to have outlived its usefulness. In either case individuals may choose to drop out of the organization or even to dismantle

it, turning their energies to other alternative organizations. Where members' paramount commitment is to higher-order movement goals, then provisional attitudes toward particular organizations, and somewhat transitory associations with them may follow.

Second, a provisional orientation may be rooted in the very act of creating a cooperative organization from scratch. The act of creating an organization where none existed before carries with it the implicit recognition that alternatives, as well as established organizations, can in fact also be dissolved.

Third, the provisional orientation may be a reflection of the values of the counterculture from which many collectives have sprung. Observers of the counterculture have depicted it as very present-oriented (Cavan, 1972; Hall 1978). Many participants seem to expect and desire an accelerated pace of social, psychological, and physical change in their lives, and this feeling is generalized to their relationships to organizations. They expect them to be relatively temporary.

History too provides examples of worker cooperatives with relatively short time horizons. During the nineteenth century, workers would sometimes create their own cooperative workplaces as a strategy to counter employer lockouts in a labor dispute. By setting up their own firms, craftsmen hoped to strengthen their bargaining powers and to get their employer to take them back under more favorable terms. Here the cooperative was conceived as a temporary solution to a problem, and its disappearance meant, in effect, that the workers had been successful in their efforts (Rothschild-Whitt, 1979b).

Provisional attitudes carry with them both positive and negative consequences for the organization. Evidence from our study of grass-roots collectives suggests that a *transitory orientation makes organizational maintenance and goal diffusion less likely,* but (as will be discussed) organizational *dilemmas* also arise from such an orientation.

In the Weber-Michels model of organizations, inertia and apathy among members create a situation in which a few are able to take control (i.e., oligarchization) and seek to perpetuate the organization as an end in itself. Much evidence shows that when an organization is either unable to achieve its original goal or when it fully accomplishes its goal, it will attempt to maintain itself by creating new, more diffuse goals (Gusfield, 1955; Sills, 1957).

A transitory orientation breaks these patterns. In the face of membership apathy or inability to move toward its original goals, a collective may simply disband. In the more unusual case of complete accomplishment of its original goals, a collectivist organization would also be more likely to dissolve than to create diffuse new goals.

Organizational ephemerality is expected, and even preferred in some cir-

cumstances, in collectivist organizations. For instance, the Food Co-op periodically encountered inadequate attendance at general membership meetings. This was reason enough, even in a time of growth and expansion, to provoke the most serious discussion of the age-old problem of member apathy. On such an occasion, Daniel, a highly respected staff member and founder of the co-op, urged that its by-laws be amended to include the following:

If we do not get a quorum for three general membership meetings in a row, then the board should be required to start procedures for the dissolution of the co-op.

To this proposal, another member quickly retorted:

Don't you think you're being a bit radical, I mean drastic? After all, the co-op still provides a valuable service, even if its members don't all participate enough.

Daniel replied:

I don't consider this a radical proposal. After all, we started the [Food Co-op] to be a community owned and controlled economic institution. If its members don't care enough about it to come to periodic meetings, then control will naturally fall in the hands of the few interested people. If and when that happens, the [Food Co-op] will be nothing more than a cheap Safeway . . . and it would be better to close down than to continue without real member participation.

Ray, a newly elected member of the board objected vehemently:

No, this is coercive. If some people don't want to participate, you can't make them conform to your ideals of participation. . . . You can't make people come to meetings if they don't want to.

This led to a heated two-hour discussion of how to maximize participation, with the major point of consensus summarized by Anne, a member of the board and founder of the co-op:

It's true that [the proposed by-law] would give the board a strong incentive for informing the members about the general meetings and encouraging their participation. Otherwise, it *is* too easy for an organization to become controlled by a small group, and we have to guard against that. . . . And it's true that the board didn't try very hard to get the membership out to this meeting. We'll have to do a lot more information spreading for the next one . . . because from the beginning [the co-op] has been based on the premise of members' control. Without that, we wouldn't be a true co-op. . . . They [the members] . . . have the last and final say about everything. The board always knew that it could impose the 2 percent surcharge. But since this represents a major policy change from selling at cost, we wanted to let the members decide.

The decision reached by the board was to refrain from immediate implementation of the 2 percent surcharge and to accept a time delay in the interest of getting wider participation at a subsequent general membership meeting.

The problem of membership apathy is not unique to the Food Co-op. It is

a perennial problem in all mutual-benefit associations (Blau and Scott, 1962, pp. 45–49) and is often a prelude to oligarchization (Michels, 1962). What is novel in the situation described above is the *response* to the problem of apathy. Although it could have been used as a justification for concentrating more power in the hands of board and staff members in the name of speeding up the decision-making process, it was not. Instead, members considered this an occasion to remind themselves that rank-and-file participation was central to the co-op's original goals – and that should the co-op be extensively eroded, it would be left without a legitimate basis for existence. The founders would rather dissolve the co-op than let it operate, however successful in financial terms, as "a cheap Safeway."

A comparable preference for self-dissolution over goal displacement was voiced at the Free Clinic. Fearing that their revenue-sharing grant would not be renewed for the coming fiscal year, staff members at the clinic began to explore the prospect of becoming a fee-for-servic agency, no longer free to patients. At a 1½-hour meeting on the subject, Nancy, the staff financial adviser, described how the budget would have to be altered. The six full-time staff positions would probably have to be cut to three, and counseling and health education programs would have to go in order to make room to hire doctors for more hours, as well as a secretary. Regarding the latter, Leah argued:

> Fee-for-service will mean billing and billing means paperwork. But no one can be a full-time paper-pusher without starting to hate the job. It's not fair to lay that on anyone.

Responding to the need for more doctor hours, Andrea pointed out:

> If we go fee-for-service we'll need to have the doctor's signature on lots of stuff and we'll need many more clinic hours to raise enough money to run on. We can't expect this level of commitment from a doctor who is a volunteer or is outside our decision-making structure, as is now the case. We'll need at least one doctor to be a part of the collective. And we must consider the impact that may have on our collective structure. The doctor might have a tendency to dominate decision making, to think his professionalism makes him superior.

After much despairing talk about how the clinic might be reorganized to accommodate fee-for-service, Rita pinpointed the tone of the meeting:

> What I feel is a death knell in the air. The more we think about the negative changes fee-for-service would bring, the more it seems pointless to continue at all. If we drastically change our collective structure – having doctors running things and cutting our staff size, I have to wonder why we should even continue. At that point we're no longer a collective. We're no longer a free medical service.

Leah poignantly agreed:

> I don't want the [Free Clinic] to go on a heart-lung machine. There's such a thing as letting a good thing die, of dying with dignity. . . . That's what I want for the [Free Clinic].

Rita continued on the same track:

> When the fiber is dead, you realize that the fiber was not the essence. . . . We must realize that it is the spirit that the [Free Clinic] represents that is important and that lives in all of us and that will be reflected in what we do with our lives. The organizational shell is not so important a thing.

Nancy was strident, but alone, in her objection to the growing "death knell in the air":

> I don't give up so easily . . . I'm a fighter. I'll put morals aside and finagle and weasel and do anything to keep the [Free Clinic] going. I'm not going to give up just because there are obstacles.

At this point, Nancy's motives were questioned by several of the staff members. She was accused of wanting to maintain the clinic because it represented a secure job to her. The meeting ended on that note of hostility.

Two points need to be made. First, when faced with an apparent inability to accomplish organizational goals – the clinic could not provide a free medical service if its grant was not renewed – most of the paid staff members would rather have dissolved the organization than water down its original goals. This is in sharp contrast with the many conventional cases where inability to accomplish goals is met with staff efforts to maintain the organization and their jobs (see, e.g., Gusfield, 1955; Messinger, 1955; Helfgot, 1974).

Second, this illustration sheds light on the dilemma arising from a provisional or transitory orientation. It is of course difficult for members to know at exactly what point an organization has outlived its usefulness. At what point is self-dissolution a needless sacrifice of an organization, a premature giving up of a difficult, but worthwhile struggle for survival? In concrete instances, the line between a sensible disbanding and a premature giving up may be devilishly difficult to perceive.

In the case described above, Nancy, the lone "fighter," proved to be correct. The clinic did manage to get the bulk of its grant proposals funded, it was able to maintain its counseling and health education components, it was able to sustain all of its staff positions, and it was able to continue to provide free medical care to financially needy patients. The uncertainty of the situation did shock the staff into realizing that "outside grants cannot last forever," and several months later they instituted measures designed to move toward greater financial self-sufficiency. They hired two doctors for longer hours and integrated them more into the collective decision-making structure. More doctor

hours allowed the clinic to stay open for more hours each week and allowed for a greater number of patients to be served. A sliding scale, fee-for-service system was instituted that still allowed many, though not all, patients to be treated for no charge, and the clinic began to charge those patients who had state or private insurance benefits to cover their medical expenses. Had the staff members who wanted the clinic to "die with dignity" moved immediately to dismantle the clinic, it would have proved premature. Their predictions about the loss of grant support proved false, and substantial goal displacement did not occur.

Another case in point is that of an artists' collective studied by Etzkowitz and Raiken (1982). Here two warring factions evolved: those who put organizational permanence above all else (the "Survivors") and those who put democratic process and other goals first (the "Innovators"). In this case, the latter prevailed, leading *not* to the end of the organization, but to different organizational choices and actions that laid the groundwork for the *continuation* of the organization on a more participatory footing (Etzkowitz and Raiken, 1982, pp. 29–31). This case, too, suggests that things are not always as they may seem to participants, and choices about organizational survival and goals often take place in a context of considerable ambiguity.

Sometimes external circumstances make it easier for people in the organization to tell whether it is time to disband. "Jane," a feminist abortion collective that provided thousands of safe abortions to women when the procedure was still illegal in the United States, was a cohesive and thriving organization as long as the members perceived an urgent public need for their service. After the Supreme Court decision in 1973 legalized abortion, the members of Jane saw that the need for safe and clean abortions could be met in hospitals, and, however regretfully, they decided to disband (Bart, 1981, p. 24).

Given the amount of energy that it takes to launch a new organization, participants have reason to be cautious about dissolving an ongoing enterprise. Their choices to support or to dismantle the organization are fraught with uncertainty. To the extent that participants wish to lower the chance that their original goals will be abandoned and that oligarchic control will develop, they raise the risk of needlessly ending the organization. To the extent that participants are more cautious about self-dissolution, they increase the risks of oligarchization and goal displacement.

The point here is not that a provisional orientation is in itself either good or bad, nor, as others have argued, that it will soon pervade many bureaucratic organizations (Bennis and Slater, 1968). Rather, it carries with it important consequences for the organization: at just those times that other organizations

would displace goals and develop an oligarchy, these organizations may opt for self-dissolution. The question of organizational life or death, however, is set in a context of ambiguity and dilemma.

This expectation that the organization as a whole may be more or less fleeting holds as well for the programs and operations within it. Members tend to regard organizational operations as experimental or tentative. Procedures and rules often are seen by members as ad hoc and flexible. Programs and operations are experimental, and if they don't work, they are altered. The sentiment that all operations and programs in an organization ought to be tentative militates against the usual ritualization of rule use that turns means into ends.

Concerning rules and procedures, the following quotation is typical:

Don't worry, if there are major objections to our new ID card system, we'll drop it. All our policies and procedures are experiments, in the sense that if they don't work, we change them – fast.

A provisional, experimental attitude toward rules and procedures may have considerable adaptive value for the organization. As Biggart (1977) points out in her study of the reorganization of the U.S. Post Office, old methods must be dismantled before new ones can take their place. In this way a transitory attitude provides a barrier against ossification and the ritualization of rule use. Nancy expressed this value in her advice to a new trainee at the Free Clinic:

I told her to start out using the systems I've worked out, but that when she feels comfortable enough in the job, to go ahead and modify my procedures [of accounting, recordkeeping, statistics, etc.]. It wasn't the hand of God that wrote those procedures down. We did what needed to be done at the time. I told her that as times change, she should change the procedures. . . . We're always in need of creative, new ways of doing things.

Staff members themselves, as well as programs and procedures, are often viewed as relatively short-term elements in these organizations. In the groups we studied, when we asked members, "How long is a long time to be here?" they tended to reply nine months to a year. Anything over two years was often considered "too long":

[Sally] probably shouldn't be staying here any longer. Not that she isn't good at what she does, it's just that the [Free Clinic] needs the enthusiasm of new people and fresh ideas.

Sense of time duration is, after all, relative, as noted by a staff member at the *Community News:*

Four years out of my 26-year life is a long, long time. . . . I just feel like I've outgrown the paper and want to go on and do something else.

Reinharz (1984, p. 37) cites the case of Ozone House, an alternative counseling service in Ann Arbor, Michigan, that provides emergency food, temporary foster care, and other services. In part, she attributes the organization's long existence – 15 years so far – to the fact that the collective has an explicit rule that no staff member may work there more than one year.

In bureaucracies, people strive to develop careers. They are provided a long-term ladder that they hope to climb. In collectivist organizations, as noted in Chapter 3, there is no hierarchy of positions and so there is no ladder to ascend. The lack of the possibility of career advancement contributes to people's short time expectations in these groups. For example, Bart notes (1981, p. 34) that the illegal nature of "Jane" meant that staff members could not use it to further their careers. This fact fostered a present-time orientation in contrast with the future/careerist orientation of bureaucracy.

In sum, bureaucracies are characterized by an orientation toward career-building, the future, and permanence. These are actually quite special attitudes toward time. Though they go without question in bureaucratic society, they are generally not shared by people in collectivist organizations. Equally plausible is an orientation toward the present, a view of the organization itself as provisional, or of one's own commitment to the organization as shorter term and contingent on the organization's ability to fulfill higher-order needs. Such a provisional attitude appears, logically and empirically, to help guard against oligarchization, rigidification of rule use, and goal displacement.

Mutual and self-criticism

The process of mutual and self-criticism is another internal feature that appears to support the egalitarian and participatory character of collectivist organizations. Collectives in as diverse societies as America, China (Hinton, 1966), and the Basque region of Spain (Johnson and Whyte, 1977) have tried to create settings that encourage constructive self-criticism. This often includes assessment of one's own behavior, of others in the group, and of the organization as a whole.

Where it is a systematic and accepted process, criticism helps to level these inequalities. Thus, *a regular and sanctioned process of mutual and self-criticism reduces tendencies toward oligarchization*. The leveling effect of one criticism session may be quite visible, but short-lived. However, when criticism sessions are institutionalized, the knowledge that one is subject to group criticism helps to curb the assertion of individual power.

Although criticism sessions seem to have positive, latent consequences for

the organization, the participants themselves are likely to justify mutual criticism in terms of its presumed benefits for the *individual* member:

> We need a place to give feedback. . . . If a person is not doing well at their job there should be a time for criticism, so they can grow into their job. A collective should help people become better, more able people. (*Community News*)

This call for group criticism was amplified by other staff members at the *News:*

> There's a need for criticism, but there's also a need for praise between us. Too often we forget the praise side of criticism. . . . People need to feel appreciated when they do a good job on something.

> Writing for the [*News*] is like writing for a void. There's no feedback at all from staff.

Ironically, however, no regular criticism sessions were instituted at the newspaper. Repeated requests for them were acceded to in principle at general staff meetings, but blocked in practice by claims that "there's not enough time for another meeting." However, in a private interview with an informally recognized leader at the paper, other bases for resistance appeared:

> Everyone who really puts work in on that paper gets my positive respect day in and day out. Not just a mechanical thing, but a true deep emotional love. . . . They work their fucking asses off. I feel so strongly about those people and they know it. And I need the same positive reinforcement from them. The whole question of not praising and not criticizing enough are from those people who don't have that respect. For the people who are carrying out their responsibilities, criticism/self-criticism exists. It's such a natural thing, it isn't even criticism. It's discussion.

The feeling that mutual and self-criticism should remain an informal and supposedly "natural" process, rather than being made into a regular and sanctioned one, was also expressed by a leader at the Free Clinic:

> Some groups find it necessary to institute sanctioned spaces where they can criticize each other because they don't dare do it otherwise. . . . The [Free Clinic] has a very healthy emotional climate. People feel free to express their anger and emotions on the spot. They needn't let them accumulate while waiting for a meeting.

Even though they may appear to agree with the principle, leaders in collectives often resist instituting regular and public forums for criticism. This is interesting for a number of reasons. First, regardless of leaders' vocabularies of motives, we found that organizations without a regular and agreed upon process for criticism were subject to explosive and sometimes destructive bouts of criticism unbound by any rules of fair play. Alternative organizations that did institute regular forums for criticism appeared to receive more considered and constructive forms of criticism. Secondly, leaders may object –

often covertly – to regular criticism sessions because they fear that they themselves may become the chief targets of the criticism.

Let us illustrate what the absence of a formal criticism system led to at the *Community News*. One member commented bitterly on the role of a former member, Tom:

> Mostly it was very insidious, behind-the-back stuff, or slap-across-the-face insults. That was all left to [Tom]. He was very good at that sort of thing. . . . [Tom] had an incredibly sharp wit and he would turn other people's ideas into jokes with his wit. Sometimes he would go on and on compiling one crack on top of another until the idea was lost somewhere in his snide humor.

And, concerning the "healthy emotional climate" said to prevail at the Free Clinic, the following example provides a counterpoint:

> [Leah] was up-front about her anger with [Sally] for not helping on the winter events mailing. . . . When [Sally] started crying, Leah accused her of trying to manipulate the situation.

Asked about the outcome, Leah said that she "felt great about it. We worked out each of our requirements for personal space, so we won't intrude on each other's space in the future." But for Sally, this event, and others like it, took on another meaning, as expressed in a private interview: "I've learned not to cross [Leah's] path. . . . Her personality has a lot of power over me. I'll do anything to avoid her wrath."

These two examples suggest that the buildup of destructive hostility and the growth of unequal influence may occur more readily in groups without a formal and sanctioned process of criticism. Those organizations that do institute a regular forum for criticism tend to collectively negotiate informal rules of fair play that define what sorts of criticism are legitimate. Social pressures are brought to bear on those who violate this sense of fairness.

People at the Law Collective arrived at the following written guidelines for their criticism meetings:

> We feel criticism should be carefully considered and thought out before it is voiced. We try to be gentle and objective, rather than abrasive, and we attempt to avoid personal attacks. In receiving criticism, we strive to consider each statement carefully, regardless of its nature or source. (Documents)

Members of the Law Collective spoke of working on their ability to criticize effectively, and thought that over the years they had made progress on this count. They would not, however, let any outsiders observe their criticism meetings because of their personal nature.

The only collective in this study in which sanctioned group criticism was observed was at the Food Co-op. There, the collective – in this case, of board members – agreed to have self-criticism sessions after each meeting. Some of

the criticism was directed at the group as a whole, rather than at individual members:

> Our energy has been too scattered tonight trying to do a meeting and inventory at the same time. In the future we should separate them and be more focused on each.

When an individual's ideas were subject to criticism, great pains were taken to critique and reject that person's proposal without rejecting the person. Criticisms were often balanced with praise. For instance, David, an active member of the co-op, came to a board meeting to propose picture identification cards for co-op usage. All of the costs of the cards had been figured, and the thoroughness of the proposal suggested that David had invested a fair amount of time and energy in the idea. "Picture IDs turn my stomach" was the quick response of a non-board member who happened to be present. Although board members, too, did not favor David's idea, they were obviously embarrassed by this tactless violation of their norms of how criticism should be presented. Board members tried to repair possible damage to David's ego by carefully explaining their objections to picture ID cards in terms of the co-op's honor system, technical difficulties with implementation, costs, and so forth. They ended by commending David on the benefit concert he was planning for the co-op and by stressing all the good things they had been hearing about it. Although the board firmly rejected his ID plan, they attempted to support him personally.

A similar sense of fair play and ego protection was voiced by a lawyer at the Law Collective:

> If it's criticism, it's not shattering. If people are laid low by it, it's trashing. There's a difference.

At the Free Clinic and the *Community News,* when criticism erupted it was not contained in special sessions set aside for that purpose, it generally was not thought out beforehand, it was not bound by collective rules of fairness, and it therefore tended to be harsh and global. Personal attacks and bitterness often were the result.

In sum, where the process of criticism is collectively sanctioned, it may serve a constructive function for the organization. By making the leaders or core members publicly and legitimately subject to members' criticisms, such forums tend to reduce the inequalities of influence and to check potential abuses of power. As a result, informal leaders generally resist calls to institute criticism forums. The tendency of leaders to become a major target of criticism in collectivist organizations may be related to the consistent research finding that task leaders are often not particularly well-liked members of groups (Bales, 1950).

But, feelings toward leaders may run deeper than that. Collectives have

contradictory feelings about leadership. They need informal leaders for a variety of reasons. However, the very presence of leaders signifies that inequalities in influence exist in an organization where such inequalities are not freely admitted. Prominent leadership is antithetical to the ethos of egalitarian control. If an individual, by virtue of some set of personal qualities, holds extraordinary sway with the group and is able to manipulate the "consensual" outcome of the decision-making process, this influence undercuts the basis of collective legitimacy. In the extreme, strong, individualistic leadership can render the consensual decision-making process a sham.

Collectivist organizations are therefore intensely ambivalent concerning leadership. On the one hand, they recognize the need for leaders or core members:

> You always need a few people to take up the slack. If that doesn't happen, if no one comes in to fill the void, the Food Co-op has big trouble.

On the other hand, they prefer to deny the existence of leaders:

> Everyone is equal at the Paper!

> The whole theory of the collective is that you don't have a leader. You have leaders, the whole group. Everyone is strong and aggressive.

What follows from this ambivalence is an extremely cautious view of the very kind of people who, in conventional organizations, would be regarded as strong, effective, or even charismatic leaders.[1] As in any organization, articulate, talented, inventive people may have magnetic appeal in co-ops, but to the extent that they can single-handedly influence the outcome of decisions, they are seen to threaten collective control over the organization. People in collectivist organizations continuously seek to reconcile the reality of individual differences with the ideal of collective control. According to a *Community News* member:

> There *is* a need for leadership. But that leadership must come from everyone; it must be mass leadership. If the leaders are individuals, that is a major flaw in the collective. Leadership by a clique is a perversion of the collective process. . . . We need an editor who can make people meet deadlines, but that position rotates, and who is the editor is chosen by the collective. The *basis* of anyone's authority is the collective, and authority can be taken away by the collective.

Despite serious attempts to articulate the meaning of collectivism, as in the above statement, and sincere attempts to abide by the collective will, individuals do sometimes assert personal authority. This generally is seen as illegitimate by the group, and if the violation is flagrant, the group may try to reassert the legitimacy of collective authority over individual authority. One of the prime mechanisms for this purpose is the process of group criticism.

This point is best illustrated through an extended example. At a regular staff meeting of the *Community News,* two influential members of the paper, Jake and Karl, came in for sharp criticism. Apparently, they had taken aside a reporter, recently assigned to city hall, and informally advised him on how they thought city hall should be covered. They suggested he treat gingerly certain progressives on the city council whose election the paper had supported. Even when in the reporter's view such progressives appeared inarticulate, stupid, or unprepared at council meetings, he was advised to "go easy" on them, for they had "good politics." The reporter objected. He wanted to "cover city hall as it is, showing fools to be fools." Members of the paper all lined up on one side: that of the reporter. One by one, they took the opportunity to remind Jake and Karl that:

Reporters have the right to write stories as they see them. . . . People of influence shouldn't pressure reporters to take their perspectives.

With that said, the tone of the meeting quickly changed. Members singled out perhaps the most influential person at the *News,* Karl, the managing editor, and launched a general attack on him. All aspects of his personality were suddenly fair game: One person rebuked him for being "unapproachable," another for being "intimidating." He was accused of "lacking trust in staff judgments" and of "guarding expertise jealously." Jake, a partner in the misdeed involving the advice to the reporter, eluded all personal criticism, probably because he was not a very powerful person at the paper.

Members then affirmed the following general principle:

Advice giving will and must informally go on, but reporters should assume this advice is from equals. They can throw it out if they want. The only [individual] authority that exists must be specifically delegated by the group.

With that action completed, the meeting took still another tack. Implicit criticism was directed at a new target, Clark, the city editor. This was not for any wrongdoing in this particular case – indeed, he had been instrumental in helping the reporter to bring the case out in the open – but for *potential* abuses of power based on his position at the paper:

Decisions about the politics of stories must come out of the Thursday staff meetings, not out of some dialogue between the city editor and the reporter. . . . Editors should not be policy makers, they should just do editorial style. . . . The line between political judgments and editorial judgments is thin.

Karl, obviously shaken and near tears from the earlier severe criticism of himself, interrupted these thinly veiled warnings to Clark, to defend himself:

[P]eople seem to be assuming, as Ronnie said, that "everyone is equal" except me. That I have to be watched because I'll pressure people and exert power. . . . I con-

sider myself very principled. I've never pressured anyone to do something, I've never made a decision outside of a staff meeting, I've never engaged in intrigue, and now my principles get questioned.

This led to a third dramatic reversal in the tone of the meeting. The reporter who originally raised this case as an instance of abuse of power apologized for generating "this major misunderstanding." The member who had earlier charged Karl with intimidation now affirmed her respect for him and his contribution to the collective. Nearly every member tried to qualify and in some cases to retract their previous criticism of Karl. The earlier criticisms were now coupled with strong praise:

If [Karl] has more influence at the paper it's due to his history of showing good judgment and responsibility in everything he does. People learn from their experience here who is worth listening to and who can be relied upon.

The issue was finally closed, 3 hours and 15 minutes after it was opened, on this note of unity and warm feelings.

Group criticism was used in this instance as a tool to check the actual exercise of individual power as well as its *potential* exercise. It was aimed specifically at those who were perceived to be the most influential members of the collective. Criticism need not be as harsh as it was in this case to have a leveling effect. As argued earlier, it is in groups that have avoided instituting regular forums for criticism (such as the paper), and therefore where no collectively held rules of fairness have been negotiated, that criticism is most unbridled and negative when it does occur.

A study of free high schools by Ann Swidler (1979, p. 81) finds a similarly erratic treatment of leaders. In Swidler's words, "[L]eaders . . . are always treading a dangerous path. Although the organization may temporarily encourage them, it is always ready to turn upon them." For this reason, she finds that the free high schools she studied had a hard time filling their directorship positions.

Similarly, Sandkull's study (1982) of six worker collectives in Oregon found peer pressure and collective self-criticism to be the main means of social control in the groups. In some of the co-ops, group self-criticism was used, in effect, to inhibit the development of leadership. In other groups, neutral outside mediators, trained in group process, were called in to facilitate and teach more constructive communication processes.

A study by Reinharz (1984, pp. 51–64) illustrates the role that a social scientist may play as a consultant to these groups. In this case, Reinharz consulted for a collective bakery that, while deeply committed to democracy on ideological grounds, had no regular forum for group feedback. As a result, interpersonal resentments had simmered for a long time. As an outside con-

sultant, Dr. Reinharz was able to introduce "feelings meetings" as a part of the decision-making process. These meetings functioned in part as the equivalent of the group self-criticism sessions we have described in other organizations. Once introduced, feelings meetings became a regular event in the collective, continuing for years with perfect attendance. Through these sessions, the group was able to clarify its goals, priorities, and organizational boundaries. Moreover, members were able to come to better decisions to put the co-op on a more secure footing, and – particularly – to open lines of communications and improve interpersonal relations in the group. The collective attributes its survival (10 years so far) and its success as a democracy to the innovation of the feelings meetings.

Contemporary collectivist enterprises often use criticism/reevaluation sessions of one type or another – consciously or not – to level inequalities of influence, to express their ambivalent feelings about leaders, and to nip in the bud leadership ambitions. In contrast, many historic communes used a more extreme version of public confession and self-criticism, but made their leaders exempt from such criticism. Where leaders are immune from criticism, as in some of the nineteenth century communes described by Kanter (1972a, p. 119), group criticism cannot check the assertion of power by charismatic leaders, nor is it intended to do so. In such cases, it serves other purposes.

Whether we are observing the extreme versions of self-scrutiny and public confession evidenced in Kanter's (1972a, pp. 106–107) nineteenth century communities and in the Bruderhof (Zablocki, 1971), or the milder process of reevaluation in the light of group ideology and goals that can be seen in the contemporary collectives (Swidler, 1979, pp. 92–95), public criticism functions as a powerful mechanism to regulate group behavior and to build commitment to organizational goals.

To emphasize, group criticism or reevaluation is often used in a context of ambivalence and dilemma. When the group both needs and resents the target of their criticism, it must be delicate in finding fault. It must walk a fine line between criticizing enough so as to level "undue" influence (enough to reassert collective authority over individual authority) but not so much as to lose a deeply committed and competent member.

Limits to size and alternative growth patterns

The face-to-face relationships and directly democratic forms that characterize *the collectivist organization probably cannot be maintained if the organization grows beyond a certain size.*

Rousseau (1950, p. 65) wrote that the upper limit for participatory democ-

racy would be groups in which "each citizen can with ease know all the rest." Weber (1968, pp. 280–290), too, acknowledged the importance of small size for democratic organization. Recent empirical work offers general support for this proposition. In a study of the effects of industrial plant size, Ingham (1970) finds that increasing size is associated with lowered cohesion of the work group, less worker satisfaction, and reduced identification with the plant. Rosner (1981, pp. 32–35) in a study of kibbutz settings, discovers that participation in both the plant and the community declines as the assembly size increases. Confirming Ingham's findings, Rosner also reports that the negative relation between size and participation is most pronounced where the motivation of members is based on identification with system goals rather than on individual material rewards, a circumstance with clear relevance for collectivist organizations of the kind we have studied.

Although it is certainly reasonable to suppose that some large number would be too cumbersome for democratic groups, it may be impossible to determine a particular threshold beyond which democratic control yields to oligarchic control. Over the centuries philosophers and political theorists have posited an upper limit, yet no consistent number has emerged from the record. Likewise, our own study suggests no particular cutoff point, since our sample of organizations is too small and their size varies widely – from approximately 10 members (at the Law Collective) to more than 100 members (at the Free Clinic) to 1,100 members (at the Food Co-op). This leads us to believe that there may be no single cutoff point concerning size but only a curve of diminishing returns – a slow erosion of democracy rather than a sudden break.

Researchers of cooperatives in communist societies have made similar observations. Kowalak (1981), having examined 30 years of postwar experience with worker cooperatives in Poland, concludes, "[D]emocracy is inversely proportional to the size of the cooperative, and it seems to be a rule in spite of several experiments being made to avoid the consequences of that rule." He therefore urges that cooperatives not exceed a size compatible with a general meeting of all of the members. Limits to cooperative size seem also to be acknowledged in China, where the mean cooperative firm size is 78, while the average size of state-owned firms is 850 (Lockett, 1981). Rather than speculate about an upper limit, it seems to us that a more fruitful question would be whether an *optimal* size exists for democratic enterprises, and if so, what it might be.

We decided to put the question of optimal size to the members themselves. Our survey turned up some interesting results. Of those who say that there *is* an optimal size for collectives in *general,* almost all locate this size at the

number of persons currently in their particular collective, give or take a few. That is, beliefs about an optimal size are quite consistent within groups, and quite disparate among groups. This may merely reflect positive bias toward one's own group. On the other hand, it may suggest that the actual optimum size for each collective is contingent upon a variety of other organizational factors, such as technology, and thus may not be generalizable. Indeed, most of the members themselves (55 percent at the Food Co-op, 71.5 percent at the Free Clinic, and 67 percent at the *Community News*) believe that there is no single optimal size for collectivist organizations in general. The best size for a collective newspaper may be quite different from the best size for a health clinic. Chickering (1972, pp. 241–227) for instance, proposes that the ideal size of organizations varies according to the task they seek to accomplish. For each organization, the main criterion would be, in Chickering's view, the avoidance of "redundancy" of personnel, that is having more people than are needed to do the job.

Complex as the issue of optimum size may be, there are undoubtedly some limits to size beyond which the familial and collectivist nature of alternative organizations is undone. In practice, members of collectives do indeed act as if they believe that size makes a difference. They often place size limits on their organizations and search for novel alternatives to conventional patterns of organizational growth.

Growth may be inhibited in direct and self-conscious ways. The following quotation from Rita, who six weeks earlier had taken over from Edward the full-time post of health education coordinator at the Free Clinic, reflects this direct approach:

> Now that I've got the health ed section organized and I'm almost done training these 20 new volunteers, I find myself making things to do in my job. Like the women's center project – things that aren't really required by the job itself. . . . Parkinson's Law sets in and the job keeps expanding to fill the time I have for it. Edward exaggerated the time demands of the job. Health ed never did require a full-time person. . . . Once I finish training the new volunteers, I'd like to cut down to half-time. I believe that the job can be done better and without all these needless elaborations and diversions as a half-time slot.

Here Rita was asking not only that she be cut to half-time and half-salary, but that the job position itself be permanently cut to a half-time one. Her request was granted. This sort of admission that less money and slots are needed to do a job would be rare indeed in a conventional, growth-seeking bureaucracy.

Sometimes, by self-consciously limiting the usual pattern of organizational growth (gaining more clients and personnel), collectives develop creative al-

ternatives to actual growth. Freedom High, unable to attract any Chicano students and unable to absorb more students than it already had, decided to form a coalition with a Chicano community cultural center. This coalition promised to broaden the school's resource base (library, art room, and so forth were now shared and enlarged) and to give its students some measure of contact with the Chicano community. Freedom High, then, acquired some of the benefits of growth without actually growing.

Building a wider network of cooperative relationships with other small, collectivist organizations is one substitute for growth that some organizations utilize. This federative principle grows out of anarchist thought (Kropotkin, 1902). Some of the most successful instances of cooperative development such as the Mondragon system in Spain attribute a good part of their success to having built a federation of mutually supportive cooperative firms and auxiliary organizations such as cooperative banks, schools, and research and development enterprises (Johnson and Whyte, 1977).

The spin-off of new, autonomous collectivist organizations is another alternative to growth. At the *Community News,* for instance, some of the staff envisioned taking about half of the collective and creating a second collectively run newspaper in another city when this one became "stable enough." Expansion in the form of a larger paper or a larger staff was not contemplated.

The Food Co-op, another case in point, planned to double its store size of 1,400 square feet because:

> We already have way too many members for the size store we have. . . . Twenty-eight hundred feet would be a good size for a store – large enough to allow for a good selection of foods and certain economies of scale, but still small enough to be a real community store.

But after this initial expansion, they envisioned no more, preferring instead to "start wholly new and independent co-ops with the additional people who want to be members." In fact, the Food Co-op did help start several other food co-ops, deliberately curtailing membership growth beyond 1,100. A separate study of food co-ops found that more than 20 percent were created as spin-offs from larger co-ops (Nagy, 1980).

The *Community News* and the Food Co-op were not unique in their concept of spinning-off parallel, collectivist organizations as an alternative to internal growth. Schumacher (1973) describes a cooperatively owned manufacturing firm in Britain, the Scott Bader Company, which requires in its by-laws the spin-off of new, autonomous cooperatives when it reaches 350 in size. Kanter (1972a, pp. 227–231) points to a similar phenomenon in some of the nineteenth-century communal ventures, such as those of the Hutterites. Johnson and Whyte (1977) too have observed spin-offs in the Mondragon system of

workers' cooperatives in Spain. The search for alternatives to organizational growth may be part of a broader reevaluation of large size in modern society (see, e.g., Molotch, 1976; Appelbaum, 1976; Schumacher, 1973; Sale, 1980).

Members of cooperatives generally do not view small size as a problem. Indeed, as just indicated, they often feel that it contains a number of advantages. Nevertheless, the limited size of such organizations may reduce their impact on the surrounding society and their value as demonstrations of alternative organizational principles. Mainstream attitudes and conventional organizations value large size far more than do people in co-ops. All other things being equal, a large organization will have more real political and economic clout than a small one. A cultural bias toward large size also makes it relatively easy for skeptics to dismiss collectives and cooperatives as trivial organizations espousing principles that would never work in large organizations. In choosing a democratically manageable size, therefore, co-ops may have to face the dilemma of trading off a degree of impact on the larger society.

Homogeneity

Consensus, an essential component of collectivist decision making, may require from the outset substantial homogeneity among members. Participants must bring to the process similar life experiences, outlooks and values if they are to arrive at agreements. The absence of a fundamental similarity in values makes reaching and abiding by a consensus much more difficult.

Bureaucracy may not require much homogeneity, partly because it does not need the moral commitment of its employees. Since it depends chiefly on remunerative incentives to motivate work and since in the end it can command obedience to authority, it is able to unite the energies of diverse people toward organizational goals. But, in collectives where the primary incentives for participation are based on shared purposes and values and where the subordinate-superordinate relation has been delegitimated, moral commitment becomes necessary. Unified action is possible only if individuals substantially agree with the goals and processes of the collective. This implies a level of homogeneity in terms of values unaccustomed and unnecessary in bureaucracy.[2]

Anyone who has participated in democratic meetings can appreciate how disruptive the contrarian can be. The holdout, while perhaps not swaying the group's eventual decision, can surely protract the decision-making process and generate frustration. In observing consumer cooperatives, Gamson and Levin (1980, pp. 9–10) find that, when an admissions policy is overly open,

common cultural values tend to be lacking and thus problems arise for the organization. For this reason, as cited in Chapter 3, many worker collectives tend to seek new members who agree from the start with their fundamental values. The Law Collective, for example, after initially selecting new staff who appeared to have congruent political values, additionally instituted a six-month probationary period to make sure that members' assessments proved correct before further committing themselves to the individual.

Consequently, collectivist organizations tend to attract a homogeneous population. In the cooperatives reported in our study (Rothschild-Whitt, 1976a) as well as in those surveyed by Crain (1978) members were disproportionately from economically and educationally advantaged backgrounds. As described in Chapter 2, they also tend to draw members with similar political views, experiences, and identification with social movements.

Homogeneity has also been a salient feature of worker cooperatives in the past. In the 1920s, Italian immigrants in San Francisco came together to form garbage collection co-ops. These cooperatives have endured over the years, have paid their worker-owners well, have succeeded in dignifying what would otherwise be considered "dirty work," and have continued to provide high-quality service at low price. Today several have become large waste-management conglomerates. The point is that these cooperatives originally came together out of economic necessity and the glue of a common cultural background. This generated feelings of trust and a sense of being social equals – the basis for any cooperative. Decades later, when new workers who wished to join the cooperatives were from other racial and minority groups, the Italian founder-owners felt no sense of kinship with them, and in addition had little desire to dilute the value of their own equity. Therefore they did not offer ownership rights to the new workers. As a result, the co-op ironically devolved into a two-class system of owner-members on the one hand, and hired workers on the other (Perry, 1978; Russell, 1982b). Similar ethnic and cultural bonds, in the case of Soviet Jewish immigrants, have united new taxi driver cooperatives in Los Angeles (Russell, 1982b).

In other successful examples of cooperative development, such as the Mondragon cooperatives, cultural homogeneity has played an equally important role in knitting the groups together. The Mondragon cooperatives have not devolved into a two-class system over time, perhaps because they are all Basques (Johnson and Whyte, 1977). In the kibbutzim cooperatives in Israel, members' common Jewish identity bonds the groups together, but separates them from their Arab hired laborers (Ben-Ner, 1982).

Efforts to develop democratic workplaces often run into difficulties if they have a very heterogenous work force. For example, the president and owner of IGP, a multimillion dollar insurance firm, decided to give half of the stock

to the workers and to set up democratic committees of workers to manage the firm. Some workers, now worker-owners, took to the new system as an opportunity and a challenge; for others, democratic decision making brought more headaches and responsibilities than they wanted. These attitudes toward democracy mirror the range of attitudes one can expect to find in the larger society. The contemporary cooperatives often avoid confronting such a range by selectively recruiting and attracting only individuals who desire participation and influence in their work. Selectivity is necessary since participatory values and habits of behavior are not widespread in our society. Bernstein (1976) suggests that a democratic ideology is essential in cooperative workplaces.

Though homogeneity eases consensus decision making and promotes group cohesion and friendship, it does present the organization with a dilemma. At a minimum, it narrows the membership base of the collective and it makes it less representative of the surrounding community. Participants in the contemporary collectives often regret the restricted nature of their constituency, especially in the light of their desire to be integrated into the community and to change it in some fashion. In the extreme, as in the garbage collection co-ops in San Francisco, the natural desire for homogeneity may exclude "outsiders" and lead to the degeneration of the co-op form.

The overrepresentation of young adults with upper middle-class origins in contemporary collectives in the United States probably results from the changing values, social movement orientations, reduced employment prospects for this group, and other factors detailed earlier. In addition, certain organizational features of collectives may have the consequence, however unintended, of further narrowing the social base. Specifically, the compensation system of the contemporary collectives, with its formally egalitarian pay principles in the context of low pay levels, may have different meanings to different individuals.

In those worker cooperatives where the pay is low and much less than it would be in comparable jobs in the outside world, collectives may have a difficult time getting and keeping members of working class origin. As a working class staff member of the *Community News* ruefully observed:

> You can't ask people who have no option of a rich family to fall back on to make a long-term commitment to the [*News*], if you can't commit the paper to supporting those people and their families.

In a private interview, another working class staff member, soon to leave the *News,* lamented:

> Part of the reason I've been bummed out at the paper is a class thing. I think a lot of the people there have a lot more life options that I do. And I think those options account for their rather haphazard attitude about things. I think growing up rich

makes a real difference. It allows people to be dilettantes and not to take things that seriously. At first I thought it was California, that people here don't take things seriously. But that's not it. . . . Being from a rich family lets people take their future for granted. Money isn't that important because there's always mom and dad.

In short, working class members of alternative organizations often feel that the members from more privileged backgrounds, while willing to live a very simple life-style now, can always fall back on their parents for money should times get rough, whereas those of more modest means have no such insurance available. This may lead committed working class members to leave collectives in search of more secure, if less inspired, work. In group discussions, this private uneasiness of working class members was never publicly acknowledged in the observed organizations.

The meager pay levels that characterize the contemporary collectives appear to stem from two causes. First, collectives sometimes generate little surplus to distribute among their members. The problem of undercapitalization in cooperatives has long been recognized (Blumberg, 1973). Second, on a more subtle level, cooperatives may continue to pay meager salaries even when they can afford more, because they fear the development of "careerism." People in collectives expect work there to be a labor of love. Some suspect that should the work become too well remunerated, it will degenerate – from their point of view – into merely a career, with attachment to one's position taking priority over organizational goals. For this reason, collectivists often want to avoid the sorts of economic incentives that might encourage people to seek a career in them.[3]

Lean salaries may assure the organization that its workers are committed by nonmonetary values, but for some members it may also mean that the collective is not a viable place of employment. The egalitarian nature of the compensation system (which is an essential feature of the collectivist organization) and the typically low pay levels (which are *not* an inherent feature) combine to take an uneven toll on members.

At the *Community News,* for example, staff members (whose average education level was a B.A.) earned an average of $160.00 per month in November of 1974. Three months earlier they had averaged $125.00 per month. Salaries at the paper were determined by the collective in plenary meetings and given out "to each according to his need." Some staff at the *News* (including some of its most helpful people) were paid nothing; the highest pay was $300.00 per month to a person with a family. Regardless of pay, full-time staff were expected to work a 40- to 60-hour week.[4]

At the Free Clinic, salaries were also collectively determined. However, here it was decided that they would each take salaries of $500.00 per month

(in 1974) and that each would be paid equally. This meant that all full-time staff members, whether a 17-year-old doing secretarial work or a Ph.D. coordinating the health education program, were paid equally (the average education was a B.A.). Later, unable to find the funds with which to raise salaries and to compensate for what they perceived to be inadequate pay for the job at hand, the staff collectively decided to lower their expectations of work hours from 40 to 28–35 hours per week and to add a number of paid vacations for themselves.

In both organizations, the salary was augmented by the nonmonetary "fringe benefits" of working in a collectivist organization, namely more autonomy and control than could ordinarily be attained in a bureaucratic or even in a professional organization.

The consequences of these two financial situations differed decidedly. At the *Community News* some of the staff lived at a less than adequate standard and a number of capable and dedicated members felt compelled to leave in search of greener pastures. Two of them soon found jobs in journalism for more than $800 per month.

At the *Community News,* staff generally made about 18–25 percent of the salary they could draw at comparable, but established, journalism jobs. At the Free Clinic, some staff people made about 50 percent of what they would draw at comparable nursing or counseling jobs for which they were qualified. The equality principle by which salaries were distributed meant that others such as secretaries made as much as 83–100 percent of what they would be paid in comparable outside jobs.

In the context of a society that grants highly unequal pay, the more or less equal compensation within collectives brings great differentials in relative sacrifice. This implies a kind of de facto inequity built into the equity principle of collectives. Most cooperative organizations tend to pay people according to both their skills and their needs, while strictly limiting the amount of differential allowed. In our observation, it is among those who perceive that they are making the greatest sacrifice by remaining in the co-op that attrition is highest.

The data on social composition in the contemporary collectivist organizations are consistent in this study and in others: Members tend to be drawn from relatively privileged backgrounds and do not have family responsibilities. Members commonly feel uneasy about their restricted social base. In the abortion collective studied by Bart (1981, pp. 25–26), members worried over their homogeneity, but Bart concludes that it was probably a "blessing in disguise," enabling members to stick together and to achieve consensus.

This outlines an important and as yet unresolved dilemma that many col-

lectives face. If pay scales are "too high," there is the risk of engendering careerism, with the attendant problem of organizational maintenance. The need to ensure a value-committed staff suggests relatively low and equal salaries, and the need for cohesion and consensus may require substantial homogeneity, but these structural facts seem to preclude many people, particularly of working class origin, from joining collectives – hardly an ideal situation for those who are trying to broaden the base of a social movement.

Dependence on internal support base

Those people with which an organization comes into regular, direct contact – that is, its members, customers, and clients – constitute that organization's internal support base. Their support, moral and financial, of the organization is crucial. We hypothesize that *the more a collective organization depends on its internal support base, the more likely it is that democratic ideals will be maintained. In addition, there is less chance of displacement of original goals.*[5] Conversely, when a collective acquires an external base of financial support (such as a foundation or government grant) its leaders tend to lose interest in the sentiments and goals of its members and clients, and thereby the likelihood of goal displacement is increased.

When collectivist organizations come to extensively rely on external financing, members typically find themselves more and more caught up in seeking such funding. In turn, they often shape the character of the organization to suit funding agencies. The Free Clinic, for example, depended on external sources for 83 percent of its budget. As a consequence, paid staff at the clinic reported spending an average of three-quarters of their time seeking continued outside revenue. After writing grant proposals and cultivating the sensitivities of those officials who award financial grants, they had little time left to attend to the volunteers and clients. In fact, when the position of health education coordinator opened up (a job whose formal responsibilities required writing health education pamphlets, speaking on health-related topics, and training and organizing a group of volunteers to do the same), the skills that the staff sought in a replacement had mostly to do with one's willingness to "hobnob with politicians," and the "ability to impress government types." Moreover, in the thick of grantsmanship, two of the staff coordinators temporarily suspended their "component meetings," the only formal arena for decision-making input that the volunteers had.

Dependence on external financial support at the Free Clinic resulted in a decline in participation levels and appeared to reduce leaders' sensitivity to volunteers' interests. It also led to more direct forms of goal displacement.

As part of the health education program at the Free Clinic, pamphlets aimed at "demystifying" health care (regarding drugs, herpes, venereal disease, and so on) were produced for public distribution. One day an important county official charged that one pamphlet that described plainly how to qualify for California's Medi-Cal program was "too political." Staff members thought that the pamphlet was valuable for patients, and its widespread use seemed to confirm that assessment. Fearing, however, that continued grant support might be jeopardized, the staff removed the pamphlet from the shelves of the clinic.

These sorts of compromises are difficult to avoid when the organization cannot pay its rent without external help. Such cooptations result not only from direct outside pressures to alter the course of the organization. Members themselves begin to monitor and structure the activities of the organization in terms of their presumed acceptability to money-rendering outsiders.

External agencies need not threaten to withdraw funds from a collective to achieve accommodations. Not uncommonly, the agency will insist on certain conditions before funding will be considered. Money often comes with strings attached. Foundations generally choose which part of a budget they wish to fund, if they choose to give at all. One private foundation decided, among many possibilities, to fund a children's clinic. This was not considered a high priority by the staff at the Free Clinic because there was another children's clinic in town, and the community need for a second one was perceived to be slight. Yet, as one staff member explained: "We're not going to let the $5,000 go to waste. We'll sure as hell do a children's clinic now." That this source of funding made the Free Clinic susceptible to an external ordering of priorities, not subject to collective control, was lamented, but not challenged, by a staff member at the clinic:

> What's "in" right now is pediatrics and geriatrics. Nothing for anybody in between. And what these foundations support goes in and out with the fads. What can you do?

In another instance, a private foundation granted the Free Clinic funds for capital expenditures. This led the staff to create pseudoneeds for an elaborate typewriter, acoustic ceiling, and a photocopy machine where other more pressing health care needs existed. Again, clinic members could not challenge the right of the foundation to earmark funds in this way.

Similarly, in recent cases where workers faced with corporate shutdowns and the prospect of personal unemployment have been able to buy their firms, they have been heavily dependent on external sources of financing. They have had to go to private banks and government agencies for loans. Banks have often insisted, as a condition for loaning the capital, that the fledging worker-owned enterprise have "responsible management." In some cases, bank-

approved managers had to be installed, with the result that the new worker-owners participated little in managerial decision making. Additionally, the worker-owners have tended to look positively upon the bank or other agency that aided the survival of their firm. These worker-owned organizations are still quite young, and it is too soon to tell what the long-term effects of this external dependency will be.[6] The situation may also be modified by the creation in 1980 of the National Consumer Cooperative Bank, which now loans money to cooperative enterprises.

The proposal of a direct relationship between dependence on external support and lessened regard for members' and clients' goals needs to be qualified. Some sources of external funding appear to be less potentially cooptative than others. The Food Co-op made a point of pursuing grants that appeared to be unconditional. For instance, they sought grants from two wealthy young heirs whose philanthropy was considered to be "radical." Like the Food Co-op, the *Community News* was mainly self-sufficient, but when it did seek grants, it sought them from rich individuals who were known for liberal causes. In fact, a new type of alternative institution has appeared, namely, the "alternative foundation."[7]

In the case of the Free Clinic, the only organization we studied that depended heavily on external funding, we saw an alternative organization that spent much of its staff time seeking continued outside revenue, compromised member participation, watered down some of its projects to accommodate outside pressures, and forfeited some measure of collective control over priorities to external foundations.

The converse of this process seems to hold as well. In service organizations that depend on the goodwill of their members and clients for financial support, as did the Free School and the Food Co-op, leaders tend to remain much more responsive to the goals and sentiments of the membership. We found in a survey of the general membership of the Food Co-op that 74 percent of the members considered their elected board of directors to be either "very" or "reasonably" responsive to their needs. In contrast, only 29 percent of the volunteer-members at the Free Clinic believed that their board was either "very" or "reasonably" responsive.

Likewise, in her study of "Jane," Bart notes that because there was a fee charged for the abortion service, leaders remained sensitive to client needs and expended no time in searching for grants (1981, pp. 20–22). A recent survey of 236 community self-help organizations also confirms our hypothesis. Milofsky and Romo (1981) find that such organizations tend to receive funds from a single type of source, rather than from some random combination of sources. They argue that is true because the sort of organization that

is able to attract a specific source of money may be unsuited to attracting an alternative source. The authors find that the most participatory organizations are those that rely on internal funding from their clients and customers.

Two of the most impressive international examples of successful cooperative development are built on internal financing. The founders of the Mondragon system of worker cooperatives in Spain recognized that independence from the state and from private banks would be crucial. They set up their own internal bank through which the co-ops' earnings could be retained and used to sponsor further cooperative development. Mondragon owes its exponential growth from a handful of people to more than 18,000 worker-owners in 25 years to this internal financing mechanism (Logan, 1981; Johnson and Whyte, 1977). Similarly, a system of 12 self-managed enterprises has enjoyed rapid growth in the Netherlands by developing its own internal bank to provide investment capital (Rothschild-Whitt, 1981).

However, in worker collectives, or more broadly in social movement organizations – especially where the clients have low incomes – the organization is faced with a difficult choice of whether to turn to internal or external funding. For example, in his study of 132 community and tenants-rights organizations, Lawson (1981) finds that without external funding these organizations have few resources and must depend on volunteer and part-time labor. When some of these community organizations do receive government funding, morale rises and staff tend to work very hard, providing services (now on salary) that they had formerly provided for no money. After a time, new staff enter, not out of deeply held conviction, but because they need a job, and the organization begins to lose its commitment and focus. In some cases, where government agencies prefer to fund multi-objective organizations, community organizations may find themselves completely shifting in focus and goals from, for example, tenants rights to youth unemployment. Further, the lessened commitment of the second generation of staff may set the organization on a course of steady grantsmanship and lowered volunteer participation (Lawson, 1981; Helfgot, 1974). In groups that rely on volunteer work, such as the Free Clinic, the external grants with which professional staff members are paid undercut the willingness of people to volunteer. Indeed, Lawson (1981: 23) concludes that "the vast majority of externally funded social movement organizations [which have developed a professional paid staff] no longer have either volunteer workers or members as such."

An intensive case study of a women's health center in New England by Sandy Morgan reveals some of the organizational consequences of accepting government funding. As at the Free Clinic, government grants here, too, brought with them project guidelines. In effect, this meant that outside gov-

ernment agencies shaped the internal priorities and programs of the collective as, for example, by curtailing funds for abortion services. In addition, government grants required specific forms to be filled out. In time, the health center ceased soliciting client feedback, concentrating instead on Comprehensive Employment and Training Act (CETA) forms and review.[8]

The dilemma for collectives posed by outside sources of funding is easy to appreciate. When such organizations lead a rather hand-to-mouth financial existence, the possibility of obtaining a foundation or government grant holds considerable attraction for members. Although there may be some reluctance to pursue outside funding because of what such a fostered dependency might do to the organization, some cooperatives eagerly seek external funding.[9] As in the case of the Free Clinic, these organizations are likely to discover that dependency on external money is a two-edged sword. External funds may allow the organization to provide a free or below-cost service to clients, but funding possibilities also seduce the energies of leaders, diverting them from devotion to organizational democracy, and in some cases, from the original aims of the organization.

Technology and the diffusion of knowledge

The egalitarian and participatory ideals of the collectivist organization probably cannot be realized where great differences exist in members' abilities to perform organizational tasks. Put more specifically, *collectivist forms of organization are undermined to the extent that the knowledge and skills needed to perform the organization's tasks* (be they medical knowledge, legal know-how, or whatever) *are unevenly distributed.*

Diffusion of knowledge and skills, crucial as it may be for effective collective control, is difficult to accomplish. It seems to require that *one* of the two following technological conditions obtain.

Either:

1. Tasks involved in the administration of the collective organization must be relatively simple so that everyone readily knows how to do them, or they must involve a relatively undeveloped technology applied in relatively non-routine situations (Perrow, 1970, pp. 75–85). An example is Freedom High. Since knowledge about the teaching process is more of an art than a science, and since every student at Freedom High was supposed to be treated uniquely, the issue of knowledge diffusion was not prominent. There was no systematic body of reliable knowledge to be communicated or monopolized. The technology employed in most organizations can be exclusively held, however.

Or:

2. If the members of the collective are of sufficiently homogeneous ability and interest to be able to learn the skills involved fairly rapidly, then the technology involved in the organization may be relatively sophisticated and may be applied in more uniform circumstances. In cooperatives that employ a more sophisticated technology, the distribution of knowledge becomes problematic, because scientific, specialized knowledge does lend itself to monopolization. Democratic organizations that utilize a more sophisticated or routine technology must therefore institute a systematic process of knowledge diffusion or risk defeating their egalitarian and collectivist principles.[10]

Sociologists of organizations have long understood that the holding of "official secrets," that is, the monopolization of knowledge, is a prime source of power in bureaucracy (Weber, 1968; Crozier, 1964). It is also an important source of power in organizations in which power differentials are not freely admitted. When knowledge inequalities are not acknowledged by the participants, as in the case below, the consequences are most frustrating and painful for members.

In another collectivist medical clinic, which we will call Southside Clinic, angry conflicts between the doctors and the paramedical staff erupted frequently. Members were confounded by the paradox of doctors who endorsed egalitarian principles in concept, but who seemed in practice to usurp decision-making power. In response, members charged the doctors with being guilty of "elitism," "authoritarianism," and "professionalism." One staff member at the Free Clinic who was in contact with the paramedical personnel at Southside Clinic, described Southside from her perspective:

> Sure they [the doctors] will let you have a collective, they'll let you talk things out, as long as you end up agreeing with them. But the minute you don't, it doesn't take long before they remind you of who's bringing in the bread and whose skills are really needed. That's the situation at [Southside] and you have no idea how toxic it is. [Southside] exists not because a collective got together and found two doctors to help them, but because two doctors decided they wanted to be hip, to come white, and they found a group to help them. . . . They have the power ultimately and it can never be a true collective.

Another staff member at the Free Clinic agreed with this analysis of Southside Clinic. She did not believe, however, that the Free Clinic would follow the same route, were it to allow a doctor "into the collective":

> You're right – a doctor may have a tendency to dominate the decision making. But [Andy] is a real possibility. *He's not like that*. He's said that he'd like to be more active in the clinic. (Emphasis added)

As the above quotations indicate, participants themselves tend to interpret conflicts as *personality* defects in the people involved. We believe, however, that the conflicts are often *structural* – and not psychological – in origin. The persistent conflicts at Southside Clinic were not due to the doctors being particularly authoritarian. Like all other members of Southside, the doctors endorsed nonelitist, egalitarian principles, yet the distribution of relevant knowledge was vastly unequal. Such disparities in knowledge are structural features of the organization, not a matter of personal idiosyncrasies, and they severely undercut the likelihood of developing or maintaining a directly democratic form of organization.

Collectivist organizations that employ a more sophisticated technology in their operations must focus on the process of knowledge diffusion if they are to avoid the tension and inequality that riddled the medical clinic described above. Toward this end, some of the organizations we observed devoted a great deal of energy to cultivating in their members a general knowledge about overall operations of the organization instead of specialized expertise. This was accomplished primarily through extensive job rotation, task sharing, and most broadly, by attempts to "demystify" normally exclusive or esoteric bodies of knowledge. Members use the word *demystification* to refer to efforts to simplify, explicate, and make available to the membership at large formerly exclusive knowledge.

The process of knowledge diffusion may take a variety of forms. When an attorney at the Law Collective, for example, was asked in an interview how the collective deals with inequalities in influence that arise as a result of differences in knowledge, he responded:

> The first thing has been to read some of the basic things that Chairman Mao has written . . . analyzing professionalism as a contradiction. . . . The main problem with professionalism is with the attitude of attorneys. That's the main thing that has to change. Attorneys have to take legal workers more seriously. The other aspect is the legal workers learning more, developing their skills. That's the rising aspect.

In practice, the Law Collective could not divide the labor completely evenly because the law prohibits nonattorneys from performing such tasks as appearing in court, giving legal advice, and visiting people in prison. All other tasks, however, were assigned without regard to professional certification. As members explained, "legal workers have to learn from doing." No regular classes were instituted to teach the newer legal workers the law, but occasionally one of the experienced attorneys would give a "shop talk" on one aspect of the law (e.g., on judicial procedures, sentencing procedures). In spite of this lack of systematic law seminars or courses, people at the Law Collective were very proud of the extent to which their legal workers, people with little or no previous law training, learned to fulfill most of the functions

commonly reserved for practicing attorneys. They happily reported the case of one legal worker who, after being a member of the Law Collective for only a month, wrote a writ of mandate to the California Supreme Court that succeeded in overturning local residency requirements for holding city office.

Personal learning through job rotation and team work is found in many innovative workplaces. In Sweden, some automobile plants use work teams and job rotation in the assembly of cars (Gyllenhammar, 1977). A unique auto repair collective in Washington, D.C., provides the means for members to teach themselves the skills of auto mechanics while repairing cars in the shop (*Syracuse Herald-American,* June 18, 1978, p. 13). Member-owners of a collective bakery in Oregon combine a number of tasks on the job (e.g., baking, bookkeeping, purchasing); at a collectivist construction firm in the same state, special efforts are made to teach building trades skills to women (Sandkull, 1982).

A particularly comprehensive and dramatic example of knowledge diffusion is found in the Mondragon worker cooperatives. For many years they have operated their own schools to train young people in technical skills they can later use in the cooperatives, and to teach cooperative values (Logan, 1981; Johnson and Whyte, 1977).

The diffusion of knowledge and skills is widely lauded and practiced in cooperatives because it is a precondition for the diffusion of influence. It is intriguing to note, however, that there are instances, too, in which knowledge diffusion comes about in more bureaucratic settings in an unplanned manner, and with no prior ideological commitment. This suggests that at least some of the features of collectivist organization, particularly the process of knowledge diffusion, may generalize to special subunits of service bureaucracies.[11]

Staff members of collectivist enterprises may try to demystify expertise not only for their worker-members, but for their clients as well. The Free Clinic, for instance, had a large number of "patient advocates," trained volunteers whose job it was to clarify medical knowledge to patients. The explicit aim was to maximize patients' knowledge about healing so that they could better care for themselves and thereby reduce their continued dependence on doctors. In the same vein, the Law Collective wrote:

> We view our clients as our brothers and sisters. We attempt to demystify the law so that when they come to us for help, they not only have their specific problems resolved, but they also learn something about the operation of the "system" and what they might do the next time a problem arises.

The Law Collective tried to implement this goal through encounters with individual clients and through efforts to help organize a "people's law school."[12]

The laywomen in the abortion collective (Bart, 1982) reportedly performed

some 11,000 successful (though illegal) abortions without the aid of physicians and with an excellent record for safety. After the members learned the necessary medical procedures, they dismissed the physicians. Bart argues that demystification helped the organization to develop and retain its egalitarian form and to provide a high-quality service.

Law or medicine, or any relevant base of knowledge, may be demystified to clients not only as a result of verbal efforts, but also by removing conventional symbols of authority from the organizational setting. For example, the Free Clinic – like most free health clinics in this respect – painted the walls vivid colors, brought noninstitutional furniture into the waiting room, oriented all of its personnel to convey a casual, first-name atmosphere, and replaced the standard white smock with blue jeans and the like for doctors, nurses, lab technicians, and patient advocates alike. These changes represent more than mere style; they have symbolic meaning. They remove important cultural symbols of authority from the setting in which alternative services are to be provided. The removal of culturally accepted props of authority signals to clients that the doctors and other medical personnel are human, and accordingly, that their professional judgments are open to question. The Milgram (1973) experiments on obedience suggest the potency of symbols of professional authority. In effect, the removal of symbols of authority may subtly encourage clients and members to respect the lessons of their own experience over the judgments of an authority figure.

In collectivist organizations that do not have clients, as the *Community News,* the demystification or diffusion of knowledge is directed at the worker-owners. Here the technology is relatively sophisticated and routine, and task sharing is the main method of skills diffusion.

At the *News,* job assignments often combine seemingly unrelated tasks such as 20 hours on editing, 10 on writing, and 10 on production. Such task distributions reflect collectivist principles, meaning that no one is stuck doing tedious work full-time, and no one is allowed to do choice work full-time. Jobs considered to be the most boring or undesirable, such as the production tasks of layout, pasteup, and so on are shared by many, from editors to advertising people. No one at the *News* argued that having people change their job activity practically every other day would be the most efficient way to put out a paper, but they did argue that task sharing was an equitable system that allowed all members to gain experience and knowledge in all aspects of the paper.

The sharing of knowledge may be further accomplished through job rotation. Job rotation at the *News* is planned and comprehensive. It arises out of principle. In reality, however, systematic job rotations are a difficult course to

follow. In planning, they are time-consuming, and in implementation, they present a great deal of change for the organization to absorb. Sometimes preparations for a coming rotation have to be laid well in advance, as when people are assigned to train for a particular job by apprenticing under its incumbent. Because job rotations require periodic retraining, they take a good deal of time. In many co-ops, the training of members in new jobs thus may be given inadequate attention, with the result that the performance of newly acquired tasks may be poor (Gamson and Levin, 1980: 30).

Plans for the first rotation at the *Community News* began months prior to its implementation, and discussion related to it (e.g., who should do what) absorbed countless hours of formal staff meeting time. This expenditure of time and energy is only comprehensible if one understands the extent to which members of the *News* prize democratic control. If we fail to identify this ultimate goal of job rotation, we miss its essential meaning to the members, and it may appear to consume time and energy far out of proportion to its organizational utility.

Job rotation at the *News* was not without its rough edges. The story of Ann and John, with which we began Chapter 1, is a case in point. Ann was a good photographer who greatly enjoyed her work at the *News*. She agreed in principle with the idea of job rotation, but didn't want to give up photography. She agonized over it, but could not think of any other job at the paper that she would want to do. Encouragement and appeals by others didn't seem to help resolve her quandary. Finally, after hours of meetings of the collective, she was assigned a new job in advertising. It didn't work out. Shortly after she started her new assignment, she unhappily confided in an interview:

> Rotation is a neat thing and I agree with the reasons we're committed to it and all that, but in practice I'm a casualty of the rotation system. . . . I can't go on much longer in the advertising section. Going to look for a waitressing job soon. I'm terrible at advertising, just terrible . . . I go into stores and ask managers to take out ads and try to act like it doesn't matter to me when they say no. No sooner than I'm out of the store, I start crying. . . . I know it's my turn to do other work besides writing, but I just don't know if I'm willing to go through six miserable months trying to sell ads.

Mismatched as Ann was in her new job, she did not question the legitimacy or the morality of the collective's assignment of her new tasks. And neither did anyone else, as reflected in this comment by one of Ann's closest friends at the *News:*

> People know [Ann's] unhappy in advertising, but she *did* offer to switch to advertising in the beginning and she hasn't suggested another job at the paper that she'd be willing to do. No one, not even her best friends could accept it if she came out and

said "Look, all I ever want to do at the paper is to write and do photography." A switch could be worked out if she would show some willingness to learn another task. But she can't ask for special privileges. . . . She has to be willing to rotate and to do the more fun jobs as well as some of the more tedious jobs, like everyone else.

Soon thereafter, Ann left the *Community News* for good. John, happy with his new job, remained as did the rest of the members. The staff of the *News* are willing to accept occasional casualties of the rotation system such as Ann because they see rotation as instrumental in the process of knowledge diffusion, and knowledge diffusion gets at the heart of collective control over the organization. Sophisticated technologies lend themselves to monopolization, and if such monopolies of knowledge are not diffused or demystified, members fear that democratic control will yield to oligarchic control. This fear is often implied in the justifications that members give for having rotation and task-sharing systems:

Expertise can become a hammer, a jealousy. If it is allowed to grow and grow, it can give some people an undue amount of power and influence in the group. . . . They can block other people from learning their expertise.

Part of the reason of having a collective is to grow and learn things. . . . People learn by doing. Rotation is part of the process of helping people to learn. The drawback to it is after a while people get very good at the tasks they've been doing. When you rotate tasks, it takes new people longer to do things and they may not be done as well as with the old people, at least until the new people grow into the new jobs.

Serious efforts, such as those at the *Community News,* to demystify knowledge also raise an important *dilemma* for the collectivist organization. Substantial diffusion of knowledge seems to entail a loss in organizational efficiency and productivity, at least in the short run. This by no means went unnoticed by staff at the *News:*

We know that ——— won't be as good a writer as ——— was, and that ——— won't be that good at selling ads at first, . . . but people get tired of what they're doing after a while. . . . You can't keep a person on a job as alienating as advertising forever. . . . We think that the long-term benefits of everyone understanding all aspects of the paper, and the kind of equality that comes from that, outweigh the short-run inefficiencies that are involved.

Clearly, people at the *News* realize that by collectively rotating tasks, some productivity might be lost. Nevertheless, all things considered, they are willing to trade off a certain measure of efficiency in order to try to make sure that knowledge about particular operations at the *News* will not be monopolized.

The Law Collective also takes pains to share or to rotate the most menial tasks, such as reception and cleanup. Like those at the *News,* members try to

avoid relying on bureaucratic criteria for dividing the labor and instead try to focus on who stands to learn the most from a particular case assignment. They assign cases to pairs rather than individuals so that members can educate each other in specific areas of law. Although they do pay serious attention to the demystification process, they tend, unlike the *Community News,* to gloss over the problems associated with it:

> Our biggest fear of an office without traditional divisions of labor was that work would not be done as quickly or as well as necessary. Since attorneys do their own typing, some individual pleadings take twice as long to prepare. But having non-attorneys in the office who also prepare pleadings helps to offset this time loss. The overall effect, we feel, is that everyone develops his or her legal skills, and as a group we become more effective and efficient than if the more menial jobs were left to the nonlawyers.

Whether or not an organization candidly faces the issues involved, a structural dilemma persists. Cooperatives employing relatively developed forms of technology must pay serious attention to sharing knowledge or risk eventual control by the experts. Sharing tasks, rotating jobs, creating apprenticeship systems and other means of demystifying knowledge enhance personal learning and perhaps member satisfaction, but time spent learning is time away from the production tasks of the organization. To the extent that tasks are distributed by criteria other than who is most experienced or talented for the job at hand, some measure of organizational productivity is sacrificed. This is a weighty dilemma for the cooperative that wishes to accomplish its tasks as expeditiously as possible while sustaining an egalitarian organization.

Nevertheless, this observation must be qualified on three counts. First, these organizations have a theoretic goal of universal competence of members in the tasks of the organization. If an organization actually approached universal competence, then it would not have to continue to invest large amounts of time and energy in the learning process, and it might be expected to be exceedingly flexible and productive. Second, in service-oriented organizations it is extremely difficult to gauge efficiency. It makes more sense to assess the *quality* of the services provided and of the decisions taken by the group. The benefits to morale of knowledge sharing may well improve the quality of services and decisions. Third, maximizing efficiency is, of course, not the most important goal to collectivist organizations.

In recent years, evidence has been mounting that small-scale, decentralized, participatory, and labor-intensive organizations may be just as productive and efficient – and by some criteria, more so – as large-scale, hierarchical, and capital-intensive modes of organization. This evidence calls into

question a widely accepted tenet of economics and organizations theory.[13] Our data unfortunately, do not permit us to add to this line of argument. In the organizations we studied, schemes that broke down the conventional patterns of differentiation did strengthen collective control, but they also entailed at least short-term losses in organizational efficiency. It is possible that these losses represent only the startup costs of any new arrangement and that they would subside in the long run. In the absence of longer-term observations, conclusions cannot be drawn. However, given the rapid turnover of personnel in collectivist organizations and the fact that rotation systems are not one-time occurrences but are repeated periodically, it is difficult to see how the "startup costs" could ever disappear.

All organizations – democratic ones notwithstanding – encounter a "free rider" problem, the tendency for individuals to avoid taking on added responsibilities or costs where the benefits of such action would accrue to everyone (Olson, 1971). Collectivist organizations may be especially prone to this problem precisely because their egalitarian structure means that individuals will receive few extra rewards for whatever extra work they may do.

Members of collectives seldom, if ever, speak in terms of a "free rider problem," but they do wrestle with the problem in more concrete terms, asking frequently what they should do about "people who don't carry their weight" in the organization. Calhoun (1980) maintains that to have a full flow of all information to everyone in an organization would be prohibitively expensive in terms of transaction costs, making democratic organizations impractical.

However, collectives may operate in such a manner as to challenge Calhoun's assumptions. Abell argues that members may *democratically* decide to permit a certain level of differentiation of information, a decision consistent with democratic principles (Abell, 1981). On the basis of empirical observation, Mansbridge (1980) notes that consensual democratic groups need not share all information and decisions when there is a commonality of interests. Our observations, too, indicate that it is not equally important that all kinds of information be shared, and members realize this. Thus, a partial solution to the free rider problem may be built into the very structure of consensus decision making. The time consumed by the decision-making process has costs, but in the process of giving input and reaching common understandings, loyalty to the group and commitment to carrying out its objectives are being built.

Data from this study generally support the Weberian notion that the decisive reason for the advance of bureaucracy has been its purely technical su-

periority, its efficiency vis-à-vis all other forms of organization in history. Weber states this proposition clearly:

> Precision, speed, unambiguity, knowledge of the files, continuity, discretion, unity, strict subordination, reduction of friction and of material and personal costs – these are raised to the optimum point in the strictly bureaucratic administration, and especially in its monocratic form. (Weber, 1946, p. 214)

For several decades, this proposition has been challenged by researchers in the Human Relations school of organizations. They have tried to demonstrate that more participatory organizations oriented to human relations stimulate greater worker satisfaction and thereby produce goods and services more efficiently.

In contrast, serious and thoroughgoing resistance to bureaucracy in collectives is based on other grounds. Although the Weberian adherent would defend bureaucracy as the most efficient and unambiguous mode of organization, the Human Relationist might retort, "No, more *satisfying* modes of organization are more productive and efficient." On the other hand, members of cooperatives might reply, "Yes, perhaps bureaucracies are more efficient, but who *wants* to maximize mere efficiency, precision, speed, continuity, and so on, anyway?" Where Human Relationists have attempted to challenge the *scientific* basis of the Weberian proposition, members of alternative institutions have generally challenged its *moral* basis.

Yet Weber might not have been so surprised had he seen the alternative, collectivist organizations develop in modern America. As he wrote: "[I]t is primarily the capitalist market economy which demands that the official business of the administration be discharged precisely, unambiguously, continuously, and with as much speed as possible" (Weber, 1946, p. 215). Since people in collectivist organizations generally have little interest in the success of the capitalist market economy, it follows that they would have low regard for its rationalistic basis.

Further, Weber might not have been altogether unprepared for a political strategy of building collectivist organizations. He recognized that parallel, autonomous, small-scale organizations, improbable as they seemed to him, were the *only* theoretical alternative to bureaucratic domination in modern society. This follows logically from his belief that once bureaucracy is firmly established, it makes a fundamental change in the structure of authority impossible, replacing it with mere changes in who controls the bureaucratic apparatus. Thus, from a Weberian perspective, the Marxist proletarian class strategy of seizing the capitalists' bureaucratic machinery is possible, but is not revolutionary. It is only a coup d'etat (Weber, 1968, pp. 987–989). The

later Marxian stage, "the withering away of the state," is revolutionary for Weber, but utopian. Thus, the creation of parallel, small-scale organizations alongside of bureaucratic monoliths is the only limited, but plausible, way "to escape the influence of the existing bureaucratic apparatus" to which Weber alludes (1968, p. 224).

The dialectics of demystification

The process of rationalization has long appeared, much as Weber described it, to be progressive. We wish to consider a counterhypothesis: that the process of rationalization is dialectical in nature, that it is inherently self-destructive.

Although the word *demystification* does not appear in many modern dictionaries, it has become a favorite and frequently used term in collectivist organizations. This word perhaps more than any other distinguishes the ethos of collectivist organizations from that of bureaucratic organizations. Demystification was defined earlier as the process whereby formerly exclusive, obscure, or esoteric bodies of knowledge are simplified, explicated, and made available to the membership at large. In its essence, demystification is the opposite of specialization and professionalization. Where experts and professionals seek licenses to hoard or at least get paid for their knowledge, collectivists would give it away. Central to their purpose is the breakdown of the division of labor and pretense of expertise. In effect, demystification reinforces egalitarian, democratic control over the organization, just as the subdivision of labor enhances managerial control over the workplace (Braverman, 1974).

In their everyday practices, people in collectives are insisting that much of what passes for expertise – not all – can be opened up and taught to any interested party, short-circuiting the usual years of training and certification. At first glance, they seem to be acting in a profoundly antirational manner. On another level, however, their efforts to extend knowledge to everyone suggests that they are taking rationality most seriously. The urge to demystify the world is at the core of rationalization (Weber, 1968; Whitehead, 1925).

A dialectical conception of the process of rationalization would lead us to the hypothesis that continued extension of the logic of rationality – the demand for the demystification of experts' knowledge – will produce the transcendence of specialized knowledge, the *ultimate* demystification.

In short form, this argument might go as follows. Weber was quite correct in seeing that rationality would entail the demystification of all domains of life. Therefore, to maximize formal rationality, bureaucracies would have to

be arranged hierarchically, with authority based on the technical expertise of the officeholder. Consistent with Weber's intent, this implies that the superior has come to a more demystified understanding of some relevant part of the world, and is therefore more technically competent than the subordinate. However, Weber did not follow the demystification process to its logical extreme.

The logical conclusion of the demystification process is equal knowledge: the complete diffusion of knowledge. That is, in the extreme, everyone would have the same demystified understanding of the world. There would be no need for "doctors" because everyone could doctor themselves, no need for "teachers" for everyone could teach themselves, no need for "sociologists" because everyone would possess the "sociological imagination," and so forth. The extension of the process of demystification to these *theoretical* extremes would undercut the very basis of rational authority, namely, superior knowledge. That is, if all members were equally competent in the knowledge and skills relevant to the operations of an organization, there would be no rational basis for hierarchical authority.

Although it is not plausible that the historical process of ever-expanding demystification of the world will go to such lengths as to completely eliminate the basis for the division of labor and functional specialization, it is important to note that this is the direction in which further demystification takes us. Thus, although it is standard fare in social science to predict ever-greater differentiation, specialization, and professionalization in modern society, there is reason to suspect that these processes cannot continue ad infinitum. As people come to perceive their work as being subdivided to the point of absurdity, they may recoil and begin to build bridges between subspecialties. Some may even choose the route embarked upon by the collectivist organizations described herein: demystification, functional generalization, and the diffusion of hitherto exclusive knowledge.

5. External conditions that facilitate collectivist–democratic organizations

The ability of a collectivist organization to achieve its participatory-democratic aspirations is conditioned not only by factors internal to the organization, but also by factors in its environment. Parallel to Chapter 4, this chapter argues that certain external conditions facilitate collectivist–democratic modes of organization, that their absence tends to undermine such forms, and that choices regarding these conditions generate important organizational dilemmas.

Oppositional services and values

Collectivist enterprises are usually created because their founders see some important social need that is unfilled by conventional businesses or public agencies. They perceive an opportunity to provide a social benefit at the same time they are creating an organization embodying their democratic ideals. Thus collectives often emerge from a two-pronged critique of mainstream society: a critique of the internal structure of rule-bound hierarchical organizations, and a critique of the failure of these organizations to meet social needs. Since the new collective tries to avoid direct competition with larger, more resource-rich organizations, it can hope to build its own market and niche. Alternative enterprises do best when they are able to ferret out a market that mainstream organizations cannot or will not enter because the product requires handmade or custom production, because public agencies fail to provide a required service, because mainstream businesses do not perceive the market for the alternative service, or because the service is illegal, as in the case of the abortion collective. Like any small business, collectivist firms must foresee potential markets and must pay the start-up costs of innovation. Thus, innovation and a burning sense of mission are often born of social criticism. *Both an oppositional stance vis-à-vis established institutions and the provision of qualitatively different products or services therefore tend to go hand in hand in collectivist enterprises. Both factors tend to unify and sustain democratic organization.*[1]

Our findings suggest that democratic organizations are likely to succumb to more merely cost-efficient modes of organization if they produce goods or services that are similar to or competitive with those produced by bureaucratic enterprises. The integrity of the collectivist organizational form is most readily maintained where the product or service of the organization is qualitatively different from that produced by dominant organizations.

All of the organizations in this study tried to provide services or goods that were different in quality from those provided by established institutions. The Free Clinic, for instance, was committed to preventive medicine, free medical care, understanding by patients of the healing process, and a wholistic approach to mental and physical health – principles that it assumed the "straight" medical delivery system did not share. Likewise, Freedom High assumed that its loose structure, its focus on learning outside the classroom, its attention to affective development, its highly critical perspective on social and economic institutions, and its measure of student control over the schooling process were anathema to the public school system. The *Community News* tried to select and present the news from a left perspective with the avowed purpose of liberalizing the local political climate. The Food Co-op sold food at cost, tried to carry only wholesome foods, educated its member-customers about nutrition, and supported other community-owned economic organizations, all of which distinguished it from privately owned, profit-based food stores. The Law Collective, in the cases it sought or avoided for social, political, or ethical reasons, defined itself as an alternative legal service. Similarly, the successful worker collectives studied by Sandkull (1982) all had found special market niches, providing quality goods or services not readily available elsewhere, such as a tofu factory, a natural foods bakery, a low-cost home construction and repair service, and a wholesale distributor of bulk natural foods.

Members' perceptions of being oppositional and therefore the target of outside harassment serve to solidify these groups and to justify their existence as "alternative institutions" in an otherwise bureaucratic society. This sense of opposition, buttressed by the perception of external harassment, is found generally in collectives.

An oppositional stance vis-à-vis established institutions does not permeate all cooperative organizations, however. In cases where worker-ownership has emerged as an attempt to save jobs in a plant closure situation, workers may feel grateful to the banks and government agencies for putting up the loan capital that allowed their new enterprise to get off the ground. This sense of positive identification with established institutions may inhibit, at least for a time, the development of a sense of group cohesion. Without a strong sense

of group cohesion, these new groups of worker-owners may not be motivated to challenge traditional managerial prerogatives or to insist on participation rights for themselves.[2]

In many other cases cooperatives have felt discriminated against by banks, and this was part of the justification for creating the National Consumer Cooperative Bank in 1980. For example, when Hubbard & Company, a utility pole hardware manufacturer, went out of business, its 100 workers sought to buy it. The banks that held the company's debt rejected the plan, reportedly because they felt worker ownership to be unworkable. In another case of a privately held firm that was closing down, Yellow Cab of Oakland, California, its workers found that their loan applications to purchase the company were refused, whereas the offer of a private buyer was accepted (Kepp, 1981, p. 30).

By and large collectives expect little support from established institutions, and in fact may be subjected to harassment. Freedom High had a multitude of fire and building code violations charged against its storefront location. Even in locations where code violations are commonly neglected, they were often carefully enforced against urban free schools (Kozol, 1972). In another case, when the local university student government voted to grant the Food Co-op start-up money, the university administration took the unprecedented stance of trying to disallow the expenditure. The co-op was required to demonstrate to the administration that the co-op would be taking business largely from a supermarket a few miles from campus and not from the small food stores in the community adjacent to campus. One co-op founder charged "the university administration didn't like the idea of the co-op underselling local capitalist grocers and maybe putting them out of business." True or not, the university chancellor was widely known to have personal economic interests in the local community.

Feminist health centers around the country have so often felt they were the targets of harassment by the medical establishment that they have created a national network of women's health centers, WATCH (Women Acting Together to Combat Harassment) to protect themselves. In a famous case, the founder of the Los Angeles Women's Health Center, Carol Downer, was arrested and charged with practicing medicine without a license. Her defense, based on a woman's right to examine her own body, was successful in court (Reinharz, 1983, p. 46).

Usually discrimination against collectives is more subtle. The Franchise Tax Board in California refused to grant the Food Co-op the tax-exempt status it requested. Although the co-op was able to show that no member or owner received any profit or inurement from the co-op, the Tax Board maintained

that the members received an indirect inurement in the form of lower food prices; hence the co-op was denied a nonprofit tax-exempt status. Board members of the co-op saw this as a form of blatant political harassment:

It [the letter of denial] was definitely written by a crew cut Bircher type who doesn't like co-ops competing with capitalist businesses.

The Food Co-op also had to pay unusually high prices to its wholesale food distributor. This is a problem faced by many food co-ops because of the relatively small volume of food they order (Giese, 1974; Zwerdling, 1975) and this is the reason co-ops are now developing in the wholesale area too. Members of the staff and board at the co-op strongly suspected that their particular wholesale distributor overcharged them because he objected to their politics. They also believed that since he knew they sold the food at cost he felt he could get by with charging them more. Similarly, workers at TRICOFIL, a worker-owned clothing manufacturer in Quebec, felt that certain large retail stores were boycotting their goods because the stores were politically hostile to worker co-ops.[3]

Whether or not charges of harassment are warranted, collectives often use opposition to mainstream institutions as a strategy. The collective's sense of mission and its internal unity are strengthened by pointing to the perceived reactionary, inept, or obsolete character of established institutions. Students and staff at Freedom High, for instance, often talked about the archaic systems and competitive values of the public school in order to positively justify the existence of their alternative. The Free Clinic did likewise with respect to the local health delivery system. Both groups ridiculed the most inadequate aspects of mainstream institutions, precisely because such aspects, used as a foil, made the alternative organization appear that much more progressive and needed.

As an illustration, the Free Clinic spent half of a staff meeting deriding "Concerned Parents," a Mormon-based group that opposed the continuation of all sex education classes in the local public school system. The Free Clinic in conjunction with Planned Parenthood had an extensive sex education speakers' program in the public high schools to protect. The ostensible purpose of the discussion at the Free Clinic's staff meeting was to choose someone to attend the Board of Education meeting who could extoll the merits of the clinic's sex education program. However, most of the time was spent simply burlesquing the antiquated attitudes of Concerned Parents. Concerned Parents made a good target for attack precisely because the issue they represented was so culturally symbolic. Lines could be clearly drawn. By focusing on this group, whether or not it represented a real threat to the clinic's sex

education program, staff at the clinic enhanced their sense of unity and mission.

This illustrates the general function that an oppositional stance serves in collectivist organizations. It strengthens the groups' sense of solidarity and purpose. In the extreme, an organization may be so oppositional in the services it provides as to be illegal, as in the case cited earlier of "Jane," the abortion collective. There is little doubt from Bart's observations (1981, pp. 17–19) of this group, that the confidentiality required by such a setup helped knit the group tightly together.

If collectivist–democratic organizations benefit from an oppositional stance vis-à-vis established institutions, then we should expect that *the introduction of reforms in the dominant, target institution, along the lines that the collectivist organization pioneered, would weaken the once-oppositional organization*.

This expectation is strongly supported by our research. For instance, during Freedom High's third year of operation, the liberalization of some of the local public school programs attracted many Freedom High students back to the public system and undercut the justification for having a free school. Likewise, at the elementary school level, once the local public school opened its own "alternative school" (consisting of three classrooms), the elementary free school folded. This has indeed been the fate of many free schools around the nation, as public schools have increasingly incorporated the reforms of the free school movement. This change may represent a success for the free school movement, if not for Freedom High itself, since it signifies that the oppositional organization has now achieved the acceptance of some of its social criticisms and the institutionalization of some of its reforms.

Rape crisis centers are another illustrative case. They started in the mid-1970s very much as grass-roots collectives, based on a harsh criticism of the lack of needed shelters and other supports for women who are victimized by violence. In addition to providing a service, the centers worked together, creating a network to rally support for legislative change. By 1980, rape crisis centers were mandated in the Mental Health System Act passed by Congress, and since then many states have passed laws to fund spouse abuse and rape centers (typically through a marriage license surcharge). This means that the centers, originally a self-help alternative service with a burning sense of mission and criticism of the mainstream's insensitivity to the problem, now have been fully integrated into the publicly funded Community Mental Health Centers, be that good or bad.[4]

The converse of this principle seems to operate as well. That is, collectives may flourish because of the character of the dominant organization they op-

pose. The *Community News* enjoyed a more rapid expansion of its circulation than it had projected at its inception. In three years the paper grew to a paid readership of 10,000 and a circulation of 22,000 – in a time when many community newspapers were declining and folding.[5] At least in part, this success was attributable to the very conservative cast of the dominant local newspaper, which gave the *News* the opportunity to reach the large group of moderate and liberal readers in the area. The dominant newspaper did not respond, leaving the *News* with its own audience and niche.

Again, it is clear that collectivist organizations thrive where they provide a product or a service that is qualitatively different from that provided by mainstream enterprises. They must fill a social need that is, for one reason or another, neglected by conventional businesses or public agencies. An oppositional stance reinforces the collective's sense of unity and mission, and the alternative products or services that it provides may secure for it a market niche. However, should the dominant target institution change to accommodate the alternative organization's purposes, then the once-oppositional organization may find itself without a niche or a rationale for existence.

This is a dilemma for collectivist organizations. They see an unfilled social need to which they want to respond. Often they would like to have some widespread effect on society. But if they reach a mass market with their qualitatively different product or service, that (now-proven) product or service may be taken over – and often watered down – by dominant organizations, undercutting the alternative organization's basis for existence. In this way, alternative organizations may sometimes have the ironic effect of strengthening dominant organizations by absorbing the risks and costs of experimentation and innovation for them.

Supportive professional base

The collectivist organization is facilitated too by having a supportive professional base in its community. The local environment most favorable to the development of participatory-democratic organizations would combine a vulnerable target institution, which leaves significant local needs unmet, with a large and supportive professional population.

The relationship between the collectivist organization and its relevant professional community can go awry in two directions.

First, rather than support, professionals may oppose or even actively harass a collective. A feminist women's health clinic in Florida found local doctors threatening to withhold hospital privileges, intimidating the physicians who worked for the collective, and filing unfounded complaints with the local

Board of Medical Examiners. The staff members of the clinic, suspecting both sexism and economic motives (the clinic charged much lower fees than private physicians), finally filed a federal court suit charging certain private physicians with conspiracy to restrain trade in violation of the Sherman Antitrust Act (see the *Poverty Law Report* by the Southern Poverty Law Center, summer 1976, p. 6).

Second, at the opposite extreme, professionals may smother the organization with overzealous support. At the Milkwood Cooperative in England, professionals and academics seized on the idea of setting up a cooperative that would hire and train young unemployed people to repair wooden pallets, utilize experienced carpenters to train the young people, and do it all in democratically managed cooperative form. To make this possible, they received a government grant. The outside professionals and university people founded the co-op, but in time they were supposed to fade away. In reality, they became the management committee for the co-op.

The result was organizational failure. The skilled carpenters were frustrated by the dull, unimaginative work, and felt that the co-op form robbed them of the authority they needed to effectively control and train the young workers. The trainees also hated the boring work and saw the cooperative form as an invitation to malinger. The experiment lasted a total of 19 months. One researcher, with wry British understatement, summed up the experience:

> Their experience, admittedly short, of working in a co-operative structure does not seem to have inspired those who worked in the enterprise with the philosophy of working co-operatively. It might be true to say that they regarded it as something of a joke, part of a game which academics play. . . . On leaving Milkwood the carpenters were unanimous in their condemnation. . . . As for furthering the cooperative cause, Milkwood cannot be said to have done the movement a service. (Rhoades, 1981, p. 36)

The overextended role of outside professionals was not the only reason for Milkwood's difficulties. Other factors included the heterogeneous work force, the dependence on outside funding, the conflicting goals of the various parties, and a dubious commitment to democracy on the part of participants. Yet Milkwood does clearly indicate the limits of professional support. Although outside professional have a valuable role to play in offering their services when requested, they cannot create co-ops out of whole cloth for others to work in. By their very nature, worker co-ops must be grass roots in origin, created by those who would work in them. Paternalistic ventures such as Milkwood, founded and managed by well-meaning outside professionals, represent a contradiction: By imposing their conception of "cooperation" on the enterprise, outsiders can only deny self-determination, the fundamental characteristic of organizational democracy.

The collectivist organizations we studied did not suffer from professional hostility, or overzealousness. They were fortunate to find in their community supportive professionals appropriate to their needs. The geographical location of these organizations, having a very attractive physical setting and climate, attracts more than its share of service workers of every profession. It is also the site of a university campus that adds greatly to the local professional population.

Professionals contributed to the maintenance of the collectivist organizations we studied in a variety of ways. For example, sympathetic professors set up a special course as a conduit through which university students could be channeled into community organizations for course credit. Without the steady supply from this course of well-educated volunteer teachers, Freedom High probably could not have existed for long. The Free Clinic recruited volunteer doctors mainly from the ranks of residents and marginally employed doctors in town. This feat would be more difficult in a city where doctors are encumbered, or blessed, by a higher ratio of patients per doctor. Another study of free clinics (Taylor, 1976) found the support of a liberal professional community important to the survival of free clinics. The Food Co-op got off the ground with seed money from a nearby university. Later, it was allowed the free use of the largest auditorium on campus for five consecutive academic quarters, and each use of the auditorium translated into a $500 to $2000 fund-raiser for the co-op. A sympathetic accountant did the Food Co-op's books for no fee at tax time. The *Community News* enlisted the free talents of several professors who wrote regular columns or special features.

In short, sympathetic professionals contribute to the development of collectivist organizations in a variety of direct and indirect ways.[6] Similarly, Stern and Hammer (1978) find that professionals who are willing to donate their time and services have helped many worker-owned firms get off the ground. A collectivist organization located in a town without a base of relevant professional support, such as a free school in a town without a surplus of teachers, or a free medical clinic in a town without many doctors, is likely to find existence more tenuous.

We have made the case that collectivist organizations are facilitated by a sense of opposition toward mainstream institutions and, simultaneously, they benefit from the support of professionals who are often employed in such mainstream organizations. Thus, the last two external conditions we have posed are paradoxically related. The need to maintain an oppositional stance toward established institutions, coupled with the need to attract professional support *from* established institutions, tends to put the cooperative organization in a bind. This may produce ambivalent attitudes and inconsistent organizational behaviors. Although these two needs place the alternative organi-

zation squarely on the horns of a dilemma, they are not, in the last analysis, irreconcilable.

Consider the case of the *Community News*. The conflict was experienced by members at the paper as a desire to be "radical" (thereby pleasing oneself and one's reference group) versus a desire to be "respectable" (thereby appealing to a broader constituency). To the extent that the *News* was seen as radical in form and content, members could find personal justification and a sense of purpose in their work, but to the extent that it was seen as respectable, members believed they would attract broader community support in the form of readership and advertising. Staff members at the paper felt genuine conflicts about this:

> I want the paper to be respectable. This will give us the wide circulation and success we want and will mean more money for staff. But I fear that too much respectableness could defeat our purpose of being an alternative. . . . There's a dilemma between being a daring alternative paper and having a wide constituency. I think we could be more daring.

Later in the same meeting, another staff member replied:

> We've done some misleading titles that slap adversaries in the face. We've been too daring at times. . . . We need to establish more contact and possible rapport with adversaries. We should try to educate people we write about and interview, not just view them as objects of exposure.

The quest for respectability was sometimes treated at the *News* as a mere source of cooptation. In reality, it was more than that. The paper's original ambition was to reach and alter the attitudes of people who did not consider themselves "radical" and who would not be likely to read an alternative newspaper. *News* staffers hoped to forge a political coalition of all shades and types of progressives, and to use the paper to mobilize such people around progressive local issues. Hence, the achievement of the paper's original social change goals required a certain measure of respectability in the community. The *News* was started as an instrument of change, and in order to reach the unconvinced, it had to appear legitimate.

The dilemma arises, we would argue, from the paper's dual need to maintain an oppositional stance toward established institutions while it tried to attract support, readership, and advertising from those institutions.

The way in which the *Community News* experienced this dilemma was, in some respects, unique among the collectivist organizations in this study. First, the paper provided a product, not a service. As such, it was not client based. Client-based organizations can cultivate, if they wish, a rather homogeneous clientele, but it is in the very nature of a newspaper – especially one with social change goals – that it must appeal to a relatively heterogeneous popu-

lation. Hence, the "respectability" issue would seem to be of more practical urgency for the *News*.

In a characteristic alternative service organization, such as the Free Clinic, the dilemma was reflected in the clinic's inconsistent posture toward its board of directors. In the early years of the Clinic, the staff sought out the most respectable, prestigious community leaders in the field of health for its board. Over time, board members came to question the less traditional services the clinic provided – they wanted to reduce the health education and counseling components – and to oppose the collectivist-egalitarian structure of the clinic, preferring instead a hierarchical structure with a director at the top. These differences led to a long series of battles between staff and board. Staff came to refer to certain members of the board as the "old guard," and to defend vehemently the oppositional practices of the clinic:

> We have a choice. We can do health care the board's way, and be no different than any public health agency and be guaranteed to last forever. Or we can stick to our vision and be a true alternative and probably not last that long. . . . What makes the [Free Clinic] different than any public health bureau is our patient advocates, our staff being a collective, health ed and peer counseling, preventive medicine, our lack of professionalism. That's what I'm here for. The old guard wants us to give up just the things that make this place different than any other place to work.

These struggles finally culminated in the "old guard's" resignation from the board of directors, an event that received considerable local media coverage.

Taking the long view, a former staff member of the clinic commented:

> There's no reason to have to lose those members of the board like this. It doesn't look good for the clinic and was totally unnecessary. They could have been phased into a Community Advisory Board and been innocuous there . . . but instead the staff let the old guard go along on the board, and then created a clash. They made those board members feel like their contributions were useless and unwanted for three years. That's not even true. There was a time when the staff really needed that board. . . . They made an important contribution to the clinic's acceptance in the community . . . made it seem more reliable. Now they create this clash and kick them off. It was unnecessary.

We would argue that the staff's inconsistent posture toward its board at the Free Clinic reflected essentially the same dilemma that the *Community News* faced. The clinic had to walk a tightrope. To some elements of its support base, particularly health professionals, government officials and foundations, it had to appear respectable and reliable. A prestigious board was instrumental in cultivating this image. To other elements of its support base, namely to paid staff, volunteers, and patients, it had to appear to be in opposition to mainstream medicine. It came to be extremely difficult, if not impossible, for the staff to maintain an image of struggling against "the system" when their

board consisted of shining representatives of that system. Conspicuously removing the "old guard" from the board helped to correct that imbalance, and served to reassert the oppositional quality of the clinic. A half-year later – and without those prestigious members of the board – the clinic found it could still impress the County Board of Supervisors, the County Health Department, and various private foundations enough to be trusted with grants. At the same time, it managed to convince patients and volunteers that it was radical enough to be trusted with alternative services. Indeed, in a survey study of this question, O'Sullivan (1977) finds that health collectives that are more tied to mainstream institutions do *not* attract more clients, but they are able to recruit more volunteers.

This same dilemma was reexpressed in numerous incidents at the Free Clinic. On one occasion, the Free Clinic was offered space in a new neighborhood medical building built by the County Health Department. Such a move offered the clinic an opportunity to leave its crowded, though colorful and homey, quarters for a much larger and improved medical facility. Some staff favored the move on the grounds that the new facility would make the clinic more respectable and would attract sorely needed physicians to work at the clinic, and its larger size would allow the clinic to serve more clients. Others among the staff argued that any addition in clients permitted by the larger size of the new facility would be more than offset by its barren quality and thus it would be indistinguishable from any conventional medical setting. Further, it was argued that because the new medical facility was across the street from a police station:

> women wanting abortions, people dropping reds, lots of people would be paranoid. Rational or not, lots of patients wouldn't come to the [Free Clinic]. They'd be afraid it was bugged or that they were being watched entering.

This situation put the clinic in a quandary. If the clinic moved into the new medical building, it would have greatly improved facilities, making it more appealing to professionals, but perhaps suspect to another necessary part of its support base – its patients and volunteers. Again we see a case in which the need of the Free Clinic to present an oppositional image (particularly to its clients, volunteers, and staff) was in conflict with its need to present a respectable image (particularly to its professional supporters). Eventually, the clinic settled the issue by deciding not to move to the new building.

In the above example the conflict was between the desires of clients, volunteers, and staff members on the one hand, and professional supporters on the other. In other instances, clients may line up on the side of the professionals, pushing for greater respectability and less "alternativeness." For instance, in its early days the Free Clinic tried to appeal to three client groups – blacks,

Chicanos, and youth culture. Staff soon learned, as one of the founders of the clinic explained:

> that these three groups do not interrelate too well, that we really should make a selection as to whom we were going to serve and go for that wholeheartedly. We selected the counterculture. . . . The counterculture group is very casual . . . barefoot, casual, wild colors . . . and the Chicano and the black have grown up with the image of medical care as being – if it's too casual it's second class, and they want the best because they think they've never gotten the best. They want the doctor in the white coat, everything spic and span, just right.

Thus, another clinic was opened later, one with white walls and so forth, and it attracted a tremendous minority clientele. This reserved for the Free Clinic the types of clients who would prefer its oppositional image.

When a collectivist organization is trying to broaden its constituency (of clients or customers), as the *Community News* did or the Free Clinic in its early days, customer demands may push it toward less alternativeness. For example, in her study of free clinics, Taylor (1976) found that black clients wanted "real" doctors and a clean modern clinic, and Katz (1975) discusses the same tension between alternativeness and minority community control in schools.

Not all collectivist organizations make outreach efforts to broaden the base of their clientele or customers. Freedom High appealed only to its "natural" constituency of affluent, white, hip students, as did the Free Clinic after its initial experimentation with a broader clientele. The members of the Food Co-op were also an entirely self-selected group. Thus, it appears that whether the clients of an alternative organization push it toward an oppositional stance or a harmonious stance vis-à-vis established institutions depends on the social base from which clients are drawn.

In sum, collectivist organizations are bolstered by an oppositional stance toward established institutions, but they also depend to some extent on professional support from established institutions. Sometimes these two needs are not in tension, as when the alternative attracts the support of professionals who are themselves critical of established institutions. Often, however, these two needs are in conflict. In such circumstances, the alternative must engage in a delicate balancing act, seeking creative solutions, and being careful not to ignore any element in its support base.

Social movement orientation

The preceding two conditions (opposition to target institutions, and support of professionals) refer to the relationship between the alternative organization

and its "straight" environment, that is, the dominant institutions and the persons who live and work in them. However, the collectivist organization may be surrounded too by many other social movement organizations. The way in which these alternative organizations relate to one another is often crucial, as is the nature of the connection between individual organizations and the larger social movement of which each is a part.

All of the organizations in this study may be classified as social movement organizations because they are oriented toward goals of social and personal change and because participation in them is motivated by values, friendships, and material incentives, in that order (Zald and Ash, 1966, p. 329).

We hypothesize that *the more a collectivist organization remains identified with and oriented toward the broader social movement that spawned it, the less likely it is to experience goal displacement.*

There is a tendency widely reported in the literature (Zald and Ash, 1966) for the founding generation of members of social movement organizations to be attached to the wider social movement, but for the second generation of members to be much more oriented toward the goals and services of only that particular organization. Earlier we described such a transition at Freedom High. The rise of an organizationally bound attitude among the new generation of staff members leads them to see their own futures as tied to the life and success of the organization, not to the movement. Hence, they become more likely to pursue organizational maintenance as an end in itself. Members narrow their sights toward providing a good service vis-à-vis other organizations in the same arena, but they may lose the larger vision of social-historical change out of which their organization was born.

This conservatism of organizational purpose, although not unusual, does not appear to be inevitable. Organizations can maintain a movement orientation into the second generation and beyond. The broader visions of the movement can provide an ideological anchor, enabling the organization to resist displacement of goals over time. At the *Community News* for example, the second generation of staff members continued to practice advocacy journalism and to press for local social change.

The organizations in this study all began with the aim of helping to create an entire "alternative community" or "cooperative sector." The possibility of building a mutually supportive network of community-controlled organizations depends on each group maintaining a movement orientation over an organizationally bound one. Members reflect their identification with "The Movement" not only by providing support services to its people and to new movement organizations, but also by dropping out of alternative organizations they see as no longer contributing to the broader goals of the movement,

and by joining other organizations that do. The latter phenomena often appear in case studies as instances of individuals getting "burned out," but our comparative study reveals a different picture. For example, many of the founders of Freedom High dropped out at the end of its second year of operation, convinced, as one of them put it, that:

> providing a groovy education to upper middle-class kids isn't the most revolutionary activity in the world.

Although at the time the "burned out" interpretation was used to account for this exodus, these same people show up later as committed members of the *Community News,* the Free Clinic, and the Law Collective. Their first allegiance was to "The Movement," not to any particular organization that currently housed them.[7]

Other researchers have observed a tendency for the personnel of social movement organizations to flow back and forth among various movement organizations, government agencies, and professional schools, all of which are devoted to a single set of policy issues (McCarthy and Zald, 1973). Our own data reinforce this observation. For example, a Freedom High staff member went on to study "confluent education" in a graduate school of education, and showed up later administering a publicly financed "open classroom" project. There were also numerous instances of staff in one movement organization who later became part of other movement organizations with seemingly disparate concerns, such as one person who switched from a free school, to an ecology action organization, to a free medical clinic.

The flow of personnel from one type of movement organization to another can be understood by developing a "movements-within-a-movement" perspective. That is, at least for some participants, the free school movement, the ecology movement, the free clinic movement, and the like are all considered subsidiary to "The Movement." For such people, these submovements are unified and assume importance only in relation to the broader social movement from which they sprang. If we employ a movements-within-a-movement perspective, we take seriously the words of some of the participants that they are part of "one struggle with many fronts." When collectivist organizations are viewed as entirely unconnected organizations, the rapid ebb and flow of personnel in and out of them is thought to reflect a fickleness of commitment, or burn-out. However, to begin to see them, as many of the participants do, as subsidiary movements within an overarching movement is to recognize a basic coherence and consistency in the actions of individual participants.

Alternative organizations give evidence of an identification with broader

movements in a variety of ways. Foremost, they usually provide services to other members of the "alternative community" and to other local movement organizations. People at the *Community News* sometimes sought medical attention at the Free Clinic, some of the volunteers at the Free Clinic bought their food at the Food Co-op, staff at the Law Collective sent their children to Freedom High, and so on. Since the cooperative community is an important reference group to most members of cooperative enterprises, members are often sensitive to the needs of individuals in that community. Some collectivist organizations go further. They try to provide direct support for other community-based alternative organizations. The Food Co-op, for instance, chose to deposit its assets from membership shares in a cooperatively owned credit union rather than in a privately owned bank. It likewise entered into a joint program with a nearby medical clinic to raise consciousness about nutrition. The *Community News* published a lot of feature stories on community-based organizations in town, thereby giving newly created collectivist organizations some public exposure and giving the public, or would-be clients, a chance to learn about alternative services.

The linkages at the community level between different types of collectivist enterprises do not stop with the support services they provide each other. They are also linked through shared boards of directors. As with corporations, such interlocks allow the heads (or in this case, the activists) of the constitutive organizations to be only a few "steps" from each other, thus easing communication, coordination, and collaborative action among them (Sonquist and Koenig, 1975; Whitt, 1982; Mariolis, 1975).

Consider the following illustration. One night, in the metropolitan area we studied, a group of community activists got together to launch an ad hoc ecology action organization. Its single purpose was to organize a large-scale national conference on energy. The "energy crisis" had recently become a national public event, and the conference was intended to raise public awareness of the politics of the oil-energy industry and of environmentally sound alternative sources of energy. The conference was a natural rallying point for the many environmentalist groups in town. However, bringing it off would take money – and time was short. A city council meeting was approaching in a few days and the newly formed group felt that it could win financial support from the city for the project if it could get the endorsements of respected, ongoing community and environmentalist organizations. The group brainstormed a list of 18 relevant organizations. The people present divided up the task of soliciting endorsements: "I'll call ———, he can give us the Sierra Club"; and so on. Finally the group seized on the name of a friend – Paul Q. Citizen – who, as it turned out, sat on 9 of the needed boards. One phone call

to Paul is all that it took to deliver 9 of the 18 targeted endorsements. The rest was downhill. Group 10, after all, could be assured that groups 1 through 9 had already endorsed the project, and so on. The new group got the needed endorsements, it got city funding, and the conference became a reality.

As this example suggests, the number of informal, interpersonal ties among community activists is typically quite impressive. Many activists are involved in a number of community organizations at the same time, and, as pointed out earlier, many flow from one collectivist organization to another over time. Thus, when the board of directors at the Food Co-op tried to change their weekly meeting night in order to accommodate a new member, they could not do it: Virtually every other night of the week some members were obliged to attend the regular meetings of other community organizations.

On a formal level, people such as Paul Q. Citizen serve on the boards of directors of a variety of different movement organizations. Such people become pivotal for movement activity in a given locale. Unlike corporate interlocks, movement interlocks are still in a nascent state; they are developing at the community or regional level, not as often at the national level; and of course they have only a tiny sliver of the resources that corporate boards have. Even so, movement interlocks, like corporate ones, have the overriding advantage of allowing seemingly diverse organizations the possibility of effectively coordinating and uniting their action. Just as interlocks among corporate leaders help to integrate the economy by putting general corporate interests over the interests of particular organizations (Useem, 1984), the growing network of interlocks among community movement organizations reflects, and helps to create, a general movement-orientation over an organizationally bound one.

Not all signs, however, point to the prevalence of a general movement orientation in collectivist organizations. People in such organizations may also have a strong identification with the particular submovement of which they are a part, such as free schools, food co-ops, free clinics, and so on.

During the year when field observations were made of the Food Co-op, the co-op aided the creation and development of four young food co-ops within a 100-mile radius; it sent a member more than 1,500 miles away to learn firsthand about the history and operation of a successful and long-standing consumer co-op in Canada; and it sent several representatives to the inaugural meeting of the Western Region of Co-ops. The birth of the Western Region of Co-ops in 1974, dedicated to seeking solutions to common co-op problems, reflected the proliferation of food co-ops that was taking place in the West. It had counterparts in other parts of the country, and in other institu-

tional domains, as in the free school movement and the free clinic movement. Freedom High, the *Community News,* and the Free Clinic also participated in the regional and national conferences of their respective submovements.

Of all these submovements, the free school movement was the earliest to develop. In 1967, free schools were just beginning: There were 30 in the United States. By 1973, more than 800 "outside-the-system" free schools existed, not counting the innumerable open classrooms and alternative schools within public school systems that were by then operating (*New Schools Exchange Newsletter,* 1967, 1973). As early as 1971, when there were some 350 documented free schools in the United States, 24 regional associations or networks had already been set up to act as clearinghouses of information and as placement services for interested students, parents, and teachers (Graubard, 1972). Many regional organizations produced newsletters and held conferences for free school people. Eight new periodicals of national scope were by then being published in North America, all aimed at criticizing the dominant educational system and working to create an alternative one. Freedom High staff were oriented to the larger free school movement, as was evidenced by their subscription to regional and national free school newsletters, by their attendance at regional free school conferences, and by their visits to other free schools.

The burgeoning of free clinics and feminist health clinics also turned into a well-developed submovement. The Haight-Ashbury Free Clinic in San Francisco was the first, having opened its doors in 1967. By 1971, there were 95 known free clinics in California alone.[8] Nationally, there were an estimated 340 free clinics by 1972. As with food cooperatives and free schools, free clinics created regional and national associations to help solve some of their shared problems. In 1970, the first regional association was founded in Southern California. As noted earlier, the rape crisis centers that proliferated in the mid- to late-1970s also created regional and national associations that lobbied for public monies and legislative reforms. In our own study, the Free Clinic sent representatives to the California regional conferences. The Free Clinic's identification with the free clinic movement was also demonstrated in the pride it took in rapidly filling up with its patient referrals two new neighborhood clinics in town.

The *Community News* and the Law Collective did not have such well-developed submovements with which they could identify, but in both cases such submovements were at least on the horizon. The *News* actively participated in the first national conference of alternative newspapers held in Boulder, Colorado, in 1975. At that meeting, attention began to be focused on the future possibility of sharing exceptional feature stories, articles, or columns

from alternative newspapers, and even of creating an alternative wire news service. The *Community News* traded subscriptions with a large number of alternative newspapers and magazines, and regularly reviewed the contents of the alternative press for items it might want to use. Although it is clear that the *News* was oriented to the alternative media, the alternative press was not yet as plentiful or as developed as a social movement as the previously discussed submovements.

The Law Collective retained an active membership in the National Lawyers' Guild, a reform-oriented organization of lawyers founded in the 1930s. Through the guild, it supported the People's Law School, "the only radical law school in the country" (*New York Times,* October 16, 1975, p. 40). It was in touch with many other legal collectives through the guild. Despite this impressive beginning, the legal field is yet to generate the number of legal collectives needed to sustain more submovement activity.

The collectivist organizations in this study all identified with their respective subsidiary movements, but they were also more generally oriented to an overarching movement. We have introduced a "movements-within-a-movement" perspective with which to understand this dual orientation. Empirically, though, commitment to these two sides may not be equal.

A general movement orientation, as discussed earlier, has the virtue of militating against the cooptation of goals. If, however, the members of a collective are preoccupied with general movement goals to the exclusion of particular submovement or organizational goals, then they will fail to accomplish organizational ends and will be left with an organizational shell of little utility to anyone. If, on the other hand, a strong organizational orientation precludes a general movement orientation, then the organization may provide a good service vis-à-vis other organizations in the same domain, but it will probably lose its sense of social-historical purpose. Organizational maintenance as an end in itself, with its conservatizing influence, will then become a more likely prospect. Thus, the collectivist organization is faced with yet another dilemma: It must strike a delicate balance between being oriented enough toward the general movement to maintain its original sense of purpose and meaning, while being oriented enough toward particular organizational goals to be able to provide a useful service to its clients, members, or customers.

This balance is not easily maintained, and some of the organizations in this study appeared to lose their equilibrium. Freedom High, for example, was started by staff who had an overriding commitment to the larger social change movement, but their dedication to education per se was thin. In the last analysis, few of them believed a liberating education alone to be an especially vital source of societal change.

Freedom High had been politically inspired, and it was able to run for at least two years on the enthusiasm of political currents external to the school. Such enthusiasm allowed staff at the school to largely neglect its pedagogical purposes. During the first two years of its operation, the activist students who had helped found the school reported enjoying it, but, even by the end of the first year, the chorus "the school isn't academically rigorous enough" became a familiar one. It was the students who were voicing this concern. Two formerly active students, in follow-up interviews about their past experiences, said they deeply valued their experience at Freedom High, but they also assessed its major weaknesses:

> I remember the fun of building a geodesic dome. Still it took a lot longer than it should have. And I enjoyed learning Chinese, for a while anyway. But the class only lasted a short while, and you can't learn Chinese in a few weeks no matter how smart you are. . . . Some of the bad . . . was the initial trauma of not learning much in the way of academics – and not being able to bring something home to my parents.

> The bad thing about [Freedom High] was that we never really got anything together. Most of the stuff we started we never finished.

Another student reported that students' needs were sometimes ignored for the social-political concerns of the staff.

In short, if staff were to effectively combine the larger movement goals that guided the creation of Freedom High with an organizational orientation, they would have had to believe that high school education makes a difference, that it is at least a sensible route to social change. They were not convinced. Hence, there was little commitment to developing pedagogical skills. The school expended a good deal of effort expressing its political values (in its relationship to its environment and in generating a very high level of internal participation and democracy) but very little effort on the substance of schooling – the curriculum. As a result, the school could not progress toward distinctly educational goals. When the original staff left (largely at the end of the second year and completely by the end of the third year) it was not because the school was not effective educationally (though, in the main, it was not), but because the school as an institution was not viewed as an effective tool for social change. As one departing staff member explained:

> We learned that it will be a very long time before this "new man" we of the revolution biz are always talking about can ever exist.

The first generation of staff members at Freedom High held allegiance primarily to "The Movement," not to Freedom High itself. After they left the school, some became committed members of the Law Collective, the Free Clinic, and the *Community News*. Others became involved in ecology action

organizations and in organizing industrial workers in a nearby city. If we adopt a movements-within-a-movement perspective, this sort of ebb and flow of personnel in and out of seemingly disparate movement organizations takes on a new coherence. Nevertheless, we also witnessed the painful consequences that an extreme movement orientation had for Freedom High as a school.

The case of Freedom High is unusual in this respect. Members of a collectivist organization are seldom so preoccupied with general movement goals that they neglect particular organizational goals. More common is the case of the Free Clinic, where an organizational orientation comes to overwhelm a movement orientation.

From the first day of field observations at the Free Clinic, evidence of an organizationally bound orientation was apparent. After an initial introduction, one staff member boasted that the Free Clinic was probably the best free medical clinic on the entire West Coast. Another staffer disagreed: "No, the —— Clinic up in Oregon is probably better." Although people at Freedom High would not dream of rating their school vis-à-vis other free schools, this was a highly relevant consideration to staff at the Free Clinic. Where Freedom High sought out black and Chicano activist groups and other community organizations for cooperative relationships, the Free Clinic tended to look upon other medical clinics as potential competitors in an informally recognized pecking order of free clinics. Where Freedom High, the Food Co-op, and the *Community News* aggressively aided the development of other community-based organizations, members of the Free Clinic did not even consider nonmedical collectivist organizations to be a salient part of their reference group, and they felt no special obligation to the rest of the local alternative community. For instance, when the clinic needed a printer to do its formal annual report, a number of possible printers were discussed. Finally, the *Community News*, which also does outside printing jobs, was chosen, but not as a statement of support for an allied movement organization, nor as a matter of reciprocity for the several favorable and timely articles that the paper had done on the clinic. The paper was chosen simply because it was considered to be the best printer in town for the money. No mention of movement politics was ever made in the course of group discussion at the clinic.

Allegiance to organizational (in this case, medical) goals to the exclusion of broader movement goals was characteristic of the Free Clinic. In consequence, it meant that the clinic was more open to cooptation of its original social change goals, as when the Medi-Cal pamphlet was removed from circulation at the behest of an influential county official. It also meant that the clinic was less sensitive to sustaining a participatory-democratic form; for

example, components meetings, the only direct input of volunteers into decision making, were suspended when some staff got too busy for them. As one staff member put it,

> I wish we *did* do things so political as to endanger our financial support. But the truth is, we don't.

On the positive side, however, commitment to organizational goals over movement goals helped to ensure that a high quality and variety of medical services were provided to the clients.

In other words, both extremes – the dominance of a general movement orientation as well as the dominance of an organizationally bound one – entail important trade-offs. Some collectivist organizations are able to resolve this dilemma. They do this by skillfully integrating the particular goals of their respective organizations with the more general goals of "The Movement." This is possible when members believe that the particular goods or services that their organizations can provide, however modest they may be in terms of quantity, are nonetheless a vital tool for societal change. In the case of the Law Collective, for instance, one staff member put it this way:

> Some of our early people weren't into legal work. . . . They left. All but one stayed with the movement though – in other ways, other places. Now we see doing good legal work as primary. Without it, we can't be effective politically.

In short, collectivist members must maintain an active concern for their organization if they are to provide a quality product or service, but they must maintain an identification with the general movement that first joined them if they are to contribute to its larger purposes. Some of the most successful cases of cooperative development have managed to maintain both. The Mondragon system has developed in a 25-year time span from a handful of people to more than 18,000 worker-owners in upwards of 80 cooperatives, at least partly because it found a way to respect the autonomy of the local co-op while allowing each co-op to gain the advantages of confederation. Using a similar balance, the Breman system of self-managed firms in Holland has grown from 2 enterprises with 150 workers to 12 firms with 600 workers in less than 10 years. At both Mondragon and Breman members are concerned with the well-being of their firms while also being concerned with the development of a cooperative sector (Logan, 1981; Johnson and Whyte, 1977; Rothschild-Whitt, 1981). In each, they have found practical ways to join hands with the other co-ops to buy supplies, obtain credit, develop markets, and train future workers. There is little question that confederations of this sort are to the mutual benefit of cooperatives, reflecting and helping to sustain their commitment to the goals of a larger movement.

Facilitating conditions: summary

Chapters 4 and 5 have presented, in propositional form, nine structural conditions that facilitate collectivist–democratic modes of organization. We have argued that the absence of each of them tends to undermine democratic forms, and each is shown to generate important organizational dilemmas.

Three further points are worth underscoring here. First, these are *not* definitional criteria, but *conditions* that facilitate democracy. They are not to be confused with necessary traits that would define organizational democracy (see Bernstein, 1976). Some or all of these facilitating conditions may not be present, and organizational democracy therefore suffers proportionally.

Second, many other outcomes or dependent variables could reasonably have been used instead of organizational democracy to assess these organizations, as, for example member satisfaction, organizational longevity, or quality of goods or services provided. We chose the level of organizational democracy because that is a central, defining characteristic of collectives, representing their most dearly held goal. Factors that promote a participatory-democratic form, such as a movement orientation, do not necessarily promote – and may even inhibit – the achievement of other possible organizational goals, such as the production of high-quality services or satisfied members. If the achievement of an egalitarian organization were their only priority, there would be little ambiguity, but collectivist organizations, like other organizations, often have multiple and conflicting goals. Under such circumstances, hard choices and dilemmas are inescapable.

Inescapable, but as we have also tried to show, not necessarily fatal for the collective or its goals. As is clear from our own case studies, and from the work of other researchers, some collectivist enterprises have been notably successful in building and retaining a democratic form while getting the job done very well.

Third, this list of conditions is not exhaustive. It is our hope that further studies of directly democratic organizations will uncover additional facilitating conditions. For instance, we did not examine the independent effect of democratic ideology because all of the organizations in this study contained many individuals who had strong democratic convictions. Other cooperative organizations might vary considerably on this score. We would certainly suspect that, as Bernstein (1976) and Gamson and Levin (1980) point out, variations in the level of democratic consciousness would have substantial bearing on the ability of an organization to function democratically. This issue may have special salience in employee-owned organizations where individuals have not previously developed participatory expectations and habits of

Table 5.1 *Summary of conditions and dilemmas*

Conditions	Effect of presence	Effect of absence	Organizational dilemmas
1. Provisional orientation	Militates against rigidification of rules and procedures, oligarchization, goal diffusion and organizational maintenance as an end-in-itself.	Permanence/careerist orientation promotes goal displacement (of all types), organizational maintenance, and oligarchization (especially when faced with member apathy).	If orientation is too transitory, then the organization may dissolve itself prematurely; the energy it took to build it will be needlessly wasted and useful social services ceased. If orientation is too permanent, then the organization may outlive its usefulness; it may displace its goals, rigidify its procedures and develop an oligarchy.
2. Mutual and self-criticism	As a regular and sanctioned process, criticism tends to level inequalities in influence and to curb individual assertions of power. It may also increase group morale and productivity.	Without a sanctioned process, criticisms tend to be illconsidered, totalistic, and unbound by any collective rules of fairness.	Ambivalence toward leadership: both need and resent informal leaders. Criticism levels undue influence and reasserts collective authority over individual authority, but "too much" risks losing the target of the criticism, often a valuable member.
3. Limits to size and internal growth	Deliberate limits to size and growth support familial and collectivist forms.	With unlimited growth, their personal and directly-democratic nature tends to be undone.	Large enough to obtain certain economies of scale (where relevant) but small enough to maintain collective control and sense of community.
4. Homogeneity	Common ethnic and/or cultural bonds foster group cohesion and ease consensus decision making.	Without the glue of a common background and values, groups may lack the level of trust and solidarity necessary for cooperation and consensual decisions.	The desire for homogeneity restricts the social base of the membership, making it less representative of the larger community, and in the extreme, the exclusion of "outsiders" may lead to the degeneration of the cooperative form.

5. Dependence on internal support base	Dependence on members and clients supports their participation and militates against goal displacement.	Dependence on external base of support (e.g., outside grants) tends to lower responsiveness to members and clients and to increase the likelihood of goal displacement and co-optation.	External financial support permits the organization to perform its services at less (or no) cost to clients, but it lowers participation of and responsiveness to members and clients and raises chance of goal co-optation.
6. Diffusion of knowledge and technology	If organizational technology is sophisticated then substantial knowledge diffusion is crucial to avoid monopolization of knowledge and oligarchization.	If technology is undeveloped and nonroutine, it does not lend itself to monopolization. But if technology is quite developed and concerted efforts to diffuse knowledge are not undertaken, then some individuals will be in a position to assert power and assume oligarchic control.	Where technology is developed, the processes of knowledge diffusion and demystification fortify collective control against oligarchization. But knowledge diffusion (e.g., job rotation, task sharing) may entail sacrifices in organizational efficiency and productivity.
7. Oppositional services and values	Heighten sense of purpose and solidifies group. Justifies collectivist forms and resistance to cost-efficiency criteria.	Undercuts sense of purpose. If mainstream institution is similar, undercuts justification for having an "alternative."	If stance toward established institutions is "too oppositional," then cannot get needed professional support from them. If stance is not oppositional enough, it may lose the allegiance of its members, volunteers, staff, or clients.
8. Supportive professional base	Channels human and financial resources from established organizations to alternatives; an indirect subsidy.	Lack of needed professional staff and volunteers. Loss of broad community respectability and support.	As above, most appeal to different elements in its support base. An oppositional image tends to garner the trust of volunteers, staff, clients, and members, but a respectable image tends to enlist the trust of relevant professionals.

Table 5.1 *Summary of conditions and dilemmas (cont.)*

Conditions	Effect of presence	Effect of absence	Organizational dilemmas
9. Social movement orientation	Lessens the likelihood of goal displacement, especially cooptation of goals.	Organization boundedness increases likeliness of organizational maintenance for its own sake, lowers levels of participation, and loss of original social change goals.	If oriented only to the general movement it may fail to accomplish organizational ends (i.e., providing worthwhile goods or services). If oriented only to the particular organization, it may lose its sense of social–historical purpose (conservatism of goals). Balance needed: particularistic service goals of the organization must be integrated with the general social change goals of the movement.

behavior. Zwerdling (1976) observed lack of experience with democracy as a problem at International Group Plans, a Washington, D.C., worker-run insurance company; and researchers in the New Systems of Work and Participation Program at Cornell University have also reported it at the various employee-owned workplaces they studied. Echoing the beliefs of the classical theorists of participatory democracy, the findings suggest that education for participation may indeed be an important part of any effective democratization process.

For the readers' convenience, Table 5.1 summarizes the past two chapters. It adumbrates the organizational effects of each of the proposed conditions, as well as the probable effects of their absence. The dilemmas they raise for collectivist organizations are also listed in abbreviated form.

Part III

The significance of democratic organizations

6. Democracy and individual satisfaction

Anarchist writer Murray Bookchin once began a speech:

> You may have changes in the economy, you may have changes in who rules, you may have changes in who rules what and in who rules who, but there is no revolution without freedom, and there is no freedom without individuals controlling the conditions of their lives.[1]

Although Bookchin was not speaking of collectivist organizations per se, this quotation distills the essence of the meaning of collectivist organizations. They represent one part of a larger drive to recreate human-scale, decentralized institutions in the community and in the workplace. To the extent that they democratize organizations, they return the locus of control to the individual.

The concern of collectives with participation and democratic control makes them highly unusual as economic organizations. Profit ultimately justifies the existence of most economic organizations in a capitalist society. Corporations may speak of corporate responsibility, they may donate to philanthropic causes, and show concern for the morale of their employees, but in the end, success is judged on the basis of profit and growth. Although cooperative organizations must generate enough surplus to pay their worker-members a livable income, profit maximization is not the main point.

Benchmarks for assessing the success of capitalist organizations are well-established but for alternative organizations they are ambiguous and multifarious. Collectivist enterprises assess themselves in terms of how well they are practicing their democratic ideals, the quality of the products or services they are providing, their ability to provide alternative places of employment, the satisfaction of their members, and – most ambiguous of all – their contribution to larger societal change. The exact priority given to these different goals varies among collectivist organizations and within a single organization over time. This, of course, makes an evaluation of their success difficult. At a minimum, alternative benchmarks must be used.

So far, this work has focused on what many collective members would no doubt regard as the most central benchmark: the creation and maintenance of

organizational democracy. We have tried to identify structural factors that improve the chances for internal democracy. Accordingly, we have focused on the organization as our unit of analysis. In this final part of the book, however, we depart from the organizational level of analysis to look both above and below that level. We look above the organization to assess the future of collectives in general (Chapter 7) and to raise larger issues concerning the implications and social significance of democratic organizations (Chapters 7 and 8). We begin Part III by looking below the level of the organization to the level of the individual member, asking what meanings and consequences collectives hold for their members.

Democracy and worker expectations

The democratic process is valued by members of grass-roots cooperatives for its presumed connection to human happiness. People come to these organizations expecting the collectivist form to bring more autonomy, satisfaction, and meaning to their work. In this chapter we attempt to assess the extent to which collective work settings do in fact reduce work alienation and contribute to satisfaction.

Studies of work satisfaction constitute an enormous literature, and an area of continuing debate. Most of these studies have focused on either (1) occupational status (or prestige), (2) income, or (3) personal factors (e.g., values, sex, age, motivation, education) as major determinants of job satisfaction (e.g., Hall and Nougaim, 1968; Vroom, 1964). Other studies have looked at the effects of participation on job satisfaction (Blumberg, 1973; Frieden, 1980). Yet the *structural* attributes of *jobs* (e.g., the amount of control or autonomy workers have in the work situation, how interesting the job is) remain a relatively underdeveloped area of research. In a pathbreaking study, Kalleberg and Griffin (1978) demonstrate that the structural facts of *class* (i.e., whether incumbents own the means of production, control the labor power of others, or sell their own labor power) and *occupational position* (the job's placement in the technical division of labor, i.e., specific tasks performed, etc.) carry considerable power to explain variations in work satisfaction. Using a national labor force sample, Kalleberg and Griffin (1978) analyze the influence of class and occupational position on job satisfaction. They conclude that:

> while occupational status exerts a net influence . . . class appears to be relatively more important as a source of inequality in job satisfaction. These results support . . . Marxian arguments about the psychological "costs" associated with membership in the working class. (Kalleberg and Griffin, 1978, p. 386)

The researchers find that both the working class and the petty bourgeoisie[2] have significantly lower job satisfaction than any other class. Workers are less satisfied because they receive relatively low financial rewards and because they work in jobs that are less intrinsically rewarding (i.e., that offer less control over the product and process of work and are less interesting). Workers would register even lower levels of satisfaction if it were not for the fact that, in these jobs that provide little control and interest, workers tend to scale down what they expect from the jobs, and lower expectations mean less active dissatisfaction (Kalleberg and Griffin, 1978, pp. 389–390). The petty bourgeoisie expect more, and are relatively low in satisfaction because:

> [they] both attain and value intrinsic rewards to a greater extent than do workers, and such a valuation enhances satisfaction for this group. . . . They obtain less income than employers and top managers, however, and this depresses their level of job satisfaction. Overall, then, the petty bourgeoisie appear to balance their levels of rewards and values and report levels of satisfaction comparable to those observed for the working class. (Kalleberg and Griffin, 1978, p. 391)

No systematic studies of work satisfaction exist for cooperatives or collectives. The above findings do contain, however, several implications for what levels of work satisfaction we should expect to find in them. We argue that the structural dimension of work is particularly relevant to an assessment of work in cooperatives. As we detailed in Chapter 3, special structural features define the cooperative form, and also are the prime reason members are attracted to the democratic organization in the first place. Moreover – as we will shortly see – class factors are the appropriate dimension for assessing Marxian expectations concerning alienation.

The class position of co-op members does not fit neatly into the classification of Kalleberg and Griffin, nor into any other with which we are acquainted. Members own the means of production, they do not sell their labor power to others, and they do not directly control the work of others. They are thus similar to the petty bourgeoisie. Yet, there is a crucial difference. They are not individual entrepreneurs, but members of a *collectivity* having influence over the work of all members. Their position thus has elements of the small employer, too. Consequently, we maintain that worker-owners occupy an ambiguous class position, different from both the petty bourgeoisie and small employers because of the unique collective element. No studies have been done on the consequences of the emergent, special class position occupied by cooperators or worker-owners, a promising area for future research.

Given the structural similarities between cooperators and the petty bourgeoisie and between cooperators and small employers, we would suppose, on the one hand, that cooperative members would experience similar levels of

satisfaction. On the other hand, there are two important complicating factors that make prediction difficult. First, as a result of their values and the self-selection process, cooperators are likely to have extremely high job expectations, certainly higher than conventional workers and probably higher than the petty bourgeoisie. This would depress actual satisfaction levels. Second, collectivism in the workplace is a unique element in cooperators' experience. According to the limited research that now exists, this element should have a positive effect on satisfaction (see Blumberg, 1973; Frieden, 1980).

We do not have the data to systematically test these ideas, but we can provide some suggestive evidence. We begin not with cooperatives but with related organizations that hold out the promise of ownership and greater voice to workers.

The worker-owned firms that have arisen since 1975 using the Employee Stock Ownership Plan (ESOP) provisions in statutes of that year illustrate our point. Many of these firms emerged as urgent and often last-ditch efforts to save jobs in the face of plant shutdowns. Other types of transfers of ownership to workers have taken place at the initiative of a retiring owner of a closely held firm, who preferred, for a variety of reasons, to sell through an ESOP to his employees rather than find and sell to an outside buyer. Some ESOPs extend minority ownership to workers, whereas others extend majority ownership. In all of these types of cases, though, ESOPs represent *conversions* to worker ownership from conventional ownership, and for that reason, even in cases where they are majority worker-owned, they are very different from grass-roots collectives. The collectives are started from scratch, and are thus in a position to gather up worker–owner–members who specifically believe in the democratic process, whereas ESOP conversions to worker ownership contain the employees who happened to work in the previous conventional firm. By examining ESOP firms, therefore, we can see how the coming of worker ownership alone – without any special, preexisting participatory ideologies – affects workers' job satisfaction.

Researchers at the New Systems of Work and Participation Program at Cornell University have studied many cases of worker buy-outs of firms that otherwise would have been shut down. At the inception of many of these firms, workers cooperate closely with management and community leaders in an attempt to save their firm and their job. During this start-up phase, workers and leaders convince themselves that it is worth the risk to buy the firm, an *espirit de corps* develops among participants, and hopes run high. Workers, when asked, say they expect that ownership will bring them a greater say in the firm. Management expects that ownership will induce workers to comply more effectively with their directives. Neither side is likely

to consider any concrete changes in the style of management or methods of decision making that would bring organizational procedures in line with the new form of ownership. The old ways continue to be followed. Attention is given instead to the more pressing matters of finance, marketing, production, and so forth. As a result, several months into the new venture, the spirit of cooperation wanes. Denied an avenue for greater voice in company affairs, workers begin to feel that their ownership is without meaning. In cases where workers have held a majority ownership position, but where they do not gain effective representation and their known preferences are explicitly contradicted by management as in the case of the Vermont Asbestos Group, they are likely to become frustrated and sell their shares to private investors.

Thus, in such buy-out situations, the anticipation of the coming ownership appears to inculcate in workers a sense of "rights" to democratic participation.[3] If the newly formed worker-owned firm does not deliver any substantial participation opportunities, these raised expectations then are dashed. The transformation to worker ownership does not match workers' initial high hopes. They still have no control. In the end, this may leave workers feeling worse than before about their workplace and may render industrial relations more tense.

We do not mean to suggest that worker ownership in the absence of a high degree of worker participation is of no social benefit. It may still save or create jobs. And ongoing research indicates that workers in these sorts of ESOPs do report a greater degree of interest in the financial success of the firm, they are glad to have the ESOP, they do plan to stay with the firm longer, and they do report more pride in their work.[4] However, on measures of job satisfaction, perceived influence, or better industrial relations, the responses appear more equivocal.

Taken together, the many recent case studies of worker-owned (ESOP) firms strongly indicate that the less participation accompanies ownership, the less effective is ownership by itself in improving the quality of people's work lives. Studies by Hammer and Stern (1980) and by Long (1979) show that there is no significant difference between owners and nonowners with regard to the preferred distribution of power in the firm. In fact, according to Hammer and Stern, the more shares workers own, the more they are willing to defer to managerial decision making. French and Rosenstein (1981) in a survey of another ESOP firm find that, among blue-collar and clerical workers, the number of shares owned is insignificant in predicting the desire for influence. These results are based on firms in which workers did have voting rights, profit sharing, and rights to certain information. In other ESOP firms, where workers do not have rights to voting, profits, or information, and where

workers were not involved in the initiation of the ESOP, ownership may not only fail to improve job satisfaction, it may actually diminish it. At Crosby Valve, workers came to see little in the ownership plan besides a bonus at retirement. One study concludes that the objective limitations on participation at Crosby caused the workers to become "cynical about ownership rather than causing them to desire or demand any right to control" (Kruse, 1981, p. 115).

Not all ESOPs are without worker participation. Many firms have developed substantial avenues for workers to participate in decision making, and in these cases we anticipate that the combination of ownership and influence will have salutary effects upon workers.

To summarize, worker ownership by itself does not seem able to generate added worker allegiance or satisfaction. In certain circumstances, ownership by itself does appear to raise workers' expectations that they will have more say, or will be treated with greater respect, but in instances where it does not deliver on this promise, it disappoints. Its fuller potential is realized only when ownership is coupled with added opportunities for employee involvement in decision making and control.

This empirical finding underscores the importance of worker participation, but it does not imply that ownership is unimportant. "Quality of Work Life" (QWL) projects try to develop employee participation in the firm, while not extending ownership to workers. Their record reveals numerous cases in which the QWL project was curtailed when management decided that the program had gone far enough in the sharing of power. It then becomes evident to all concerned that any participation program that is called into being by management can be called to a halt by management.

This is not true in situations in which workers are owners of the firm. Worker ownership provides the *legal foundation* and ultimately the *motivation* for worker participation. It provides the legal foundation because voting and other participation rights can be tied to equity ownership, and then managers cannot unilaterally terminate participation. It is a motivator for the workers because they now have a claim to any productivity gains deriving from participation.

Grass-roots collectives, in contrast to ESOP and QWL firms, have a fundamental commitment to democratic participation. Though they are often cooperatively owned, democratic *control* is the defining characteristic, not a mere adjunct to ownership. Because they are started from scratch, co-ops attract and recruit people who want specifically to build democratic workplaces. Thus self-selection is a powerful filter, and co-ops do not contain the broad cross section of the working population that ESOPs have. Also, because they are created as new organizations, they can be developed at a delib-

erate pace, not in extremis as organizations that need to preserve jobs in a community. Where ESOPs often reproduce earlier styles of management and methods of decision making, co-ops inherit little from the past and are free to experiment with new forms of organization.

Less alienation, more stress

With both worker ownership and democratic control, can co-ops offer a more satisfying work life? From a Marxian standpoint, work in cooperatives should be more satisfying than in capitalist enterprises. Marx argued that under capitalism workers are not able to control either the process or the product of their labor. Since they must work as hired labor, in a workplace owned and controlled by others, their labor power is not freely given, they are unable to influence the disposition (i.e., selling, trading, storing, giving away) of the final product, and their labor merely serves to enrich an opposing class. Under such conditions, work is inherently exploitative, coercive, and alienating for the worker. In Marx's words, from *The Economic and Philosophical Manuscripts:*

> His work is not voluntary but imposed, *Forced labour.* It is not the satisfaction of a need, but only a *means* for satisfying other needs. Its alien character is clearly shown by the fact that as soon as there is no physical or other compulsion it is avoided like the plague. Finally, the alienated character of work for the worker appears in the fact that it is not his work but work for someone else, that in work he does not belong to himself but to another person. . . . It is another's activity, and a loss of his own spontaneity. (Quoted in Bottomore, 1964, pp. 169–170)

As an example, in the never-ending search for profits, capitalists impose a rigid division of labor upon the workers:

> [A]s soon as the division of labour begins, each man has a particular, exclusive sphere of activity, which is forced upon him and from which he cannot escape. He is a hunter, a fisherman, a shepherd, or a critical critic, and must remain so if he does not want to lose his means of livelihood. (*The German Ideology,* from Bottomore, 1964, p. 97)

Marx's essential thesis was that a communist society would free workers from this condition of exploitation, alienation, and rigidified work roles. Surveying the achievements of workers' cooperatives in the mid-nineteenth century, Marx wrote:

> By deed, instead of by argument, they [worker co-ops] have shown that production . . . may be carried on without the existence of a class of masters employing a class of hands . . . [and that] hired labor is but a transitory and inferior form, destined to disappear before associated labor plying its toil with a willing hand, a ready mind, and a joyous heart. (Quoted in Avineri, 1968, pp. 179–180)

It follows then, from a Marxian perspective, that where workers *do* have control over the conditions of their labor, as in a collective, work should be freely motivated, less objectively alienating, and more subjectively satisfying.

Our observations of the contemporary cooperatives, however, lend only partial support to the Marxian expectations. We argue that although collectivist workplaces *do* lower alienation, providing engaging and meaningful work to their worker-owners, expectations also are much higher, and the picture is rather mixed in terms of overall satisfaction.

Following extensive field observation in each of our collectives, written surveys were conducted in three of the groups: the Food Co-op, the Free Clinic, and the *Community News*. In contrast to the findings of Kalleberg and Griffin (1978), occupational position (as indexed by organizational membership) appeared to have no effect here. The three organizations differed little in overall response patterns, so all data have been aggregated. Staff members at each of the organizations were asked, "In general, how satisfying do you find your work at ——— to be?" Of those responding, 38 percent declared it to be "the most satisfying thing I've ever done;" 46 percent indicated that it was "very satisfying;" for 16 percent it was "fairly satisfying;" and none reported "very little satisfaction" or "not at all satisfying." Although we do not have directly comparable data, these responses seem to indicate very high levels of satisfaction compared with responses in conventional organizations (HEW, 1973) and certainly much higher than for the categories of workers and petty bourgeoisie reported by Kalleberg and Griffin (1978). The collectivist structure of the organization may account for this greater level of satisfaction among cooperators compared with that of their nearest class equivalents.

A more detailed question in our survey tried to pinpoint the organizational features with which co-op members were most or least satisfied. Of 21 listed dimensions, the ranking listed in Table 6.1 emerged (in descending order).

Table 6.1 indicates that the organizational qualities with which virtually all members are satisfied (75 percent or more) concern the overall mission and service of the organization, the autonomy it offers, and the defining attributes of the collectivist form, such as equality and collective ownership. Similarly, in the Mondragon cooperatives in Spain, surveys of worker-members show that members rate autonomy more highly than money (Logan, 1981, p. 17). Attributes of work life with which relatively few members are satisfied (50 percent or less) relate to the organization's efficiency, job pressures, money, and job security – the very things that bureaucracies are considered good at providing. Virtually no one (8 percent) feels that the operations of the organization go smoothly. This indicates that members' subjective impressions of

Table 6.1 *Satisfaction of co-op members*

Dimension	Percentage satisfied with it
Feeling of doing something worthwhile	92
Collective ownership of the workplace	92
Freedom from supervision	90
Chance to help individuals	88
Accounting to coworkers	83
Equality with coworkers	82
Appreciation from clients	82
Changing the community	73
Recognition from coworkers	73
Learning for future career	70
Opportunity to be creative	67
Opportunity to do own thing	64
Room for spontaneity	64
Sense of community	64
Chance to change society	54
Money	50
Job security	50
Freedom from high pressure	42
Efficient decision making	42
Provide high-quality product or service	33
Smooth operations	8

what their collectives are good at or not good at are consistent with the structural features and constraints of collectivist-democratic organizations we set forth in Chapter 3. *Here psychology accurately reflects organizational structure.*

One source of dissatisfaction is harder to interpret. Members gave a low ranking to the quality of goods and services provided by their organizations. Only one-third of the members were satisfied with the quality of the product or service. As our impression is that all organizations were in fact producing high quality services, this statement may simply reflect the high, near-perfectionist standards of respondents, or it might reflect members' frustration about providing quality services in what they may feel is an inefficient manner.

Other studies also have discovered positive psychological benefits of col-

lectivism. In one of the few social psychological studies of a collectivist organization, Schlesinger and Bart (1982) find that the women in "Jane" reported a deep sense of satisfaction with what they were accomplishing. The researchers attribute this to the broadened skills and competencies that developed as a result of the multifaceted jobs in the organization. Self-confidence went up and people developed a desire to learn more and to participate more. The extensive work by Menachem Rosner on the democratic general assemblies and industrial plants that make up the kibbutzim system in Israel demonstrates that a taste of participation does lead to the desire for more, and thus supports the notions of the classical theorists of participatory democracy: J. J. Rousseau, J. S. Mill and G. D. H. Cole. Having studied the directly democratic assemblies both in the community and in the workplace of the kibbutz, Rosner is able to separate the effects of participation in the community general assembly from participation in the plant assemblies. Overall, Leviatan and Rosner (1980) find that mental health and reduced alienation are more related to involvement in the *community* assembly than in the plant assembly. Rosner (1981, pp. 17–25) finds that attendance at the community assembly meetings correlates positively with members' sense of commitment, satisfaction, and influence. In the plant assembly, on the other hand, satisfaction and influence correlate negatively with attendance at the meetings. In other words, those who come to community meetings are satisfied; those who come to plant meetings are not. This is because members feel that the kibbutz community assembly has considerable influence on outcomes important to them, whereas the plant assembly is not perceived to have enough influence. This study underscores the psychological benefits of participatory democracy at the community level.

Another way to appreciate the effect of cooperative organizations on the individual is to compare, within a society, similar organizations with different ownership forms. There are only three types of ownership available in our society: private, governmental, or cooperative. Garbage collection is one of the very few industries in the United States in which examples of all three can be found. In a study of private, municipal, and cooperative garbage collection firms, researchers were able to compare the effects of each form of ownership. Overall, the cooperative firms, by allowing their worker-owners to feel like "partners" rather than "garbagemen," and by permitting them to pursue many business tasks (such as collecting accounts in addition to picking up garbage) bring pride of ownership and a sense of dignity to this usually stigmatized job (Perry, 1978). Russell, Hochner, and Perry (1979, p. 339) find that worker-owners show a statistically higher level of satisfaction than non-owning workers in the cooperative firms.

Not all research findings are favorable to workplace democracy. The garbage collection co-ops were started more than 50 years ago and today contain both worker-owners and nonmembers who have been hired to help out. Comparing the attitudes of these two groups, Hochner (1981) finds that those members of the cooperatives who perceive their organizations to be democratic are no more satisfied than members who do not. On the other hand, hired help who see their workplaces as more democratic are more satisfied with their work. Moreover, adding to job complexity has little effect on worker-owner's job satisfaction, but job complexity and higher pay do boost the hired workers' job satisfaction.

Another anomalous set of findings comes from Yugoslavia, where Obradovic (1970) finds that members of the workers' councils in self-managed firms are more alienated than nonmembers. In the United States, worker ownership has also brought some surprising findings regarding the individual. Although one might expect worker ownership to raise commitment to the firm and thereby reduce absenteeism, for example, the latter has not happened. In one employee-owned firm studied by Hammer, Landau, and Stern (1981), the number of "voluntary" absences (elective time off) went down following the conversion to employee ownership, but the number of "involuntary" absences (due to illness, for example) went *up*. Overall, the absenteeism rate did not change. Finally, Norwegian social scientist Frederik Engelstad, (personal communication), comparing a family-owned newspaper with a collectively owned paper in Norway, concludes that the collective did raise commitment and lower alienation among its members, but that it also raised their stress levels and the rate of eventual "burn-out."

How can one make sense of the mixed picture that these findings present?

People in collectives, at least in the United States, are usually very different from their counterparts in private firms or government agencies. First, collective members are a self-selected group who have sought out and have taken great pains to develop a very unconventional form of ownership and control in their work organizations. The boldness of this move suggests that they have an unusually high sense of efficacy. Such a sense of mastery, or belief in one's own ability to effectively control one's life, was evident in all of the five collectives we studied. Of those members who answered two items on the survey intended to measure sense of personal efficacy, more than 80 percent reported they "strongly agree" or "agree." Second, as noted, members of collectives want and expect a great deal from their work. Most of them have come to these organizations for idealistic reasons. They have a strong sense of mission, they yearn for work that is worthwhile, autonomous, and integrated in its execution. In short, they are, and expect to remain, highly involved in

their work. *All* of our respondents answered "fairly great" or "very great" to an item on our questionnaire, "How big an effect would you say that your involvement in ——— has on your life at the moment?" In addition to lofty standards for work institutions, people in collectives tend to be critical of society and of the institutions that comprise it. It is out of this critical spirit that their desire to change society comes.

For all of these reasons, we conclude that member–worker–owners of collectives are not easily pleased. The great expectations they bring to their work (and the atypical inducements to which they are responding) tend to counteract somewhat the overall – but still very high – levels of satisfaction they apparently derive from their work. Therefore one must be careful in interpreting satisfaction statements of cooperators and in comparing them to similar statements from conventional organizations.

Worker expectations help us better understand the puzzling findings of Hochner and Russell. The hired (non-owning) workers in the garbage collection co-ops showed a boost in their satisfaction levels when their workplaces allowed for more democracy, more multifaceted jobs, or more pay. This was not true for the worker-owners. It is likely that for the hired help (for whom this is a conventional wage-labor exchange) very little is *expected* of a garbage collection job, and thus if they get to attend a few meetings with management or receive more interesting job duties or more pay, then they are happy.

For the worker-owners, expectations are higher, and so added increments of these benefits become background factors that are accepted but do not add to satisfaction (Hochner, 1981). Similarly, Obradovic's finding that members of the workers' councils in Yugoslavia are more alienated than nonmembers may be a case of unfulfilled hopes.

High expectations and the sense of mission in collectives may lead to more intense, engaging work, but engagement exacts a price: stress. Because work in collectives is freely chosen, and because it is relatively autonomous and equitable, we, like Marx, would expect members to show little alienation. However, contra-Marx, we are led to conclude that the very features that define and give meaning to collectivist work also generate personal stress.[5]

Three structural features of collectivist organization in particular cause personal stress: collective authority, de-differentiation, and familial interpersonal relations. First, collective authority often leaves nebulous the question of what is the appropriate exercise of individual judgment. An individual may need to make an immediate job-related judgment, but feel unsure whether it is rightly an individual decision or a group decision that must wait for the next

group meeting. In bureaucracies, because authority resides in individual positions, rights and responsibilities can be more clearly delineated than in collectives. Thus, the individual in a collective may feel uncertain as to whether a decision should be delayed until the next group meeting, or acted upon, and thus run the risk of group criticism for overstepping unspoken bounds of personal authority.

Second, the effort to reduce specialization in collectives and give people a broader array of jobs to do introduces an additional form of role ambiguity. Instead of having one job, the individual may be expected to do several different tasks. The person may not know what the priority is at any given time. If one is part of a team that has been assigned to a certain task, then it may not be clear which team members are to do what. In a bureaucracy where specialization is respected and jobs are narrow, everyone (in theory) knows what they are supposed to be doing, and expectations are limited. This may be boring, but role expectations are at least clear.

The price of the stimulation and freedom implied in broad job roles is that jobs are more taxing and the organization is less predictable. Blurry, ill-defined jobs tend to make people anxious: There is always more one *could* be doing, and what one should be doing is not necessarily clear. Gamson (1981) has described the same problem in the Ann Arbor collectives she studied. The strong sense of purpose among participants only makes this problem worse. Because members are working for a cause, they are driven to take on an overload of work, leading to the familiar phenomenon of "burn-out."

This explains why Engelstad (personal communication) discovered that although the collectively owned publishing house in Norway provided less alienating work, it raised the levels of stress. Swidler finds the same pattern in her study of collectively run free schools. Because teachers did not hold legal-rational authority, they had to draw on their own reserves of charm and skills to engage the students: "[T]hey must fuel their teaching with their private lives. This process is exhausting, and the more successful teachers are at it, the more worn out they become" (Swidler, 1979, p. 71).

The demystification of knowledge and job rotation in collectives give members an opportunity to widen their competencies, and this builds self-esteem. On the other side of the picture, though, job rotations sometimes produce a mismatch – as in the case of Ann, the photographer – in which a person does not want, or is not adept at, the new job.

The desire for demystification in these groups makes illegitimate most claims to special expertise. Those who are, or could be, excellent in a job have no incentive to highly develop their skills, because they are unlikely to

receive extra rewards for outstanding work and they might even be criticized as "elitist." There is thus an incentive to downplay one's skills. In the words of one woman who had been a member of a collectivist band:

> At the beginning even though the skill levels were hopelessly unequal, the fact that you thought there was a commitment to learning things made everything seem possible. . . . It later began to be apparent that there was not this commitment; instead there was "militant amateurism." . . . In the case of the [alternative music band] one woman in the movement seriously advanced the idea that we should not put our names on the record because that would separate us from all our sisters who couldn't be on the record and weren't playing in bands. For a long time we never played on a stage because that would elevate us symbolically above The People. (Wenig, 1982/1983, p. 18)

Third, the small size and egalitarian nature of these groups and the close, personal relationships that knit them together, while aiding satisfaction, also ironically contribute to stress. Having peers in a position where they are obligated to evaluate the collective's activity means that job performance will be judged not by a boss (who in a bureaucracy one can shrug off as a fool or worse), but by people whom one knows well and cares about. This makes people work harder, but family-style performance appraisals are more anxiety laden. As we have seen in earlier examples, criticism from peers carries more sting.

A self-selection factor, too, operates in cooperatives: Co-ops often attract idealists, people who demand a strong sense of purpose from their work. Such people are probably also more prone to guilt than most. This disposes them to overburden themselves with extra responsibilities and tasks. Coupled with the aforementioned structural factors, the individual self-selection factor therefore compounds stress. In short, having a cause one believes in makes work more meaningful, and building a cooperative with close friends provides social satisfaction and a sense of community, but these same things raise the anxiety associated with possible failure and with performance evaluations. Extending ownership and control in the workplace integrates work into people's lives, making work more central to their psyches. People care more, and they take on more work. They experience stress, impairing overall levels of satisfaction. We argue that this is not a failure of particular collectivist organizations or of the individuals in them. Rather, it is part and parcel of the way collectives do things. It is the consequence of collective authority, equality, personal relations, and de-differentiation.

Collectives appear to have a difficult time dealing with this dilemma. Recognizing the unusual intensity of collectivist work, some collectives try to implement practical ways to give members an occasional respite from these

stresses. The Free Clinic, for example, initiated "mental health days," giving members the right to choose a certain number of days off explicitly to renew themselves. That workers need a certain amount of time off is illustrated by one of the worker buy-out cases studied at Cornell. There, as noted, researchers found that following the conversion to employee ownership, the overall absenteeism rate did not change, but the rate of "voluntary" absenteeism went down, whereas "involuntary" absenteeism increased (Hammer, Landau, and Stern, 1981). It seems workers change the excuses they give for being absent. For example, they say they are ill rather than they want time off. This alteration in the justifications people give for being absent may well be a response to changes in the informal culture of the workplace. With worker ownership come new group norms and attitudes that support maximum effort and allegiance. In this context, workers may feel compelled to come up with better excuses for their absences, absences that remain important to them. The data from this study then suggest that the extension of ownership to workers may not reduce the need for time off, and if our argument regarding stress is correct, it may actually increase the need.

As we have seen, collectivists expect a high level of both personal fulfillment and social utility from their work. They construct organizations designed to achieve these ends, and in practice they seem to succeed fairly well. There are positive psychological benefits deriving from collectivism, such as a generally strong sense of work satisfaction, broadened competencies, and raised self-esteem. With them, however, comes an undesirable side-effect: personal stress. Individuals will, of course, react in a variety of ways to this stress, but a substantial level of stress does appear to be endemic to collectivist organizations. Our observations lead us to suppose that in order to capture the other benefits of cooperation, many people in cooperatives decide that it is a cost worth bearing.

7. The future of cooperation

The question remains: How will cooperatives fare in the United States, and how significant a sector of the economy will they become? Will they remain relatively marginal, deviant forms of organization, rising and falling in waves as they have in earlier periods of American history? Or does the current burgeoning of cooperatives signal the beginning of a larger and more permanent cooperative sector?

We believe the future of the cooperative form of organization will depend on three main factors: whether they can be competitive in a market economy, and if so, in what specific niches; what the average life span of individual cooperatives turns out to be; and what role the state takes vis-à-vis cooperative enterprises.

Economic performance

The definitive study of the economic performance of cooperatives has yet to be done. There have been, however, a few economic studies of more limited scope that bear on this question.

The issue of how to properly measure economic performance is complex. In ordinary capitalist firms, profitability is usually used as the yardstick of success. When this yardstick is applied in the context of producer cooperatives, the results can be misleading for two reasons. First, members of cooperatives can manipulate the profit level down or up by deciding to pay themselves higher or lower wages. Abell and Mahoney (1981) have therefore argued that residual profit bears no necessary relationship to the economic viability of a cooperative enterprise, and urge instead the use of value-added per unit of labor as a better measure of economic performance. Second, although grass-roots collectives strive to earn enough surplus to provide each worker-member with a livable income, profit maximization is not their primary aim. Thus, profit becomes merely a limiting factor, with some level being necessary for survival. To evaluate collectives solely on the basis of profit would be to use the wrong criterion.

Nevertheless, the ability of a collective to generate an economic surplus by having revenues exceed costs by at least a moderate amount is important to the viability of the organization and to the attainment of its members' individual and collective goals. The issue of economic performance therefore is relevant to the future of collectives.

Some writers have argued that cooperative firms will perform worse than their privately owned counterparts in a capitalist society. Vanek (1977), an economist of self-managed firms, contends that cooperatives rely heavily on internally generated savings, that they give limited returns to share capital, and that they show a preference for consumption. The net result, argues Vanek, is that cooperatives typically will underinvest and will therefore have an unreasonably low capital-to-labor ratio. Empirical studies of worker-owned firms in countries such as Sweden do reveal a tendency to put additional earnings into higher wages rather than investment (Lindkvist and Svensson, 1981). On the other hand, studies of cooperatives in developing countries such as India, Peru, and Senegal do not support the Vanek hypothesis of high consumption and low savings (Abell and Mahoney, 1981). Moreover, in the Spanish Mondragon cooperatives, profits are credited to individual worker's accounts in a system-wide cooperative bank, leaving these collective savings to be plowed back into the co-op firms. Far from being cash-starved, Mondragon is able to retain some 85 percent of the profits for reinvestment (Johnson and Whyte, 1977). The Vanek thesis of overconsumption and underinvestment does not appear inescapable.

Others argue that co-ops, operating in a capitalist context, suffer discrimination, which impairs their ability to obtain bank credit, enter new markets, bring about advantageous relationships with suppliers, and so forth.

Co-ops do suffer from outside prejudice against the cooperative form, and there are many corroborating anecdotal accounts. However, systematic documentation of economic discrimination does not exist and would be difficult to demonstrate. If a bank declines to advance a loan to a co-op, it might be a case of pure prejudice, or it might be simply the bank's assessment of the perceived financial risk. It is no doubt easier for large, established businesses to obtain loans and to penetrate new markets, and it is possible that co-ops suffer only the same handicaps as small businesses in general. This issue is not clear.

Although we agree that external prejudice on political or ideological grounds does exist, its extent may be overstated by both theorists and participants in the field. As discussed in Chapter 5, participants in co-ops achieve a heightened sense of worth and unity from seeing themselves in opposition to mainstream institutions. Since they receive a secondary gain from the per-

ception of external slights and from the interpretation that these are prejudicial in nature, we might expect a bias in this direction. As an explanation for economic problems, we maintain that this factor is overshadowed by others. Since some evidence suggests that capital starvation is not the main reason for co-op failures (Abell and Mahoney, 1981, pp. 13–14), the bank prejudice hypothesis should recede in importance.

Of greater consequence, we suggest, are the structural biases contained in the patterns of economic concentration and in the laws of taxation and incorporation. Of importance too, is the lack of active support of co-ops on the part of the state (to be discussed later). For people outside co-ops, lack of knowledge and indifference may be a more ubiquitous problem than active hostility toward co-ops. For example, very few attorneys know anything about the rather obscure body of co-op law, and therefore co-ops often are greatly disadvantaged in finding relevant legal forms, instituting effective accounting procedures, and claiming appropriate tax entitlements.

The Webbs (1920) in an early and influential critique of cooperatives, took the view that co-ops were destined to underperform in comparison with their capitalist counterparts, not so much because of underinvestment or a hostile environment, as ineffective management. In the Webbs' view, the one-person-one-vote principle encumbers good management, slowing down decision-making processes and blocking technological innovation in the organization.

The Webbs' argument, although it has been echoed by many since, has not been subjected to empirical testing. We question, for example, the argument's implicit assumption that "good management" in a privately owned hierarchical firm is the same as "good management" in a democratic cooperative. If the characteristics outlined in Chapter 3 are in fact desired and typical in collectives and anathema in pyramidal bureaucracies, then there is no reason to expect that such a transference is possible. To the extent that the values, goals, and processes of the organization are different, principles and procedures that would constitute good management should be different. For this reason, claims that cooperatives entail "bad management" require careful scrutiny. Further, the Webbs' argument was published in 1920 when management entailed less specialized functions than it does today. The subsequent specialization of management functions makes it important to be more precise as to which functions of management might be deficient in co-ops and which might be strong.

Few systematic studies have compared the management outcomes of co-ops and those of conventional enterprises (certainly an important subject for future research). One exception is the work of Jones (1980). In a study of British footwear cooperatives, he finds that they *are* able to introduce technological change. In fact, the more participatory they are, the more techno-

logical innovations they introduce. Jones's data are confined mostly to the shoe industry, however, and he uses a number of indirect measures of relevant variables.

Our own observations convince us that co-op members are generally quite good at personnel management – related skills such as group process and human relations. They are also fairly perceptive and judicious in the choice of personnel who will be dedicated and hard working. Letting go of people when circumstances warrant it (in recessionary times or when an individual does not work out) is a difficult problem for many cooperatives. This is probably an inherent problem of co-ops. As a legal entity, the co-op form allows workers to become owner-members. Thus termination is usually reluctantly done and emotionally painful in these family-like groups. Collectives sometimes choose to build in future flexibility in the labor force by not extending ownership to new members, but of course this dilutes the co-op form and produces a two-tiered system of owners and hired help. Alternatively, members may decide in a recessionary period to cut time and pay for everyone by a certain percentage rather than lay off anyone. In either case, because of their broader ownership structure, co-ops are faced with personnel management issues that are thornier than those in ordinary firms.

In terms of financial management, Abell and others have found worse returns to capital in co-ops. Abell (1981) had argued this is the single most fundamental problem of co-op management. Partly, this may be a matter of skills that may be learned, since the members of the co-ops Abell studied (in the Third World) were barely numerate, much less literate. In response to a similarly perceived failing in co-ops in the United States, the New School for Democratic Management established a national program to teach financial and general management skills to cooperators.

Another set of skills, related to the above but analytically separable, are entrepreneurial in nature. It may be that the disadvantages of cooperatives with respect to returns to capital derive from a lack of entrepreneurial talent. This could be a built-in disadvantage of cooperation. As Abell and Mahoney (1981, p. 19) put it:

> Consider a potential entrepreneur (assumed self-interested) with either his own capital or access to loan capital and what he believes to be a marketable idea. Why should he choose to establish an IPC [Industrial Producer Co-op] when he faces (a) limited returns to his capital, (b) no guarantee of control of the enterprise, and (c) a situation where the benefits of his idea become a bounded public good within the cooperative? Surely he will rather be attracted to a partnership or traditional private firm.

In short, entrepreneurs in a capitalist society have little incentive to share with others the fruits of their ideas or control over the production process. We

believe, with Abell and Mahoney, it is this lack of attractiveness for individual entrepreneurs that contributes to the relatively low historical frequency of cooperative formation.

Some present-day cooperatives come about because the head of a private firm wants to transform it into a cooperative or democratic enterprise for essentially moral or religious reasons.[1] Other instances of worker ownership come out of the urgency of communities that stand to lose jobs because of local plant shutdowns. In these cases, some form of cooperation or employee-ownership may be tried as the solution of last resort. On the other hand, grassroots cooperatives tend to be created spontaneously by groups of people who are homogeneous in values and who bring relatively equal levels of capital and skills to the enterprise, as in our cases. Through the development of a co-op, as in the examples of the plywood and garbage collection enterprises, members can syndicate the financial risk, pull together the labor and the skills of many, build a community, and earn a livelihood. Conspicuously absent from the history of cooperatives is the solo entrepreneur who, armed with a fresh idea and the capital to make it work, gathers up a group of people to implement the idea and makes them into a co-op. Co-ops provide no incentives for the *individual* entrepreneur.

Workers' ownership and cooperation can bring certain economic advantages, particularly enhanced worker commitment. Worker stakes in the ownership of the firm and worker voice in decision making do appear to promote workers' sense of allegiance and commitment to the firm and to strengthen bonds of solidarity among the work force. For instance, in the early stages of worker buy-outs of failing firms, workers – drawing from their years of observation and experience on the shop floor – often come forward with constructive suggestions for cutting down on waste and improving quality. The move to worker ownership also causes a change in the culture of the factory. Informal norms disapproving of waste and voluntary absence begin to take root. Turnover declines, as might be expected, when employees have a financial stake in the firm. Under such conditions there is less need for supervision, and the costs of surveillance decrease.[2] The result of these processes is higher labor productivity.

Closely related to improvement in worker commitment is a strengthening of the bonds that hold the group together and contribute to its morale. In some instances, as in the case of threatened job loss, solidarity comes from the perception of a common enemy and a common circumstance. Similarly, labor-management conflicts such as lockouts or strikes sometimes precipitate the creation of a workers' cooperative. For example, the *Citizen's Voice,* a co-op newspaper in Wilkes-Barre, Pennsylvania, arose out of a conflict with

the *Times-Leader* and its new conglomerate owner, Capital Cities Communication, Inc. It was this conflict with Capital Cities that helped to unite an otherwise diverse group of workers represented by four separate unions (Keil, 1982). In other instances, workers may be bound together by ethnic and kinship ties, as was the case in many of the nineteenth-century co-ops and in the garbage collection and plywood co-ops of the 1920s. Sometimes, too, cohesion derives from the common religious, cultural, or value background of the participants, as in the case of the Israeli kibbutzim, the Mondragon co-ops, some of the co-ops Abell studied in Senegal, the Breman enterprises in Holland, and others. In many cases, of course, solidarity arises from the combination of these elements. In the contemporary collectives that are the subject of this work, members are bound by common values – though they are not religious or ethnic in origin – and by their oppositional stance toward dominant institutional arrangements.

Worker solidarity, like commitment, is of significance beyond the gains in worker satisfaction and morale that it may bring. One research team has found in its study of cooperatives in developing countries that high solidarity goes with various measures of economic success, just as low solidarity goes with economic failure (Abell and Mahoney, 1981, p. 14). This team posits that cooperatives rely on their solidarity and commitment advantages to achieve their economic performance; if these are lacking, the result is more diseconomies than in a conventional enterprise. As is apparent from the organizational features outlined in Chapter 3, a collective orientation depends on mutual trust. Internal conflict is especially disruptive precisely because of the consensual basis and personal relations that characterize these groups. Thus, compared with conventional firms, higher levels of worker commitment and solidarity are often observed in cooperative enterprises – but by the same token, they are also more necessary.

Research concerning the economic performance of worker-owned firms has begun to accumulate, and much of it shows good performance. Studies of worker-owned firms that have emerged out of impending plant closures reveal a number of cases – such as Bates Fabrics in Lewiston, Maine, and Saratoga Knitting Mills in Saratoga, New York – in which corporate earnings after the conversion are far higher than before. A survey of 30 worker-owned firms found, when controlling for size and industry, that worker-owned firms show a level of profit higher than similar conventional firms. Further, the most important single determinant of profitability is the percentage of the company's equity owned by the nonsalaried employees. As this percentage goes up, so do the company's profits (ISR, 1977, pp. 2–3).[3] Another study currently in progress by Tannenbaum and associates indicates that employee-

owned firms are neither more nor less profitable than conventional firms, but the employee-owned firms do stay in business longer (reported in Rosen et al., 1986, p. 2).

Except in the case of ESOPs, few studies have examined the economic performance of worker cooperatives. For those co-ops that have been studied, the results look generally positive. The kibbutz industries, probably the most studied cooperative system in the world, show high levels of profit per worker, high labor productivity, and a healthy return on capital investment compared with that of privately held companies in Israel (Melman, 1971, pp. 203–220; Barkai, 1977). Looking at Italian producer cooperatives, Jones (1981) finds that workers' capital stakes promote productivity in some industries, but hurt it in others. In comparing British footwear cooperatives and conventional firms in the footwear industry, he finds no difference in productivity. When examining American producer cooperatives historically, Jones finds some industries (such as plywood) in which labor productivity surpasses that of conventional enterprises, and other industries (such as barrel making) in which the cooperatives have performed worse (1980, pp. 144–145). Overall, the picture presented by Jones is a mixed one, in which the relative performance of worker co-ops depends on the industry in which they are located.

Virtually no economic studies have considered contemporary grass-roots co-ops. Much information does exist, however, on some of the cooperatives that were established during the 1920s and are still in operation today, notably the plywood and garbage collection co-ops. In these, an impressive record of performance has been demonstrated.

A study of the 15 plywood cooperatives in the Pacific Northwest finds that they average 20 to 30 percent higher productivity than conventional plywood firms (Berman, 1967, p. 189). Commensurate with this productivity, salaries in these worker-owned firms are 25 percent higher than those in privately owned plywood firms. When the Internal Revenue Service challenged some of the cooperatives for paying salaries higher than the industry average and sought to tax the excess as corporate income, the cooperatives successfully demonstrated that the higher wages were justified on the basis of productivity 25 to 60 percent greater than the plywood industry average.

The refuse collection cooperatives in San Francisco have consistently outperformed their counterparts. They provide higher-quality services at lower prices than the refuse collection firms of any other major city in the United States. At the same time, worker-owners in the cooperatives are paid an average annual income that is far higher than that paid to employees of the private refuse collection firms, which in turn is higher than wages paid to municipal garbage collection workers (Russell, 1982a). Workers in the pri-

vately owned firms are paid an average of 25 percent more than in the municipal firms, and worker-owners in the cooperatives receive 83 percent more.

In sum, a systematic study that controls for firm size, industry, and internal structure is needed before we can draw firm conclusions about the economic performance of co-ops. In the light of the available evidence, we are led to provisionally conclude that worker ownership and democratic management bring effectively higher levels of worker commitment and solidarity, and this often can be turned into a labor productivity and profitability advantage. But this economic advantage is precarious in cases where mechanisms are not established to give workers more voice in company affairs. Instances of worker ownership without avenues for democratic participation run the risk of losing the morale and therefore the economic advantage that would otherwise be inherent in this organizational form. This danger can be seen in the declining economic fortunes of several employee-owned firms (such as South Bend Lathe and the Library Bureau in Herkimer, New York) that appear to have paid insufficient attention to the participation implications of broadened ownership. In an overview of the economic research on producer cooperatives, Jones (1980, p. 147) concludes that co-ops without democratic majority control perform worse economically than those with democratic control. Likewise, Olivarius finds in a survey of 400 producer cooperatives in the United Kingdom a strong correlation between economic vitality and the degree to which decision-making procedures are democratic.[4]

The state

The degree to which the governmental machinery of the state fosters the development of cooperatives varies greatly across societies. Logically, government can take four stances toward co-ops: It may attempt to repress them, it may be indifferent and inactive, it may encourage or facilitate private cooperatives, or it may directly set up co-ops. The future strength of the co-op sector in the United States, as anywhere, will depend not only on how well co-op values fit with values in the rest of society and on how economically competitive they can be, but also on the action (or inaction) of government. It is therefore difficult to predict the future course of a cooperative sector without knowing which groups will hold political power and what their biases may be.

The U.S. government over the decades has maintained a largely indifferent stance toward co-ops, with the notable exception of the farm co-ops that did receive significant supportive legislation in the 1930s and now represent a strong part of the agricultural economy. Nonfarm producer cooperatives,

however, have been neglected. The U.S. government has shown no cognizance of worker co-ops and has taken no role in their formation. The hundreds or thousands of producer cooperatives formed during the 1970s are grass roots in origin and, given their predilection for independence and self-sufficiency, might not be receptive to state overtures even if they were forthcoming. Indeed, they likely would be suspicious that outside intervention would undermine local autonomy and control.

In this country, co-ops have for the most part been created and operated on the basis of self-initiative, with periodic help from related social movements. This is not the case in many other parts of the world. In other countries, co-op development often has been fostered by the state and supported by other major institutions such as labor unions, political parties, and private business. It is instructive to look at examples of what other nations have done to encourage cooperative enterprise, and what resulting benefits have accrued to the state.

We examine cooperative development, especially the state's role in such development, by focusing on the important cases of France, Italy, Spain, Portugal, Poland, and China. Our strategy here is to draw attention to some important examples of cooperative development around the world that are lesser known. We also wish to help bring to light a valuable recent research literature, much of which is relatively inaccessible because it is as yet unpublished or appears in publications that are not widely circulated.[5] We try to draw out of these cross-national experiences some general lessons for the United States.

Western Europe contains a surprising number of co-ops. Though the co-op movement was severely reduced during the facist period in Europe, since 1945 worker cooperatives in France have grown to more than 600 (Thornley, n.d., p. 5). In Italy, the cooperative sector has 400,000 workers (Jones, 1981, p. 6) in some 3,000 co-ops (Derrick, 1981, p. 3) and accounts for 7 percent of the gross national revenue (Thornley). In Britain co-ops declined after the war, but have spread since 1975 with the support of about 40 local Cooperative Development Agencies (CDAs). Today there are approximately 500 co-ops in the United Kingdom (Jones, 1981, p. 6; Derrick, 1981, p. 3; Cornforth, 1983). The CDAs generally derive their funds from labor-controlled city councils.

Support for cooperatives in France goes back to 1884, when a government agency was established to provide advice to co-ops; by 1938 it had capital to loan to co-ops. In recent years the governments of France, Italy, and Britain have awarded an increasing number of contracts to cooperative enterprises because of their reputation for reliability, quality, and low cost, and private

business is beginning to subcontract particular jobs to co-ops for essentially the same reasons. In Italy, the Ministry of Works and Social Services was empowered in 1971 to develop cooperatives and to set up courses to train co-op leaders. The government in Italy also has set up an agency to provide loans to co-ops through the Banca d'Italia. As in France, loans are given at favorable (i.e., subsidized) interest rates and expert advice is made available to cooperatives.

Another major underpinning of the cooperatives sector in Italy and in France has been the development over the years of cooperative federations. Italy has three such federations and France has one. These are private, self-help organizations, though taken together they have garnered the support of five of the political parties in Italy. They function to provide business expertise to co-op participants. In addition, during the 1970s consortia (Consorzi) have developed in Italy to serve centralized functions for the co-ops: acquiring raw materials, marketing finished goods, negotiating for loans and contracts, and providing consultation. The federations also play a key role in promoting and setting up new co-ops. Thornley estimates that 60 percent of new Italian co-ops are set up with federation help. Sometimes the government, private business, and labor unions call upon the federations to form worker cooperatives out of failing firms (Thornley, n.d., p. 18). In Britain there are also self-help cooperative organizations (such as the International Common Ownership Movement) intended to promote co-ops, but these have relatively meager resources compared with those in France and Italy.

Hegland (1981) reports at least 130 "alternative institutions" in Denmark, much like their grass-roots counterparts in the United States. A study of worker-owned enterprises in Sweden (Lindkvist and Svensson, 1981) located 60 such companies encompassing 2,000 workers in industrial production. In addition, the study found many co-ops in the service sector, such as architectural firms, legal coops, and engineering firms quite like those reported in this study. The authors report that, as in the United States, the Swedish co-operatives are of three types: "paternalistic" firms in which ownership is transferred from a private owner to the workers for philanthropic or idealistic reasons; "defensive" firms in which worker ownership is used as a strategy to save jobs when a plant is faced with a shutdown; and "offensive" cases in which cooperative firms are started from scratch. All three types have been expanding since the Swedish industrial minister decided to support worker-owned firms with monetary grants.

Comparing data on worker participation for Western Europe, the Stephenses (1982) argue that the strength of labor union organization and of labor party (or socialist party) incumbency in office determine the extent to

which shop-floor democracy will be realized. The degree to which a country's labor force is unionized and the resulting political power of labor appear to have direct effects on the passage of legislation favorable to workers' participation and on the development of all kinds of workers' cooperatives. Although this may not be a complete explanation for the different types and degrees of worker participation in Western Europe, liberal government policies supported by labor, often in coalition with other parties, have had a great deal to do with the spread of workers' participation or cooperation in Western Europe. In contrast, in other countries we see that workers' cooperatives have developed in spite of the policies of the nation-state, or in other cases, because of revolutionary changes in who holds state power.

One of the most extensive networks of workers' cooperatives, involving more than 20,000 worker-owners, exists today in Spain. This system, dating from 1956, developed in spite of decades of fascist rule under Franco. To understand the widespread development of cooperatives under such oppressive circumstances, we must look at the particular politics and history of the region. Mondragon, where this development has occurred, is in the Basque region of northern Spain, a region that has a long history of intense separatist and anarchist sentiments and bitter memories of the devastation of the Spanish Civil War. It is in this context that the Basque Nationalist Party, affiliated with the Basque trade union movement, first proposed the establishment of a cooperative economy in 1933. In addition to the political support provided by the regional party and trade unions, an important cultural underpinning was the rural Basque tradition of offering one's labor for community service (Logan, 1981, p. 8). Moreover, the solidarity required for workers' cooperation derives in the Basque region, as indeed we have argued it does everywhere, from the homogeneity of the participants – in this case cultural, political, and religious in origin – and from the perception of a common enemy.

As in Italy and France, federative activity has played an important role in the Mondragon system. From the beginning, Mondragon's leaders saw the need for auxiliary institutions to support the cooperative development. Today a cooperative bank, school, child care center, and technical innovation center support the co-ops' financial, ideological, social, and technological needs.

Portugal, after five decades of fascist rule, is another country that illustrates how quickly grass-roots cooperatives can develop when given the opportunity. By 1978, Portugal had 511 agricultural co-ops with some 42,000 worker-members, and an estimated 1,200 industrial co-ops with some 59,000 members. In a detailed examination, Hammond (1981) has shown that the Portuguese revolution of 1974–75 set the stage for this flowering of grass-roots cooperatives. Although the new state leadership that resulted from the

revolution was centralist and socialist in orientation, it did allow local worker control efforts to develop. In this context, decentralized neighborhood committees were formed, landless farm workers took over plantations and turned them into co-ops, state-chosen union leaders were thrown out and replaced by democratically elected ones, and workers took over industrial plants (Hammond, 1981). Thus, worker control of the factories was part of a broader movement for democratic control of all institutions. The revolution had ushered in a climate of freedom and self-determination, and this new *zeitgeist* favored the development of workers' cooperatives as well as these other forms of popular control. The revolutionary (but centralized) state did not pursue workers' control as a policy, but did accept it, and the police did defend the new worker co-ops, and the nationalized bank gave them credit.

In late 1975, a right-wing military coup took place, purging the revolutionaries from positions of power. After this date, no more factory occupations took place. Since the counterrevolution, many of the agricultural co-ops and some of the industrial co-ops have been forcibly dismantled by the government and returned to their original owners, with the aid of the military to quell worker resistance. Other co-ops have been allowed to continue to exist, probably because they are able to provide employment in a deteriorating economy (Hammond, 1981).

To summarize, we have seen in the nations of Western Europe that the more liberal capitalist states (i.e., those supported by labor parties) tend to produce legislation and policies that favor worker participation. In Mondragon, even without state support, co-ops can still develop extensively at the grass-roots level if there is enough local and regional support. Portugal demonstrates the implications for the birth and demise of co-ops, even at the grass-roots level, of revolutions and counterrevolutions in state power.

Next we turn to Eastern Europe, to see how a centralized socialist state may respond to worker cooperatives.

Providing rare information concerning cooperative development inside Poland, Kowalak (1981) reports that the first worker cooperative dates back to 1873, but that co-ops were a negligible part of the economy until the end of World War II. Following the war, many small and medium-sized factories were turned over to cooperatively organized groups of workers. In some cases this was done because the factories were partly ruined by the war; in other cases the owners feared expropriation, and so they handed over their plants and joined the co-op. Since 1945, cooperatives have gradually come to be seen as an effective and relatively inexpensive way to diminish unemployment. They have been used to create job opportunities, especially for those who would otherwise find it difficult to enter the work force: women with

young children, disabled persons, pensioners, university students, and others. A 1961 Polish law recognized and defined the rights of co-ops. By 1979, 203,000 disabled persons were employed in the worker co-ops. All together, the cooperatives employ about 800,000 people, representing 6.7 percent of the work force in Poland (Kowalak, 1981, pp. 3–4). Reports indicate that from 1965 to 1975 the cooperative enterprises had higher rates of growth and productivity than the state industrial enterprises (Kowalak, 1981, p. 19).

Since 1975 the Polish cooperatives have become more centralized. First, there were many mergers of co-ops, presumably to bring economies of scale. Such mergers, however, have led to a large organization incompatible with democracy. In addition, the central and regional state-sponsored unions (to which the co-ops must belong) have begun to usurp the cooperatives' power by making important decisions concerning credit, investment, supply, and marketing. As decisions are increasingly being taken by "persons situated above the enterprise," the autonomy and power of the local cooperatives is dwindling, and the feelings of dissatisfaction and alienation that plague conventional enterprises are beginning to spread in the cooperative enterprises. As a result of this centralization of power in Poland, the productivity advantages that the cooperatives once had have now ended (Kowalak, 1981, pp. 16–19).

At the time of the presentation of his paper (May 1981), Professor Kowalak envisioned good prospects for the development of worker cooperatives in Poland, since this was one of Solidarity's (the independent trade union) chief aims. With the subsequent reassertion of state authority and the suppression of Solidarity, there is now little information available on the status of cooperatives in Poland.

In Poland the state was quick to grasp the job-creation potential of cooperatives and promoted them in order to stimulate national economic recovery. In time, however, as workers' power grew, the state was not content with autonomous co-ops, and it pushed decision making to progressively higher levels of the system. In short, the Polish state wanted to capture the economic benefits of cooperation, but it wanted to control the organizations. Since cooperation implies the right of workers to control the process of production and the surplus, this usurpation of power by the state produced an internal crisis in the co-ops and detracted from members' motivation to work, a contradiction that many state socialist societies may face in their treatment of co-ops.

Moving outside of the Soviet bloc to yet another type of socialist state, we see in China some similarities to Poland in the treatment of cooperatives. In a detailed analysis, Lockett (1981) shows that the first interest in establishing

cooperatives in China surfaced in 1919 with a student-based movement; then during the 1920s and 1930s credit and producer co-ops were brought into existence by radical intellectuals, international relief agencies, missionaries, and peasants. Reports indicate that the credit co-ops lasted some time, but the producer co-ops were often repressed by the state. During the two decades that preceded the Chinese revolution, the development of cooperatives in China was intricately tied to the evolving struggle between the two main contending political parties: the Khoumentang (KMT) and the Chinese Communist Party (CCP). The KMT, in search of a "third way" that was neither capitalist nor communist, decided to promote cooperatives. During the 1930s, the KMT set up thousands of credit, agricultural, and producer cooperatives. Later, however, as the co-ops began to pose an economic challenge to the urban and rural property owners who represented the base of political support for the KMT, the KMT withdrew its support of the cooperatives, and in fact became antagonistic toward them. The Communist Party, on the other hand, started from a position of suspicion toward the co-ops, and gradually came to embrace them during the 1940s, urging a position of more popular control of the co-ops and more profit sharing than the KMT had allowed.

With the revolution in 1949 and the ascendancy of the CCP, cooperativization of the economy accelerated (Lockett, 1981). By 1955, about 10 percent of the handicraft work force was involved in producer cooperatives. This amounted to some 850,000 persons. During the first 12 days of 1956, however, the government attempted to convert by fiat most of the remaining handicraft enterprises to a cooperative form. Suddenly there were 3.7 million persons in approximately 100,000 handicraft cooperatives. The party found, however, that by so directly and speedily trying to set up co-ops or to intervene in privately owned enterprises to turn them into co-ops, it encountered resistance and created organizational problems. The party then turned to more indirect means of control over the economy, including tax incentives, control over wages and work conditions, prices, and contracts for purchase of products. During the period of the Great Leap Forward there was another wave of new cooperative formation, this one largely the result of local initiative – often by unemployed women – to create employment in a difficult time.

Since 1958, the distinction between cooperative enterprises (those owned and run by the workers in them) and collective enterprises (owned and run by a locality) has become blurred, and evidence concerning magnitudes is scarce and often unreliable. The line between cooperatives and collectives has become particularly blurred because the government has placed controls on wages, work conditions, and surplus in the co-ops and has appointed managers in them (Lockett, 1981). This effectively removes from cooperative

members the control they would normally expect to have over the process and product of their labor, and reduces the usual motivation for working in a co-op. In fact, it brings into question to what extent enterprises called "cooperatives" in China during this period can be considered cooperatives as that term is generally understood elsewhere.

According to a 1973 study (reported in Lockett, 1981), labor productivity was 3.2 times higher in the state industrial sector than in the cooperative sector, although the state achieved this productivity advantage on the basis of 5 times more capital per worker. In other words, co-ops have demonstrated their usefulness in China as enterprises that can create jobs with less capital.

Since around 1976 and the break with the Cultural Revolution, several changes in economic policy in China have tended to favor the development of co-ops. First, the CCP leadership has looked to co-ops as a means of creating relatively inexpensive jobs for the millions of unemployed urban youth. Second, the turn toward more consumer goods and lighter industry favors cooperative development. Third, there has been a tendency to try to loosen the reigns of control over the co-ops and the collectives and to allow them to elect their own managers and to retain more of their surplus. Recent newspaper articles have praised cooperatives' superior flexibility and job-creation potential, especially in small-scale production, in subcontracting for the larger state enterprises, and in services. This reflects a positive change in party attitudes toward co-ops, but the fundamental contradiction between the requirements of a centrally planned economy and the desire of workers in cooperatives for autonomy has not been resolved. Some members of the CCP view the cooperatives as less socialistic; others want to develop out of the mixture of cooperatives, collectives, and state enterprises some new type of social property (Lockett, 1981).

The minimal involvement of the state in cooperative development in the United States stands in marked contrast to the cases described above. As noted in Chapter 1, this country has experienced five successive waves of grass-roots cooperative formation in its history – waves that occurred without government support or repression. Cooperatives, consistent with their anarchistic roots and their values of self-sufficiency, generally have not sought state aid.

Today, however, some forms of government recognition and advancement of co-ops are emerging in the United States. In 1979 a National Consumer Cooperative Bank, modeled on the successful farm cooperative bank of decades earlier, was passed into law by Congress and signed by President Jimmy Carter. The bank itself, however, was the product of consumer activism, and as the name suggests, was intended to aid consumer co-ops. As an after-

thought, a clause was put into the legislation allowing up to 10 percent of its funds to be loaned to producer co-ops. To date only a handful of worker co-ops have received loan support from the bank. With this small exception, the government has done nothing to assist producer co-ops.

In the mid-1970s, however, the federal government did begin to give considerable support to Employee Stock Ownership Plans (ESOPs), a new form of property that in effect broadens ownership to include the workers in an enterprise. By establishing a legal form, special tax incentives, and a financing mechanism (out of future pretax earnings), the government has provided strong incentives for these sorts of transfers to worker ownership. A recent study by Rosen et. al. (1986, p. 1) locates eight thousand ESOPs, involving some 7 to 8% of the national workforce. In an estimated 10% of these ESOP firms, workers hold or will come to hold, a majority of the voting stock, but in most of these firms workers hold between 15% and 40% of the stock.

Since 1974, Congress has passed 16 laws to encourage employee ownership, and 13 states have passed 16 laws to do the same (Rosen, 1986, p. 2). Why has the U.S. government become involved in promoting ESOPs? It is always difficult to infer motivations from actions, especially collective actions such as an act of Congress, but some common themes emerge in the speeches and writing of political leaders. ESOP legislation has been politically unusual in that it has received broad bipartisan support from conservatives as well as liberals. It appears to be a phenomenon into which different people read different virtues. To the conservative, it is "People's Capitalism": a chance to revitalize our capitalist economy by broadening its ownership base. To the liberal, employee ownership means a fairer distribution of future wealth and the potential for greater worker control in the production process. To both, employee ownership offers a way to save jobs in plants that might otherwise be shut down by their corporate owners. Additionally, the connection between workers having an ownership stake in their firms and worker motivation and productivity has not been lost on policy makers. Some have argued that worker participation schemes appear to grow in times of declining corporate profits and productivity (Heckscher, 1980).

Labor unions, on the other hand, started out lukewarm to ESOPs. Naturally, trade unionists want to preserve jobs wherever possible, but they are concerned about the role trade unions would have in worker-owned organizations. Moreover, they are concerned about the financial risk to workers in the minority of ESOP cases in which the former pension plan is exchanged for shares of company stock because this practice eliminates the diversification that one might seek in a long-term pension investment. In addition, those ESOPs that have emerged out of plant closings have often entailed wage cuts

in exchange for stock, and union leaders have feared that this might have a contagion effect, depressing wages in a whole industry. Finally, many of the early ESOPs (1975–80) were initiated by management and gave workers little say in company affairs. Since 1980, however, studies show a change in trade unionists' attitudes toward worker ownership. Several new cases of worker ownership have been spearheaded by local unions and contain ample measures to ensure a voice for workers in voting rights, shop-floor participation, representation on the board of directors, access to financial information, and appointment of managers. These cases have received widespread attention in the mass media and especially in union circles. Consequently, some unions in the United States are beginning to view the ESOP as a flexible instrument that can be shaped to benefit workers, communities and unions, if they can be involved in its planning and development.[6] Even with this change in union attitudes, the extent of labor union organization in the economy is much less in the United States than in Europe, and there are no labor parties. As a result, trade unions in the United States have not been nearly as vigorous or effective in pushing for national policies encouraging workers' cooperation or participation, as they have in Europe.

Although it is within the realm of state power to encourage workers' cooperation, any prediction for the United States would have to depend on who holds state power. In a context of high unemployment, worker cooperatives offer a policy option that has barely even been examined. Evidence from nations as diverse as Italy, Spain, Poland, and China demonstrates that jobs can be created in a cooperative sector more inexpensively (with less spent on a per-job basis) than in conventional firms. This is the major reason why governments in those countries have embarked on programs to encourage worker cooperatives, and it is the chief benefit that has accrued to them. Similar potential exists in the United States. Recently, it has been shown that ESOP firms where the employees own a majority of the stock generate three times more net new jobs per year than comparable conventional firms (Rosen, 1986, p. 2).

A program of workers' ownership, or cooperativization, would have as a political asset the perception that it is a third route, neither capitalist nor socialist. As we have seen, this could mean a broader base of political and popular support for such a program. On the other hand, although both socialist and capitalist countries have found good reasons to support economic development through a program of workers' cooperation, albeit of different types, they have both faced a dilemma with it as well. Both capitalist and socialist societies want to capture the economic benefits of cooperation, but they want to control it. For the capitalist country, an economically vital co-

operative sector would compete against private enterprise; for the socialist country, it would compete against state enterprise. Because worker cooperatives are neither fish nor fowl, they can make both sides edgy. At the same time, they cannot be rejected easily by either side. We have seen most clearly in the cases of Poland and China that government efforts to deny local members control of the co-ops (by moving important decisions to the state apparatus) has had a deleterious effect on motivation and performance in the co-ops. Similarly, efforts in the United States to reduce or eliminate local member control in employee-owned firms (by not granting voting rights and other avenues for worker voice) can be expected to have the same consequence.

In the last analysis, the role of the state in promoting cooperatives is limited. The experience with worker co-ops in Spain, Portugal, Italy, Poland, China, and elsewhere suggests that their democratic character can be preserved only if they are grass-roots formations. Once co-ops exist, the state can effectively facilitate them, if it wishes, through the provision of loan capital, purchasing contracts for goods or services, seed grants, technical advice, or tax incentives. The co-ops themselves can be expected to build self-help federations to provide support services for each other. Democratic control is an inherent feature of workers' cooperation, of whatever form, and this means that external agencies such as governments, or banks, or labor unions can restrict the autonomy of the enterprise or its members only at the peril of losing many of the benefits of the cooperative form.

Life span

Another factor that will greatly affect the future growth of the cooperative sector is the average life span that individual cooperatives may expect. Logically, cooperatives may terminate in two ways: They may cease to exist, or they may be transformed into a nondemocratic form of organization, thereby losing their defining cooperative character. Experience teaches that the processes by which cooperatives (or more generally, democratic organizations) tend to die or to degenerate can be inhibited or speeded up by specific choices that their members make at critical junctures.

Earlier in this book we made the case that a provisional or transitory orientation follows from participant commitment to higher goals, and that such an orientation was characteristic of the particular population of people in our collectives. A provisional orientation generally is not characteristic of people with greater family or financial responsibilities. Certainly, members of the plywood and garbage collection co-ops of the 1920s wanted their co-op en-

terprises to last, and worker-owners in the contemporary wave of ESOPs equally desire organizational stability. Organizational continuation is important, both for individual job security and for the goods and services the organization provides to the community, and thus relatively few people are in a position to be unconcerned with organizational longevity.

The data on how well cooperatives meet this need for stability are seriously limited. Systematic research comparing the longevity of worker cooperatives and that of comparable small businesses has not been done. Only a few scraps of data are available. Jones (1979) found in his study of nineteenth century worker cooperatives a median duration of somewhat less than 10 years, although longevity varied widely between industries. More than half of the cooperatives dating from the 1920s and 1930s, particularly in the plywood industry and in the refuse collection industry, are still in operation today. Of the contemporary wave of grass-roots collectives, still less is known. One effort to make phone calls to all of the collectives listed in the 1976 San Francisco Bay Area Collective Directory found that almost one-third of the previously listed collectives had disconnected phone numbers by 1981 (Kepp, 1981, p. 31). If it is assumed that all those collectives with disconnected numbers had gone out of business, this information would indicate only a 33⅓ percent rate of dissolution over five years. This compares with an 80 percent rate of dissolution over five years for conventional small businesses in the United States, according to figures from the Small Business Administration. Although many participants in and observers of the contemporary collectives mythologize a short life span, we are aware of no general data that support this belief, and in fact what little data exist suggest that collectives are doing a good deal better than conventionally owned small businesses.

These figures on longevity do not say what specific causes may lead to the demise of cooperatives. The death of a co-op can mask a number of different phenomena.

Many co-op failures are essentially economic in origin. They are market failures. In this respect they are no different from their small business counterparts. In the view of Beatrice Webb (1921, p. 229),

> So long as the co-operative society . . . constitutes only one among other forms of production and distribution . . . the co-operative society has to maintain itself in continual rivalry with capitalist enterprise. . . . Except in so far as it can effect a genuine improvement or economy in management, every step by which it departs from the competitive standard set by its capitalist rivals results in lowering the margin between cost and price. Any wide departure, whether in the way of higher wages, shorter hours, more favourable conditions of employment, or failing to take advantage of the best terms of obtaining raw materials or of employing the most efficient processes, means failure to serve the customers on the same terms as the capitalist trader. Thus the co-operative society, if it is to continue to exist and make headway

against capitalist enterprise, cannot go beyond the currently prevailing conditions of employment.

There is, no doubt, some truth to these assertions, especially where co-ops exist in a relatively unregulated market economy. On the other hand, we join others in arguing that to the extent that worker cooperatives democratize ownership and control of the workplace, they have demonstrable advantages in terms of labor motivation, productivity, lowered waste, and higher quality. Ultimately, this often lessens the need for and cost of management supervision. Cooperatives often do best where they are providing services or goods that are not available from capitalist enterprises. For this reason, some of the contemporary co-ops have turned to customized or hard-crafted goods, and others have pioneered alternative services, such as wholistic health care, legal services for the poor, feminist counseling, or conservation and ecology-related consulting. The cooperative's oppositional ideology allows it to fill a specialized niche: to serve specific social needs that generally go unattended by conventional businesses. By providing qualitatively different goods or services, the cooperative may reduce its direct competition with capitalist enterprises.

In other circumstances, not contemplated by the Webbs, cooperatives may disband, not out of economic failure, but because they choose to do so. Some worker cooperatives in the nineteenth and twentieth centuries were formed as a result of a strike or work lockout. In these cases, the cooperative was conceived by participants as a temporary solution to put pressure on the previous employer, and the voluntary termination of the cooperative signaled not its failure but its success in convincing the employer to take the workers back under more favorable terms.

Some of the contemporary co-ops decide to dissolve, despite the fact that they are economically viable, because members do not want to see certain alterations made in the form or purpose of the organization. As discussed earlier, Free Clinic members considered this line of action, but ultimately rejected it. Where changes appear imminent and where members are at odds with the original purposes for which they joined the organization, their commitments to higher goals may win out. In such cases, they may purposefully disband the organization, an unlikely event in bureaucratic organizations.

Still other processes, having little to do with economic failure or purposeful dissolution, may lead to the demise of cooperatives. As Whyte (1978) points out, cooperatives have often faced a "catch-22 situation: over a period of years they were doomed to lose this form of ownership either because they failed or because they were successful." Cooperatives may be killed by their very success. Several causes for this have been observed. Individuals generally start a cooperative as equal partners, but with the passage of time and the

success of the firm, they see that it is not in their economic self-interest to let new members in, since the creation of new shares will only dilute the equity of the previous members. If more workers are needed to meet the demand of growing production and sales, cooperative members learn that they can often make more money by hiring additional help and expropriating the surplus value of others, rather than by extending full ownership rights to new workers. As a result (unless there is a proviso in the original charter of the co-op permitting new workers the right to membership after some probationary period, or unless the co-op is exceptionally committed to democracy for idealistic reasons), there is a tendency over time for cooperatives to hire more help and to extend fewer memberships. This leads to a serious deterioration in the cooperative form, producing a two-class system of owners and nonowners.

In addition to this strong economic incentive, cooperatives often have ethnic or cultural bonds that lead them in a similar direction (Russell, 1984, 1985). Such ties are an enormous aid in the formation stage of the co-op, helping the group to work together smoothly, but on the other side of ethnic/kinship/friendship solidarity is the possibility that it may lead to the exclusion of "outsiders," people who are seen to be different. In the end, the Italian immigrants who founded the garbage collection co-ops in San Francisco in the 1920s, like the Soviet Jewish immigrants who started the taxi co-ops in Los Angeles 50 years later, resisted the entrance of other minority group members into their co-ops, producing a sharp line of demarcation between members and hired help, the "ins" and the "outs" (Russell, 1984, 1985). This is an example of the solidarity that can be created by a "clan" structure (Ouchi, 1980).

Finally, degeneration of the cooperative form may be hastened by the influence of the outside environment. Even where cooperatives seek seclusion from the environment, communal autarky is a practical impossibility. In his study of the kibbutzim system, Ben-Ner (1982) argues that communes gradually adapt their values and their form of organization to the environment, even though adaptation may be antithetical to collectivist values and principles of organization. Using a novel empirical approach, Ben-Ner looks at "revealed preferences" implicit in kibbutzim choices for collective consumption versus private consumption. He computes a ratio of collective to private consumption in different kibbutzim in 1955 and 1965, and the data indicate a strong decommunalization of values (Ben-Ner, 1982, pp. 27–30). In order to do business with dominant economic enterprises, communes must, according to Ben-Ner, develop structures and values that are isomorphic with surrounding institutions. We would add that outside inequality forces cooperatives to pay members more unequally than they might otherwise desire in order to

attract and retain highly skilled personnel. Similarly, outside record-keeping requirements may foist off the same need on reluctant collectives. These adaptations to the environment (i.e., isomorphism) may help the cooperative to survive, but they may also lead to the degeneration of the cooperative form.[7] Organizational theorists have argued that the tendency toward isomorphism is a fact of all organizational life, but that it varies with the environment of the organization (Emery and Trist, 1965; Aldrich, 1979).

These processes of developing a closely knit "in-group" that resists taking in outsiders, of hiring help rather than offering full membership, and of gradually adapting to the values and the ways of the outside environment, are unplanned, natural, perhaps even imperceptible. But taken together, they weaken the fabric of the collective. With fewer members and fewer shares, the price of the shares is bid up. And with the decommunalization of cooperative values, emotions will no longer hold the cooperative together. Ultimately, the cooperative may lose its integrity as a cooperative, and members may become more open to private investor offers for the purchase of shares.

The *Real Paper,* for example, was ultimately defeated by its own economic success. After 2½ years of operation, this newspaper had built up an impressive circulation and was turning a substantial profit. This made it attractive to private investors who offered $325,000 for the paper. Like the *Community News* in this study, the *Real Paper* was collectively owned by its staff members. For those staffers who were fatigued, disillusioned, or otherwise ready to move on, the $325,000 proved irresistible. Hence, what began as a staff-owned and controlled alternative paper became, by virtue of its financial success, a privately owned enterprise (Kopkind, 1974, 1975). Similarly, Bernstein (1975) describes a cooperative plywood factory where pressures to sell out to a large corporation mounted for a variety of reasons, not the least of which was the attractiveness to the workers of getting big money for their share of the stock (between $20,000 and $40,000 apiece). Again, the very financial success of the cooperative enterprise makes it that much more enticing to sell into private hands, thereby eliminating altogether the collectivist basis of the organization.

To end here on a negative note would be deceptive. The point of this section has been to show the various ways in which co-ops come to an end, either by ceasing to exist or by changing into something else. Many co-ops do not die early deaths, but continue to function as effective and democratic organizations over a period of many years. Overall, as was previously noted, co-ops appear to survive considerably better than conventional small businesses, an impressive testimony to their viability. Mere survival of an organization says nothing, of course, about whether internal democracy has been preserved,

but organizational democracy can be made to work indefinitely if members are committed to the principle, if they are knowledgeable about the organizational conditions that foster its development, and if they are wiling to work at creating, maintaining, and occasionally renewing direct democracy in the workplace. To paraphrase Menachem Rosner, there are no permanent achievements. Democracy requires permanent struggle.

The proliferation and success of cooperatives depends not only on the struggle for democracy within the collectivity, but also on social forces beyond the organization's control. The future of cooperatives will hinge on the extent to which cooperatives receive state and other major institutional support, compete effectively in the marketplace, and develop the organizational capability for long-term, democratic survival.

8. Overview and conclusions

Organizations come and go. Cooperatives, in particular, have risen and fallen in at least four previous waves in American history. Only history can tell how long the current movement will last, whether it will burgeon into a substantial and lasting sector of the economy, or whether it will become a forgotten experiment.

The master trend of the twentieth century has not been toward greater democratization of organizations – quite the contrary. It has been toward growing concentration both in the economy and in government institutions, with fewer and fewer economic units having control over an increasing share of the assets. Trends of this magnitude, however, often set into motion social forces that oppose them, countertrends that eddy against the main current. The desire for self-determination is such a countertrend. As the autonomy of the individual diminishes in ever-larger organizations, and as control becomes increasingly remote, some individuals recoil. Joining with others, they try to build communities, families, and workplaces that offer autonomy and control. What all of the examples discussed in this book have in common is this simple, but profound, desire for self-initiated, self-paced, self-controlled work. Given, too, the necessity and desire for meaningful group life, individual freedom becomes a value to be maximized within the context of local, collective control.

Today, the drive for workplace democracy has forms of expression that were unknown a short time ago. In the United States there are small grassroots collectives in virtually every community, "quality of work life" efforts in many *Fortune* 500 corporations, and worker-owned ESOPs in scores of formerly private industrial plants. Similar trends exist in Europe, in addition to the legislated workers' councils and workers' representatives on boards of directors in many European nations. The breadth of the drive for workplace democracy today is indicated by its international scope. The constituency, too, is broad and various. The people involved in, for instance, quality-of-work-life programs are different from the people involved in co-ops or in ESOPs. Each group has its own conferences, newsletters, and networks.

They are each in a sense participants in separate submovements with little articulation between them, but they are responding to the same human need for voice, for some measure of control, and the same organizational need for flexibility and innovation. Democratic control gives the organization the benefits of experience and knowledge from all levels of the organization; it gives the individual the chance to be heard. The many faces that workplace democratization has assumed this past decade and the international appeal of this movement suggest that this is not a passing fad, but an enduring historical evolution.

The recent growth of democratic forms in the workplace comes as a counterpoint to the time-honored sociological belief in the inevitability of oligarchy, a belief so widespread that it led Alvin Gouldner (1955, p. 507) to comment that social scientists have become in this respect "morticians, all too eager to bury men's hopes." This book takes seriously the evidence for the fragility of democratic systems, but does not present simply another case study in which democracy yields to oligarchy and abandons its original goals. Instead, we observed grass-roots democratic organizations in order to tease out the generic organizational features of democracy and the specific conditions that aid or impede the struggle for democratic control. By taking this conditional approach, we hope both to contribute to an empirically grounded theory of organizational democracy and to support by systematic investigation people's aspirations for self-managed work.

The larger political meaning of this set of organizational phenomena will reside, we believe, in the ability or inability of new forms to demonstrate that democratically managed organizations can "work." They must show that they can accomplish the tasks that need doing in this or any society (education, health care, production, etc.) without recourse to hierarchical authority relations. To the extent that they manage to do this, they provide a concrete model of what cooperative relations to production can look like.

American values and organizational democracy

The American public may be more receptive to these new ideas and forms of organization than is generally recognized. A poll by Hart Associates asked Americans which of three types of economic systems they would prefer to work in: Only 8 percent preferred work in a government agency, and only 20 percent chose the private investor–owned company, the now dominant form. The majority, 66 percent, preferred to work in companies owned and controlled by their employees (Rifkin, 1977, pp. 45–57). In addition, Hart found that only 17 percent of the public would keep the economic system as it is,

while 37 percent would make "minor adjustments," and 41 percent favored "making a major adjustment to try things which have not been tried before." Recent Yankelovich polls show sharply declining levels of confidence in many American institutions and occupational groups.

These polls would seem to reflect substantial changes in public attitudes. They suggest an increasingly prevalent questioning of established institutions and an openness to alternative ways of organizing work. But delegitimation alone does not ensure change, particularly if it is replaced by cynicism. This is where concrete alternative models play a role. In the absence of any international models that have gained widespread acceptance in America, domestic models of democratically controlled institutions are being created. In their very form of operation, democratic workplaces serve an educative function: They begin to convince participants (and clients and onlookers) that "workers do not need bosses to get work done" (Vocations for Social Change, 1976), that hierarchical authority structures and their corresponding stratification systems may not be necessary incentives to get people to do their jobs, and that labor and capital need not be separated.

Although some members in the collectives we observed did set out to build a model, an exemplar organization, that would demonstrate a better way to organize work, countercultural visions always take root among a minority. Is there any evidence that the collectivist values found in the co-ops are diffusing throughout our society?

It is reasonable to suppose that most ordinary people are not as value-driven as co-op members. Indeed, most scholars have presumed that people turn to cooperatives or ESOPs for practical, economistic reasons, though there have been relatively few empirical studies of the actual motives of participants. A case study of a mid-1970s worker buy-out of a firm scheduled for shutdown found job-saving motives to be foremost among both blue-collar and white-collar employees (Hammer and Stern, 1980). Examining historical materials on cooperatives in the United States from 1880 to 1935, Shirom (1972) concludes they were formed by members chiefly as a defensive measure to save jobs or to avoid downward mobility, and secondarily, to achieve entrepreneurial ambitions. Totally lacking in America, he argues, were collectivist ideals, although such ideals occasionally were found among European cooperators. Greenberg, picking up Shirom's argument in a recent study of the plywood co-ops, finds that members of the cooperatives (started in the 1920s) joined the firms primarily for economistic reasons. Over time, the experience of being in the co-op reinforces in the members their original individualistic, petit-bourgeois values (Greenberg, 1981).

This view contrasts markedly with the communal and egalitarian values we

found in our study. As we argued in Chapter 1, we see participation in collectivist organizations as motivated by a *coalescence* of material and ideal interests. Participation provides a means of livelihood consistent with members' values, where conventional work situations are distasteful or not available to members. Even in conventional work settings, we expect that values favorable to cooperation are more widespread in our society than is supposed by the economistic view.

Hochner and his associates are studying a case unfolding at present wherein A & P, a national supermarket chain, closed most of its stores in Philadelphia. This put 2,000 employees out of work, but A & P agreed, in a unique contract with the United Food and Commercial Workers union, to reemploy many of these people in new "Super Fresh" markets, which would have a QWL program to involve employees in decision making. Other employees would be allowed to purchase closed A & P stores and run them independently. The Super Fresh stores and the "O&O" stores (meaning worker owned and operated) opened in late 1982 and early 1983. Because the former A & P workers have an unusual set of options open to them, the researchers are in a position to analyze differences between those workers who choose worker ownership and those who don't. Hochner and Granrose (1986) find that those workers who pledge money for worker ownership (a worker must pledge $5,000 to become a member of an O&O) *are* more interested in avoiding unemployment, and *are* more entrepreneurial than nonpledgers. However, they *also* have more collectivist/participatory values than nonpledgers. The strength of the latter factor was contrary to the researchers' initial hypotheses and came as a surprise. Hochner and Granrose (1986) conclude that "the level of collective idealism expressed by these workers points to a greater interest in workplace reform than is generally recognized among American workers."

The finding of such values among grocery workers suggests that the countercultural belief in autonomy and self-determination of the 1970s has spread far. Further evidence comes from a just-completed reanalysis of the Hart public opinion survey (Zipp, Luebke, and Landerman, 1984). The researchers find the social base of support for workplace democracy to include nearly *all* segments of the society, especially blue-collar workers and professional/technical workers. The only class that does not support the redistribution of power in the workplace are managers and owners. Further, affluence does not much diminish blue-collar workers' support of democracy. Union membership has no significant effect, and there are no significant differences by race or party, although liberals are more in favor. Young persons and females are somewhat more likely to support democratization than middle-aged persons and males. On the basis of the evidence, the authors conclude that workplace democracy

is an issue that cuts across important political cleavages in the United States. Where most surveys of political-economic attitudes in this country have focused on traditional welfare state items, using these to draw conclusions about working class conservatism, Zipp and his associates argue that political attitudes are multidimensional and that quite different conclusions are warranted on issues of workers' control.

The findings of Zipp et al. are reinforced by a recent reanalysis of the 1977 Quality of Employment Survey national probability sample of working Americans. Excluding owners and managers from their analysis, the researchers find that support for workplace participation is stronger among women than among men and stronger among nonwhites than among whites (Fenwick and Olson, 1984). Somewhat at odds with the Zipp findings, however, Fenwick and Olson find that support for workplace participation increases with education, but declines with age, occupational status, and income. Also, contrary to Zipp et al., they find that union members favor participation substantially more than nonunion workers (1984, pp. 17–18).

Finally, comparing a sample of United States workers in Indianapolis and a sample in Sweden, Haas (1980) finds both groups of workers to be extremely supportive of workplace democracy, but only the Swedes favor having a "final say" in work-related decisions.

Although the collectivist values in small grass-roots co-ops cannot, on the face of it, be assumed to be shared by broader segments of society, the above studies indicate that the counterculture may have been a leading indicator of value shifts in America. Today, even before workplace democracy has received sustained public discussion, it apparently holds natural attraction for broad segments of the working population. The current backlash against big business and big government may well speak of a desire to enhance individual autonomy and grass-roots collective control.

The desire for grass-roots, democratically controlled organizations is not limited to the American counterculture nor, indeed, to twentieth century industrial workers. In times of social upheaval, locally initiated alternative institutions have played an important role.

In revolutionary periods, alternative work organizations are required, not only for their propagandistic functions, but also to provide needed human services and products in a time when mainstream institutions are disrupted. Roberta Ash (1972, pp. 75–78) documents the vital role played by locally organized alternative institutions during the American Revolution. Dual structures, such as the Committees of Correspondence and the Minutemen, were crucial to the Revolution's success. In times of general strikes, ad hoc worker-established cooperatives are frequently organized (again, at the local

level) to provide essential services to the population (Brecher, 1972). Likewise, the "soviets" played a necessary role in revolutionary Russia, providing decentralized and effective alternative industrial, political, and social structures. Later, after consolidating his power, Lenin abandoned the soviets in favor of a more centralized model of social organization (Gurley, 1976). During the Spanish Civil War, self-managed workers' committees collectivized both industry and agriculture until they were militarily defeated (Guerin, 1970, pp. 114–143). Throughout American history workers' cooperatives have risen in times of social protest and subsided during quieter times, persistently reasserting the desire of working people to control directly the process and end products of their work. In recent revolutionary actions in Portugal, as we saw, worker cooperatives sprang up in agriculture and in industry to control production, until their development was halted by a military coup. In Poland, too, local control and self-management in the economy were a major demand of Solidarity until the movement was repressed.

It is clear that collectivist, alternative structures have been important historically. Yet, the organizations studied here are very different from the Russian soviets, the Portuguese collectives, or the alternative structures discussed by Ash. The kinds of alternative organizations discussed by Ash and Brecher were born in revolutionary times and were therefore more self-consciously and immediately revolutionary in purpose. Few of the contemporary cooperatives would claim revolutionary goals.

We must leave to history the question of what social impact the current wave of alternative institutional development in the United States will have. So far, collectivist organizations have attracted only a rather limited social base of clientele and worker-owners, and they have remained predominantly, though not entirely, in the service sector of the economy. Whether they can spread in significant ways to other segments of the economy and attract a wider social base remains to be seen.

Important, some of the ideas and lessons from the experimentation with democratic management in the collectives over the past 15 years are being applied, albeit in somewhat watered-down fashion, to the more recent developments of worker-ownership using the ESOP form and to QWL in conventional firms. However, there has been considerable reinventing of the wheel. It is not necessary to argue that the organizational practices and values of the collectives are being directly appropriated, without due credit, by some of the ESOP and worker participation programs. Indeed, we believe one of the major weaknesses of the workplace democracy movement is that most people in the ESOP and QWL movements have little awareness of collectives, or of each other's experiences. In our opinion, each movement could learn much

from the achievements and difficulties of the other two (Rothschild-Whitt, 1983). Each of these three movements appears to represent different constituencies' responses to the same spirit of the times: a growing desire by individuals to control fundamental aspects of their work lives, and a growing realization of the need for organizations that can sustain commitment, innovation, and flexibility. The grass-roots collectives may have been the earliest and purest expression of this *zeitgeist,* but what began as oppositional values and practices are now on their way to such broad acceptance in the business world and among the public at large that the collectives are perhaps in danger of losing some of their oppositional flavor.[1]

Work and play

The basic contradiction within democratic organizations is between the logic of substantive rationality, in which the democratic process is of value in itself, and instrumental rationality, which is directed toward a product. All organizations are concerted actions toward some end and, as such, must be instrumental. One may begin by making shoes because one gets pleasure from the act of creating them. But what begins as a vehicle for self-expression must also sell on a market. It is impossible to live in a market economy, dominated by exchange-for-profit maximization, and not be touched by this concern.

As sites of production, co-ops exist in a world of markets. Otherwise, they could be purely playful enterprises. In a sense, collectivist organizations may be said to be developing a coalescence of work and play in their dual commitment to process and product. Concern for product is forced upon them by the necessity of exchange in a market. Concern for process is of their own creation and is at odds with commercial values.

People have had different reasons for introducing greater participation in the workplace. In QWL it is to improve productivity and the quality of product in order to better compete for markets. It is assumed that drawing information from the widest group of employees and involving them will add to the organization's know-how and to members' commitment and satisfaction. In co-ops, the purpose of democratic control over the organization is to ensure that the product or service will be in line with one's own values, that the distribution of surplus will be more egalitarian, and that the process of work will be, in and of itself, autonomous and therefore "fun."

To oppose formal rationality principles with substantive rationality principles is perhaps only an academic and bulky way of contrasting organizational practices that are intended to achieve something in the marketplace with organizational practices that are undertaken because they are directly grati-

fying for the people involved. Possibly the collectivist organization can arise only where technological capacity is great enough to free most from toil. We can hunt in the morning, fish in the afternoon, and talk philosophy at night only when we have the technological capacity to easily sustain material existence. When work is relatively free from the press of necessity it becomes self-expressive, playful activity.

The mechanical-industrial age vastly increased humankind's capacity to reproduce material existence. Now we appear to be moving into an electronic age that vastly again increases our capacity in this respect and also alters the nature of work, from transforming things to creating and disseminating new values, services, and knowledge. This transformation perhaps will give us more freedom to merge work with play.

This is, we believe, the fundamental meaning of the organizational paradigm defined and analyzed in this book. After 50 years' elaboration and specification of the behavior and theory of bureaucracy, the experience of the grass-roots collectives suggest a wholly new model of organization that must be assessed against its alternative values and aspirations.

Awareness that the world's resources are limited and shrinking suggests that we are moving away from gargantuan size and rapid growth rates as yardsticks of success, to a concern with the social utility and quality of products or services. The move from the mechanical age of moving parts to the electronic age of integrated systems will lead us to seek organic forms of organization to replace mechanistic ones. The greater difficulty and complexity of problems faced by organizations will come to require the gathering of information from wider sectors of the organization, especially those involved in the direct production or provision of services, in turn calling for cross-department communication and collaborative relations. Ultimately, in a world of limited resources and complex questions, values and priorities assume more importance. And so democracy comes to the fore.

The pioneering experiments in alternate forms of organization may be deviant in our society, but we hold that the major shifts in our society will make them more "normal" over time. The exact forms that future organizations will take in modern society are not predictable, but we believe that the organic, collaborative, and democratic forms with which these alternatives are experimenting presage parallel changes in other organizations, which will come perhaps long after the current wave of co-ops comes and goes. It is thus important to observe these naturally occurring experiments now, to see what we can learn from them about the structures and processes of democratic management.

We have developed an empirically grounded theory of organizational de-

mocracy. The essential feature of such organizations, from the Weberian perspective, is that authority is collectively held. From the Marxian perspective they are important because they alter the social relations to production, dissolving at the organizational level the antithesis between capital and labor. Those who work also manage. The upshot of both is that the organization derives its logic and its unity from substantive values. Such values can come to the fore and be implemented only if there is democratic control of the process and product of labor. These organizations signal a shift from production for exchange value to production for use value, from a market calculus to a social utility calculus.

The major contribution of collectivist organizations to both social science and to social change is, as we see it, their development and practice of a new model of organization. We have tried to identify the organizational properties that characterize this model, the constraints on its practical realization, and the conditions enhancing members' potential for democratic control of the organization. We have pursued a dialectical analysis of organizational democracy, examining the contradictory forces within this form or acting upon it, and the potentials for reconciling such forces and synthesizing a viable form of democratic management. As a result, we have continually turned our attention to the dilemmas that collectivist organizations face in their everyday efforts to get the job done, while still retaining their democratic form.

The essence of play is that it does not have an instrumental end in mind: In itself it brings gratification. It is a triumph of process over goals. However, in our society a clear line of demarcation separates work from play. We are told not to "confuse" the two, that they are opposites. Collectivist organizations see them as related. In a sense, although they don't put it this way, members are trying to integrate the world of work with the sentiments of play. By putting process before product, they are trying to find a place for expressive impulses in an arena ordinarily reserved for instrumental activity. Of course it might be argued that even the most formalized bureaucracy cannot eliminate all traces of human emotion and expression, but the point is that these are regarded as inappropriate or misplaced in the bureaucracy. In the collectivist organization they are cultivated and sought. They are part of the way that the organization *accomplishes* its business.

Notes

Introduction

1 Our definition is similar to that used by Rob Paton in a study of worker cooperatives in England (1978, p. 2).

2 The term *producers' cooperative* is used in the legislative clauses of the National Consumer Cooperative Bank statute. For this reason it is now beginning to be picked up by some participants in the co-ops, but it is still a minority expression.

Chapter 1

1 For an analysis of how organizations that reject authority find functional alternatives to authority that allow them to maintain social control, see Swidler (1979).

2 For a fuller examination of the history of anarchist ideas and movements, see: Guerin, 1970; Woodcock, 1962; Krimerman and Perry, 1966; Benello and Roussopoulous, 1972; Bookchin, 1980.

3 Our synonymous use of the terms *authority* and *domination* (*Herrschaft*) is consistent with the interpretation of Reinhard Bendix (1962, pp. 290–292) and the translation of Weber by Roth and Wittich (Weber, 1968, p. 946), but it differs substantially from the translation by Henderson and Parsons (Weber, 1947, p. 152).

Chapter 2

1 The mean age of the general membership at the Food Co-op is 21 years, and only 8.5 percent of its membership 24 or older. The staff and board members of the co-op (who are much more likely to identify with the New Left) show a mean age of 24.5 years. At the Free Clinic the mean age is 27.5 years, and at the *Community News* it is 24.5 years.

2 If, as our evidence indicates, there is considerable continuity between activists of the 1970s and those of the 1960s, then the New Left is alive and active, albeit with an altered political direction and strategy. This does not imply that it always will be. Like others, we have argued that conditions of affluence render traditional work and accumulation values obsolete and incoherent, especially to those who are young enough and lucky enough to have grown up with the assumption of affluence. The reintroduction of conditions of apparent scarcity has caught such people unprepared. Thus, political energy may smolder and collective action may be dispersed as individuals seek to secure their personal futures. If the line of argument here is basically correct, then the appearance of scarcity can be expected to have a conservatizing influence.

Chapter 3

1 The organizational structures and processes of alternative service organizations bear considerable, though certainly not complete, resemblance to collectivist organizations in contemporary China. For specific points of comparison see Martin King Whyte (1973) and Bettleheim (1974).

2 As organizations grow beyond a certain size they are likely to find purely consensual processes of decision making inadequate, and may turn to direct voting systems. Other complex, but nevertheless democratic, work organizations may sustain direct democracy at the shop-floor level, while relying on elected representative systems at higher levels of the organization (see Edelstein and Warner, 1976).

3 Actually, Weber did recognize the possibility of directly democratic organization, but he dealt with this only incidentally as a marginal case (1968, pp. 948–952, 289–292). Although Weber's three types of legitimate domination were meant to be comprehensive, both in time and in substance, it is difficult, as Mommsen points out (1974, pp. 72–94), to find an appropriate place for modern plebiscitarian democracy in Weber's scheme. Weber did come to advocate the "plebiscitarian leader-democracy," but this was a special version of charismatic domination (Mommsen, 1974, p. 113). He did not support "democracies without leadership" (*fuhererlose Demokratien*), which try to minimize the domination of the few over the many, because organization without *Herrschaft* appeared utopian to him (1974, p. 87). Thus it is difficult to identify the acephalous organizations of this study with any of Weber's three types of authority.

4 Research in the San Francisco Bay area found that free school teachers have higher degrees from more prestigious universities than their public school counterparts (McCauley, 1971, p. 148).

5 Industrial organizations in China have implemented similar changes in the division of labor. These were considered an essential part of transforming the social relations of production. Their means for reducing the separation of intellectual work from manual work were similar to those used by the alternative work organizations reported in this paper: team work, internal education, and role rotation. For specific points of comparison see Bettleheim (1974) and Whyte (1973).

6 The eight dimensions put forth here are clearly interrelated. However, there is evidence from bureaucracies that these dimensions may be somewhat independent (Hall, 1963). That is, an organization may be highly collectivist on one dimension but not so on another, and the interrelationships between these variables may be elusive. For instance, of seven propositions offered by Hage (1965) in an axiomatic theory of organizations, six could be supported by the organizations in this study. One, however, that higher complexity produces lower centralization, was contradicted by the evidence of this study, although it has received empirical support in studies of social service bureaucracies (Hage, 1965; Hage and Aiken, 1970). Hage suggests that relationships in organizational theory may be curvilinear: When organizations approach extreme scores, the extant relationships may no longer hold or may actually be reversed. This is an important limitation to bear in mind, especially as we begin to consider organizations such as the ones in this study, which are by design extreme on all eight continua proposed in this model.

Menachem Rosner, a scholar of the Israeli kibbutzim, has argued on both theoretical and empirical grounds that the dimensions of collectivist democracy are internally cohesive (1983). In an empirical test of the internal consistency of this sort of model, Leslie Brown (1984) found that 8 out of 18 food co-ops met her criterion of having at least 8 out of her 12 characteristics of participatory democracy. In an original research proposal to the National Science Foundation, Etzkowitz and Cuzzort (1983) suggest that an alternative orga-

nization may reject hierarchical authority in favor of egalitarian decision making, while at the same time accepting the formalization of rules. This implies that the dimensions of the model presented in this chapter do not necessarily have to go together. They maintain that their model, "Type *X*," may have advantages over collectivist democracy.

7 The result of this effort was a two-tiered structure: The paper is incorporated as a general corporation and a trust, which owns all the stock in the paper. Each six months of full-time work is worth one voting share in the trust. This grants ultimate control of the paper to the staff, past and present. Immediate control is exercised by the Board of Directors of the corporation, which consists of the currently working staff. As a member of the paper said, "[T]he structure is neither graceful nor simple, but it . . . guarantees that the working staff will maintain editorial control, and makes it nearly impossible ever to sell the paper."

8 See, for example, the abortive attempt to raise capital for employee ownership at Kasanof's Bakery (*Boston Phoenix,* April 26,1977).

9 Organization-environment relations are always reciprocal. In part, the low wages, hard work, and intense personal involvement that make collectivist organizations seem so costly may be due to costs imposed by the environment. Conversely, collectivist organizations rely on goods and services produced by the surrounding bureaucratic organizations, e.g., light bulbs, fast food chains.

10 Swidler (1976) vividly describes the extent to which members of a free school will ransack their private lives to locate sources of glamour that will enhance their sense of worth and influence in the group.

11 Mansbridge (1976) observes that even the most genuinely democratic organization will accept some measure of inequality of influence in order to retain individual liberties.

Chapter 4

1 "Charismatic" individuals are not entirely absent from alternative institutions, of course. Swidler (1979) aptly describes the use of "charisma," or more broadly, of personal appeal, as a means of social control in free high schools. Charismatic authority – complete with hierarchical leadership and privileges based on rank – was evidenced in many of the successful, long-lived nineteenth-century communes examined by Kanter (1972a, pp. 117–120). However, such formalized authority would be anathema to the organizations in our study.

2 Organizations that are homogeneous in this sense probably register substantial agreement over organizational goals (or what Thompson and Tuden [1959] call "preferences about outcomes"), but register considerable disagreement about how to get there ("beliefs about causation"). In such cases, Thompson and Tuden predict that organizations will reach decisions by majority judgment. A collegium type of organization, they maintain, is best suited for solving judgmental problems. This would require all members to participate in each decision, route pertinent information about causation to each member, give each member equal influence over the final choice, require fidelity to the group's preference structure, and designate as ultimate choice the judgment of the majority. On all but the last point they correctly describe collectivist work organizations. Further, as they point out, the social science literature does not contain models of this type of organization as it does for bureaucracy (Thompson and Tuden, 1959, p. 200).

3 The effect of level of compensation, or more broadly, of the level of financial resources, on the collectivist organization requires further study. On the one hand, there is evidence from the work of Kanter (1972a, pp. 78–80) that an austere life-style contributed to commitment in nineteenth-century communes, whereas affluence diminished it. Likewise, Duberman (1972) found that economic precariousness helped to knit the community together and to

generate commitment at Black Mountain, a forerunner (1933–56) of the free school movement. People in the cooperative organizations we observed seemed to believe that a low level of financial gain would help to avert careerism. Still another possibility is that limited financial resources encourage innovation, by forcing the organization to imagine new ways of fulfilling their organizational needs that are not so costly as conventional means. What is needed is a systematic comparison of relatively poor cooperatives with quite prosperous ones, several of which have developed over the last few years, to investigate the effects of affluence on cooperative organizations.

4 The meager pay and demanding work schedule at the *Community News* bring to mind the sort of self-exploitation that owners of small businesses often practice (Galbraith, 1973). In some ways, the economic marginality of alternative enterprises is not unlike that of any small enterprise. Both seem to impose heavy doses of self-exploitation. In fact, the social change goals of collectivist organizations may provide a much stronger ideological justification for self-sacrifice than do the private profit goals of the small entrepreneur.

5 Similarly, Eisenstadt (1959) argues that direct dependence of the organization on its members and clients is a condition that facilitates the debureaucratization process. However, Eisenstadt is concerned with the debureaucratization of already bureaucratic organizations, whereas this work focuses on new organizations that have resisted bureaucratization from their inception.

6 Researchers in the New Systems of Work and Participation Program at the School of Industrial and Labor Relations, Cornell University, have investigated cases of employee-ownership: Robert Stern and Tove Hammer have studied the Library Bureau at Herkimer, New York; Janette Johannesen has examined the Vermont Asbestos Group; and Michael Gurdon the Saratoga Knitting Mill in Saratoga Springs, New York. In each of these cases, the role of the banks in limiting the participatory rights of the worker-owners has been in evidence (Rothschild-Whitt, 1979b).

7 This was reported in the *Los Angeles Times,* January 26, 1975, pt. 5, p. 5. Drawing on inherited wealth and on the profits of some countercultural enterprises, some young wealthy activists are committing their philanthropy to radical social change efforts. These radical sources of financial support are still relatively rare, and their impact on the receiving organization may be unlike that of conventional agencies of external funding.

8 Personal communication with Sandy Morgan, Department of Anthropology, University of North Carolina, Chapel Hill, 1980.

9 Though this sort of eagerness may be widespread, especially in the beginning stages of alternative organizations, it is not without exception. The Food Co-op, for instance, was offered a $6,500 grant from the student body government of a nearby university. Although this money was critical in allowing the co-op to buy the food refrigeration equipment it needed before the story could be legally opened, it did not grasp it eagerly. The co-op seriously debated the strings that might be attached to the grant and its effect on the future development of the co-op, before deciding to accept it. Sarason (1972) and Kozol (1972) have also noted the rarity of this kind of careful anticipation of future organizational problems in alternative institutions.

10 For an interesting analysis of the impact that advances in technology may have on cooperative forms of organization, see Russell, Hochner, and Perry (1977). Here the introduction of new packer trucks and centralized billing procedures in garbage collection cooperatives in San Francisco has reduced the number and complexity of jobs, carrying with it less task sharing and more division of labor. In a historical overview of this issue, Braverman (1974) shows how increasing differentiation and specialization in our society have degraded work, making jobs less wholistic, less skilled, and less equal.

11 Although the sharing of information, and therefore of influence, is value-based in collectives, it may also arise as a structural need in more conventional organizations. In a study of Children's Center, a psychiatric facility for children, Blau and Alba (1982) show that although this organization did not set out as a matter of policy or ideology to be collectivist in nature, certain units did evolve in a participatory-democratic direction. This happened because the complexity of role relations in the organization promoted interunit communication. The researchers found that individual staff members gained more influence in decision making to the extent that the *unit* they were part of was integrated into organization-wide communication networks. In other words, to the extent that individuals had access, through their units, to the information and contacts possessed by higher levels of the organization, they had more influence. The Blau and Alba research indicates that some of the general characteristics of collectivist democracy, as discussed in Chapter 3 (especially knowledge diffusion) may evolve in ordinary service bureaucracies. This is so because people's jobs require negotiating a complex network of role relations, for which they need rather broad information and contacts.

12 The Law Collective contributed to the development of the Peoples College of Law in Los Angeles. Established under the auspices of the National Lawyers Guild, Peoples College was described by the *New York Times* (October 16, 1975, p. 40) as "the only radical law school in the country." Peoples College was unique in many ways. It was run primarily by its students, had free child care, was guided by an activist philosophy, and offered courses on topics rarely considered in conventional law schools (e.g., on legal aspects of tenant-landlord relationships, police brutality, racism, immigration).

13 For a good review of this growing body of literature, with special reference to Third World countries, see Vail (1975). For industrialized society, Rueschemeyer (1977) has shown that traditional efficiency arguments for differentiation may be seriously flawed. Melman (1969), Stein (1974), and Frieden (1980) have addressed the issue of efficiency as it relates to cooperative enterprises.

Chapter 5

1 In positing an oppositional relationship between the alternative organization and its environment, this condition assumes that the society in which the alternative exists is predominantly capitalist and bureaucratic. Hence, this condition would not be expected to hold in a socialist–collectivist society.

2 This line of reasoning fits well the data from a number of case studies of worker-owned firms—Vermont Asbestos Group; Saratoga Knitting Mill; and the Library Bureau in Herkimer, New York—that were conducted at the New System of Work and Participation Program, School of Industrial and Labor Relations, Cornell University.

3 *Canadians for a Democratic Workplace Newsletter,* vol. 1, no. 4 (1977), p. 7.

4 It is not clear that the publicly funded centers will provide better services than did the grassroots organizations. In fact, the opposite might be the case. A recent study (Forman and Wadsworth, 1983) finds that many federally funded community mental health centers provide few, if any, rape-related services, despite the mandate to do so.

5 There are more than 300 "straight" community newspapers in the Los Angeles metropolitan area. Unlike the *Community News,* they are privately owned, they see their purpose as covering traditional community news (e.g., civic clubs, high school football, PTA), and what national news they do contain is simply from the established wire services. They do not do the sort of original investigation, analysis, and comment that characterizes the *News*. All of these papers face financial hardships owing to the rapidly rising costs of

newsprint and advertising cutbacks by retailers in times of recession. But in a time when the readership of these traditional community newspapers is declining, that of the *Community News* is growing. (*Los Angeles Times,* 22 June 1975, pt. I, pp. 1, 25–28).

6 For illuminating discussions of the structural bases of this sort of intellectual and professional support for social change efforts, see Flacks (1971) and Zald and McCarthy (1975).

7 It is no doubt true that many individuals do get "burned out" by their experiences in alternative institutions. Emotional drain and exhaustion may be built into the structure of collectivist organizations, as Swidler (1976) vividly describes in free schools. What we are stressing here is that these individuals often resurface in other equally committed contexts.

8 For further details, see the *Socio-Economic Report* put out by the Bureau of Research and Planning, California Medical Association, October 1971.

Chapter 6

1 Delivered at the Caucus for a New Political Science, Brown University, November 8, 1975.

2 Starting with a basic schema developed by Wright (1976), Kalleberg and Griffin (1978, p. 380) categorize class as follows: "(1) the *petty bourgeoisie,* who own the means of production but do not control the labor power of others, (2) *small employers,* who own both the means of production and control the work activity of not more than 50 employees, (3) the *working class,* who, having neither ownership nor managerial functions, simply sell their labor power in the market . . . (4) the *managerial class,* who, while denied ownership roles, do control the labor power of others . . . [and (5) the *bourgeoisie*] who own production facilities and employ more than 50 persons." Lacking a large enough sample, they do not examine the job satisfaction of the last category.

3 Although worker expectations appear to rise when workers buy out firms to avoid a shutdown, expectations do not necessarily rise in other ESOPs. For example, when a founder sets up an ESOP for estate planning/financial reasons – depending on how the ESOP is described to the workers – employees may see the ESOP as merely a retirement benefit, a good profit-sharing tool, an indication of management's good-heartedness, or a tax dodge for the company (Katherine Klien, personal communication, 1983).

4 Based on personal communication with Katherine Klein of the National Center for Employee Ownership (1983).

5 Personal stress, particularly at high levels, is generally considered unpleasant, and most people seek to avoid it. However, not all stress is undesirable. All organizations must create some level of stress in their members. Psychologists have proposed that there is probably some *optimum* level of stress, below which little motivation or activity exists, and above which severe disorganization sets in. We are here considering disruptive levels of stress.

Chapter 7

1 See, for example, the cases of the IGP insurance firm in Washington, D.C., and of Breman enterprises in Holland (Rothschild-Whitt, 1981).

2 A study by Keith Bradley of the London School of Economics concludes that the economic performance of the Mondragon co-ops is better than that of local competing companies, partly because workers are more motivated and need less supervision (cited in Logan, 1981, p. 9).

3 For a detailed review of the research on workplace democracy and productivity, see Frieden (1980).

4 Personal communication with Ann Olivarius, Oxford University, 1982.

5 For these reasons and because of limitations of space, we do not examine in this section

the cases of Yugoslavia and Israel, two of the best-known and most extensively studied examples of cooperative development. Interested readers should consult, for example, Obradovic and Dunn (1976), Arzensek (1983), Rus (1970), Leviatan and Rosner (1980), Rosner (1983).

6 For a detailed analysis of the ways owners and managers have shaped ESOPs versus the ways workers would fashion them, and for an assessment of the recent union-initiated cases of Rath in Iowa, Hyatt-Clark in New Jersey, and the "O&Os" in Philadelphia, see Rothschild-Whitt (1984).

7 Contrarily, Holleb and Abrams (1975) have argued that collectives must forsake their early structure of personal relationships, autonomy in work, undifferentiated tasks, and commitment to ideology and must develop hierarchies, and formalized interaction and lose their ideological commitment before they can develop into "consensual democracies." The latter they view as a "stable compromise between the values and rewards of communalism and the necessities of existing in the real world" (1975, p. 149). Although Holleb and Abrams regard this as a positive process through which collectives should grow, there is no evidence in our study or any other that we know of to suggest that collectives can substantially bureaucratize and then revert back to some more mature form of democracy. The "real world" to which Holleb and Abrams want collectives to adapt is, of course, a capitalist–bureaucratic world in which the dominant institutions, social relations, and attitudes are in the main antithetical to collectivist principles of organization. With Ben-Ner, we would argue that the more cooperatives or communes decommunalize their values and organizational structures, the weaker and less able to resist environmental influences they become.

Chapter 8

1 It has been estimated that one-third of the *Fortune* 500 firms are developing worker participation programs (Walton, 1979), and a study by the New York Stock Exchange (1982) speaks admiringly of such efforts. A recent QWL conference (1981) attracted 1,500 participants, with many corporations sending a contingent of 10 or 15 people as representatives. In the sessions of the meeting, the corporate presenters spoke of the "new," "innovative," and even "revolutionary" ideas they are implementing in their firms. Virtually all of these ideas were in fact modest versions of participatory democracy. None of the corporate presenters gave any indication of awareness of cooperatives, and none seemed to have considered the possible importance of worker ownership in motivating the participation they desired.

In a very different context, Rothschild-Whitt spoke about the experiences of collectives in the United States at an international conference in Brussels. She found scholars, particularly those from Eastern Europe and the Soviet Union, to be incredulous at the idea that a movement toward worker ownership of any kind could be unfolding in the United States. How could such a thing gain support, they asked, in the heartland of capitalism? Americans, more familiar with the value Americans have long placed on autonomous work, on democratic decision making, on self-reliance, and on ownership, should be much less surprised.

References

Abell, Peter. 1981. "A Note on the Theory of Democratic Organization." *Sociology,* 15:262–264.

Abell, Peter, and Nicholas Mahoney. 1981. "The Social and Economic Potential of Small-Scale Industrial Producer Cooperatives in Developing Countries." Prepared for the Overseas Development Administration, London.

Aldrich, Howard. 1979. *Organizations and Environments.* Englewood Cliffs, N.J.: Prentice-Hall.

Aldrich, Howard, and Robert Stern. 1978. "Social Structure and the Creation of Producers' Cooperatives." Paper presented at the Ninth World Congress of Sociology, Uppsala, Sweden, August.

Appelbaum, Richard. 1976. "City Size and Urban Life." *Urban Affairs Quarterly,* December, pp. 139–170.

Argyris, Chris. 1974. "Alternative Schools: A Behavioral Analysis." *Teachers College Record,* 75:429–452.

Arzensek, Vlado. 1983. "Problems of Yugoslav Self-Management." *International Yearbook of Organizational Democracy,* 1:303–313.

Ash, Roberta. 1972. *Social Movements in America.* Chicago: Markam.

Avineri, Shlomo. 1968. *The Social and Political Thought of Karl Marx.* Cambridge: Cambridge University Press.

Bales, Robert F. 1950. *Interaction Process Analysis.* Chicago: University of Chicago Press.

Barkai, H. 1977. *Growth Patterns of the Kibbutz Economy.* Amsterdam: North Holland.

Bart, Pauline. 1981. "Seizing the Means of Reproduction: An Illegal Feminist Abortion Collective." Chicago Circle, School of Medicine, University of Illinois. Photocopy.

Bates, F. C. 1970. "Power Behavior and Decentralization." In *Power in Organizations,* ed. Mayer Zald, pp. 175–176. Nashville, Tenn.: Vanderbilt University Press.

Bendix, Reinhard. 1962. *Max Weber: An Intellectual Portrait.* Garden City, N.Y.: Anchor Books.

Benello, C. George, and Dimitrios Roussopoulos, eds. 1972. *The Case for Participatory Democracy.* New York: Viking Press.

Ben-Ner, Avner. 1982. "Changing Values and Preferences in Communal Organizations: Econometric Evidence from the Experience of the Israeli Kibbutz." In *Participatory and Self-Managed Firms: Evaluating Economic Performance,* ed. Derek Jones and Jan Svejnar. Lexington, Mass.: Lexington Books.

Bennis, Warren, and Philip Slater. 1968. *The Temporary Society.* New York: Harper & Row.

Berman, Katrina. 1967. *Worker-Owner Plywood Companies: An Economic Analysis.* Pullman, Wash.: Washington State University Press.

Bernstein, Paul. 1976. "Necessary Elements for Effective Worker Participation in Decision Making." *Journal of Economic Issues,* 10:490–522.

Bettleheim, Charles. 1974. *Cultural Revolution and Industrial Organization in China*. New York: Monthly Review Press.

Biggart, Nicole. 1977. "The Creative-Destructive Process of Organizational Change: The Case of the Post Office." *Administrative Science Quarterly,* 22:410–426.

Blau, Judith, and Richard Alba. 1982. "Empowering Nets of Participation." *Administrative Science Quarterly,* 27:363–379.

Blau, Peter. 1970. "Decentralization in Bureaucracies." In *Power in Organizations,* ed. Mayer Zald, pp. 150–174. Nashville, Tenn.: Vanderbilt University Press.

Blau, Peter, and W. Richard Scott. 1962. *Formal Organizations*. San Francisco: Chandler.

Bluestone, Barry, and Bennett Harrison. 1980. "Why Corporations Close Profitable Plants." *Working Papers for a New Society,* 8:15–23.

Blumberg, Paul. 1973. *Industrial Democracy: The Sociology of Participation*. New York: Schocken Books.

Bookchin, Murray. 1980. *Post-Scarcity Anarchism*. Palo Alto, Calif.: Ramparts Press.

Bottomore, T. B., ed. 1964. *Karl Marx: Selected Writings in Sociology and Social Philosophy.* New York: McGraw-Hill.

Bowles, Samuel, and Herbert Gintis. 1976. *Schooling in Capitalist America*. New York: Basic Books.

Braverman, Harry. 1974. *Labor and Monopoly Capital: The Degradation of Work in the Twentieth Century.* New York: Monthly Review Press.

Brecher, Jeremy. 1972. *Strike!* San Francisco: Straight Arrow Books; distributed by World Publishing Co., New York.

Brown, Leslie. 1984. "Patterns of Organizational Democracy: New Wave Food Coops and the Challenge of Congruence." Sociology Department, University of Minnesota. Photocopy.

Buber, Martin. 1960. *Paths in Utopia*. Boston: Beacon Press.

Calhoun, C. J. 1980. "Democracy, Autocracy, and Intermediate Associations in Organizations: Flexibility or Unrestrained Change?" *Sociology,* 14:345–361.

Cavan, Sherri. 1972. *Hippies of the Haight*. St. Louis: New Critics Press.

Chickering, Arthur W. 1972. "How Many Make Too Many?" In *The Case for Participatory Democracy,* ed. Benello and Rossopoulos. New York: Viking Press, pp. 213–227.

Cicourel, Aaron. 1964. *Method and Measurement in Sociology.* New York: The Free Press of Glencoe.

Coleman, James. 1970. "Social Inventions." *Social Forces,* 49:163–173.

Cornforth, Chris. 1983. "Experiences in the UK." Cooperatives Research Unit, Open University, Milton Keyes, England. Photocopy.

Crain, Joyce. 1978. "A Survey of U.S. Producer Cooperatives: Summary of the Initial Phase." Center for Economic Studies, Palo Alto, Calif. Photocopy.

Crozier, Michael. 1984. *The Bureaucratic Phenomenon*. Chicago: University of Chicago Press.

Derrick, Paul. 1981. "Industrial Cooperatives and Legislation." Paper presented at First International Conference on Producer Cooperatives, Copenhagen, Denmark, June.

Duberman, Martin. 1972. *Black Mountain: An Exploration in Community,* New York: E. P. Dutton.

Eckstein, Alexander. 1977. *China's Economic Revolution*. New York: Cambridge University Press.

Edelstein, J. David. 1967. "An Organizational Theory of Union Democracy." *American Sociological Review,* 32:19–31.

Edelstein, J. David, and Malcolm Warner. 1976. *Comparative Union Democracy: Organization and Opposition in British and American Unions*. New York: Wiley.

Eisenstadt, S. N. 1959. "Bureaucracy, Bureaucratization, and Debureaucratization." *Administrative Science Quarterly,* 4:302–320.

Elden, J. Maxwell. 1976. *Democracy at Work for a More Participatory Politics: Worker Self-Management Leads to Political Efficacy.* Ph.D. diss., University of California, Los Angeles.

Ellul, Jacques. 1964. *The Technological Society.* New York: Alfred A. Knopf.

Emery, F.E., and E.L. Trist. 1965. "The Causal Texture of Organizational Environments." *Human Relations,* 18:21–32.

Engels, Friedrich. 1959. "On Authority." In *Marx and Engels: Basic Writings on Politics and Philosophy,* ed. Lewis Fever. Garden City, N.Y.: Anchor Books.

Etzioni, Amitai. 1961. *A Comparative Analysis of Complex Organizations.* New York: The Free Press of Glencoe.

Etzkowitz, Henry, and Ray Cuzzort. 1983. "Organizational Succession and Emergent Forms." Research proposal submitted to the National Science Foundation.

Etzkowitz, Henry, and Laurin Raiken. 1982. "The Transformation of Artists' Organizations: A Case Study." New York University. Photocopy.

Etzkowitz, Henry, and Gerald Schaflander. 1978. "Fight for the Sun." Department of Sociology, State University of New York, Purchase, New York. Photocopy.

Fenwick, Rudy, and Jon Olson. 1984. "Contested Terrain or Created Consensus: Why Workers Want Participation in Workplace Decision-Making." Paper presented at the annual meetings of the American Sociological Association, San Antonio, Tex.

Flacks, Richard. 1967. "The Liberated Generation: An Exploration of the Roots of Student Protest." *Journal of Social Issues,* 23:52–75.

1971. "Revolt of the Young Intelligentsia: Revolutionary Class Consciousness in Post-Scarcity America." In *The New American Revolution,* ed. Roderick Aya and Norman Miller. New York: The Free Press.

Forman, B.D., and J.C. Wadsworth. 1983. "Delivery of Rape-Related Services in CMHCs: An Initial Study." *Journal of Community Psychology,* 11:236–240.

French, J. Lawrence, and Joseph Rosenstein. 1981. "Employee Stock Ownership and Managerial Authority: A Case Study in Texas." Paper presented at the annual meeting of the Southwestern Sociological Association, March.

Frieden, Karl. 1980. *Workplace Democracy and Productivity.* Washington, D.C.: National Center for Economic Alternatives.

Galbraith, John Kenneth. 1973. *Economics and the Public Purpose.* Boston: Houghton Mifflin.

Gamson, Zelda. 1981. "Problems of Collectivist-Democratic Workplaces." Paper presented at Harvard University, June.

Gamson, Zelda, and Henry Levin. 1980. "Obstacles to the Survival of Democratic Workplaces." Center for Economic Studies, Palo Alto, Calif. Photocopy.

Gardner, Richard. 1976. *Alternative America.* Privately published.

Geise, Paula. 1974. "How the 'Political' Co-ops Were Destroyed." *North Country Anvil,* October-November, pp. 26–30.

Gillespie, David. 1981. "Adaptation and the Preservation of Collectivist-Democratic Organization." Department of Sociology, Michigan State University. Photocopy.

Glaser, Barney, and Anselm Strauss. 1967. *The Discovery of Grounded Theory: Strategies for Qualitative Research.* Chicago: Aldine.

Goodman, Richard, and Lawrence Goodman. 1976. "Some Management Issues in Temporary Systems: A Study of Professional Development and Manpower – The Theater Case." *Administrative Science Quarterly,* 21:494–501.

Gouldner, Alvin. 1954. *Patterns of Industrial Bureaucracy.* Glencoe, Ill.: The Free Press.

1955. "Metaphysical Pathos and the Theory of Bureaucracy." *American Political Science Review,* 49:496–507.

Graubard, Allen. 1972. *Free the Children.* New York: Pantheon Books.

Greenberg, Edward. 1981. "Industrial Self-Management and Political Attitudes." *The American Political Science Review,* 75:29–42.

Guerin, Daniel. 1970. *Anarchism: From Theory to Practice.* New York: Monthly Review Press.

Gurley, John G. 1976. *Challengers to Capitalism: Marx, Lenin, and Mao.* San Francisco: San Francisco Book Co.

Gusfield, Joseph. 1955. "Social Structure and Moral Reform: A Study of Women's Christian Temperance Union." *American Journal of Sociology,* 61:211–232.

Gyllenhammar, Pehr. 1977. *People at Work.* Addison-Wesley.

Haas, Ain. 1980. "Workers' Views on Self-Management: A Comparative Study of the U.S. and Sweden." In *Classes, Class Conflict, and the State,* ed. Maurice Zeitlin, pp. 276–295. Cambridge, Mass.: Winthrop.

Hage, Jerald. 1965. "An Axiomatic Theory of Organizations." *Administrative Science Quarterly,* 10:289–320.

Hage, Jerald, and Michael Aiken. 1970. *Social Change in Complex Organizations.* New York: Random House.

Hall, D., and K. Nougaim. 1968. "An Examination of Maslow's Need Hierarchy in an Organizational Setting." *Organizational Behavior and Human Performance,* 3:12–35.

Hall, John. 1978. *The Ways Out.* London: Routledge and Kegan Paul.

Hall, Richard. 1963. "The Concept of Bureaucracy: An Empirical Assessment." *American Journal of Sociology,* 69:32–40.

Hammer, Tove, and Robert Stern. 1980. "Employee Ownership: Implications for the Organizational Distribution of Power." *Academy of Management Journal,* 23:78–100.

Hammer, Tove, Jacqueline Landau, and Robert Stern. 1981. "Absenteeism When Workers Have a Voice: The Case of Employee Ownership." New York State School of Industrial and Labor Relations, Cornell University. Photocopy.

Hammond, John. 1981. "Worker Control in Portugal: The Revolution and Today." City University of New York. Photocopy.

Heckscher, Charles. 1980. "Worker Participation and Management Control." *Journal of Social Reconstruction,* 1:77–102.

Hegland, Tore. 1981. "Social Experiments and Education for Social Living." Aalborg University Center, Denmark. Photocopy.

Helfgot, Joseph. 1974. "Professional Reform Organizations and the Symbolic Representation of the Poor." *American Sociological Review,* 39:475–491.

Hendricks, Wendell. 1973. "On the Growing Edge." *Friends Journal,* June.

Henry, Jules. 1965. *Culture against Man.* New York: Vintage Books.

Hinton, William. 1966. *Fanshen: A Documentary of Revolution in a Chinese Village.* New York: Vintage Books.

Hochner, Arthur. 1981. "The Mentality of Worker Ownership: Reflections on Some Anomalous Findings." Paper presented at the First International Conference on Producer Cooperatives, Copenhagen, Denmark.

Hochner, Arthur, and Cherlyn Granrose. 1986. "Sources of Motivation to Choose Employee Ownership as an Alternative to Job Loss." *Academy of Management Journal.*

Holleb, Gordon, and Walter Abrams. 1975. *Alternatives in Community Mental Health.* Boston: Beacon Press.

Ingham, G. K. 1970. *Size of Industrial Organization and Worker Behavior.* Cambridge: Cambridge University Press.

Institute for Social Research, University of Michigan (ISR). 1977. *Employee Ownership*. Technical Assistance Project, Economic Development Administration. Washington, D.C.: U.S. Department of Commerce.

Jackall, Robert. 1976. "Workers' Self-Management and the Meaning of Work: A Study of Briarpatch Cooperative Auto Shop." Center for Economic Studies, Palo Alto, Calif. Photocopy.

Jackall, Robert, and Joyce Crain. 1984. "The Shape of the Small Worker Cooperative Movement." In *Worker Cooperatives in America*, ed. R. Jackall and W. M. Levin. Berkeley, Calif.: University of California Press.

Jenkins, J. Craig. 1977. "Radical Transformations of Organizational Goals." *Administrative Science Quarterly*, 22:568–586.

Johnson, Ana Gutierrez, and William Foote Whyte. 1977. "The Mondragon System of Worker Production Cooperatives." *Industrial and Labor Relations Review*, 31:18–30.

Jones, Derek. 1979. "Producer Cooperatives in the U.S." Department of Economics, Hamilton College. Photocopy.

1980. "Producer Cooperatives in Industrialized Western Economies." *British Journal of Industrial Relations*, 18:141–154.

1981. "Italian Producer Cooperatives, 1975–1978: Productivity and Organizational Structure." Paper presented at the First International Conference on Producer Cooperatives, Copenhagen, Denmark, June.

Kalleberg, Arne, and Larry Griffin. 1978. "Positional Sources of Inequality in Job Satisfaction." *Sociology of Work and Occupations*, 5:371–401.

Kanter, Rosabeth Moss. 1972a. *Commitment and Community*. Cambridge: Harvard University Press.

1972b. "The Organization Child: Experience Management in a Nursery School." *Sociology of Education*, 45:186–211.

1977. *Men and Women of the Corporation*. New York: Basic Books.

Kanter, Rosabeth M., and Louis Zurcher, Jr. 1973. "Alternative Institutions." A special issue of *The Journal of Applied Behavioral Science*, 9.

Katz, Michael. 1975. *Class, Bureaucracy, and Schools: The Illusion of Educational Change in America*. New York: Praeger.

Kaye, Michael. 1972. *The Teacher Was the Sea: The Story of Pacific High School*. New York: Links Books.

Keil, Thomas J. 1982. "Extraorganizational Factors in the Emergence and Stabilization of a Producer Cooperative." Paper presented at the International Sociological Association meetings, Mexico City.

Keith, Pam. 1978. "Individual and Organizational Correlates of a Temporary System." *The Journal of Applied Behavioral Science*, 14:195–203.

Keniston, Kenneth. 1968. *Young Radicals*. New York: Harcourt, Brace, and World.

Kepp, Michael. 1981. "The Berkeley Collectives: An Outside Perspective." *Communities*, 47:29–31.

Kopkind, Andrew. 1974. "Hip Deep in Capitalism: Alternative Media in Boston." *Working Papers for a New Society*, Spring, pp. 14–23. See also follow-up comment by Bernstein in *Working Papers for a New Society*, Summer 1975, pp. 2–3.

Kowalak, Tadeusz. 1981. "Work Co-operatives in Poland." Paper presented at the First International Conference on Producer Cooperatives, Copenhagen, June.

Kozol, Jonathan. 1972. *Free Schools*. Boston: Houghton-Mifflin.

Krimmerman, Leonard, and Lewis Perry, eds. 1966. *Patterns of Anarchy*. Garden City, N.Y.: Doubleday.

Kropotkin, Peter. 1902. *Mutual Aid: A Factor in Evolution*. London: McClure, Phillips.

Kruse, Douglas. 1981. "The Effects of Worker Ownership Upon Participation Desire: An ESOP Case Study." Honors B.A. thesis. Economics Department, Harvard College, Cambridge, Mass.

Lawson, Ronald. 1981. "He Who Pays the Piper: The Consequences of Their Income Sources for Social Movement Organizations." Urban Studies Department, Queens College, New York. Photocopy.

Leavitt, H. J. 1964. *Managerial Psychology.* Chicago: University of Chicago Press.

Levitan, Uri, and Menachem Rosner. 1980. *Work and Organization in Kibbutz Industry.* Norwood, Pa.: Norwood Editions.

Lindenfeld, Frank. 1982. "Problems of Power in a Free School." In *Workplace Democracy and Social Change*. Boston: Porter Sargent.

Lindkvist, Lars, and Claes Svensson. 1981. "Worker-Owned Companies in Sweden." Paper presented at the First International Conference on Producer Cooperatives, Copenhagen, Denmark, June.

Lipset, S. M., Martin Trow, and James Coleman. 1962. *Union Democracy.* New York: Anchor Books.

Litwak, Eugene. 1961. "Models of Bureaucracy Which Permit Conflict." *American Journal of Sociology,* September, pp. 177–184.

Lockett, Martin. 1981. "Producer Cooperatives in China: 1919–1981." Paper presented at the First International Conference on Producer Cooperatives, Copenhagen, Denmark, June.

Logan, Christopher. 1981. "Adapting the Mondragon Experience." Paper presented at the First International Conference on Producer Cooperatives, Copenhagen, June.

Long, Richard. 1979. "Desires for and Patterns of Worker Participation after Conversion to Employee Ownership." *Academy of Management Journal,* 22:611–617.

Mansbridge, Jane. 1973a. "Town Meeting Democracy." *Working Papers for a New Society,* 1:5–15.

1973b. "Time, Emotion, and Inequality: Three Problems of Participatory Groups." *Journal of Applied Behavioral Science,* 9:351–368.

1977. "Acceptable Inequalities." *British Journal of Political Science,* 7:321–336.

1980. *Beyond Adversary Democracy.* New York: Basic Books.

1982. "Fears of Conflict in Face-to-Face Democracies." In *Workplace Democracy and Social Change*. Boston: Porter Sargent.

Marcuse, Herbert. 1962. *Eros and Civilization*. New York: Vintage Books.

Mariolis, Peter. 1975. "Interlocking Directorates and Control of Corporations: The Theory of Bank Control." *Social Science Quarterly,* 56:425–439.

Marx, Karl. 1938. *Critique of the Gotha Program*. New York: International.

1967. *Capital*. 3 vols. (A, B, C). New York: International.

Maslow, Abraham. 1954. *Motivation and Personality.* New York: Harper & Row.

McCarthy, John, and Mayer Zald. 1973. *The Trend of Social Movements in America: Professionalization and Resource Mobilization*. Morristown, N.Y.: General Learning Press.

McCauley, Brian. 1971. *Evaluation and Authority in Radical Alternative Schools and Public Schools*. Ph.D. diss., Stanford University.

Melman, Seymour. 1969. "Industrial Efficiency under Managerial vs. Cooperative Decision Making: A Comparative Study of Manufacturing Enterprises in Israel." *Studies in Comparative Development*. Beverly Hills: Sage.

1971. "Industrial Efficiency under Managerial versus Cooperative Decision-Making." In *Self-Governing Socialism,* ed. B. Horvat et al.: International Arts and Science Press, pp. 203–220.

Merton, Robert. 1957. *Social Theory and Social Structure*. Glencoe, Ill.: The Free Press.

———. 1964. "Anomie, Anomia, and Social Interaction: Contexts of Deviant Behavior." In *Anomie and Deviant Behavior,* ed. Marshal Clinard, pp. 213–242. New York: The Free Press of Glencoe.

Messinger, Sheldon. 1955. "Organizational Transformation: A Case Study of Declining Social Movement." *American Sociological Review,* 20:3–10.

Michels, Robert. 1962. *Political Parties: A Sociological Study of the Oligarchical Tendencies of Modern Democracy.* New York: The Free Press.

Milgram, Stanley. 1973. *Obedience to Authority: An Experimental View.* New York: Harper & Row.

Mills, C. Wright. 1959. *The Sociological Imagination.* New York: Oxford University Press.

Milofsky, Carl, and Frank Romo. 1981. "The Structure of Funding Arenas for Community Self-Help Organizations." Institute for Social and Policy Studies, Yale University. Photocopy.

Moberg, David. 1979. "Experimenting with the Future: Alternative Institutions and American Socialism." In *Co-ops, Communes and Collectives,* ed. John Case and Rosemary Taylor, pp. 274–311. New York: Pantheon Books.

Molotch, Harvey. 1976. "The City as a Growth Machine: Toward a Political Economy of Place." *American Journal of Sociology,* 82:309–332.

Mommsen, Wolfgang. 1974. *The Age of Bureaucracy: Perspectives on the Political Sociology of Max Weber.* New York: Harper & Row.

Mouzelis, Nicos. 1968. *Organization and Bureaucracy: An Analysis of Modern Theories.* Chicago: Aldine.

Nagy, Michael. 1980. Personal communication (letter), Department of Sociology, Concord College, Athens, W.Va.

New York Stock Exchange. 1982. *People and Productivity: A Challenge to Corporate America.*

Obradovic, Josip. 1970. "Participation and Work Attitudes in Yugoslavia." *Industrial Relations,* 9:161–169.

Obradovic, Josip, and William N. Dunn, eds. 1976. *Workers Self-Management and Organizational Power in Yugoslavia.* Pittsburgh: University of Pittsburgh.

Olson, Mancur. 1971. *The Logic of Collective Action: Public Goods and the Theory of Groups.* Cambridge, Mass.: Harvard University Press.

O'Sullivan, E. 1977. "International Cooperation: How Effective for Grassroots Organizations?" *Group and Organizational Studies,* 2:347–358.

Ouchi, William G. 1980. "Markets, Bureaucracies and Clans," *Administrative Science Quarterly,* 25:129–141.

Palisi, Bartolomeo. 1970. "Some Suggestions about the Transitory-Permanence Dimension of Organizations." *British Journal of Sociology,* June, pp. 200–206.

Pateman, Carole. 1970. *Participation and Democratic Theory.* Cambridge: Cambridge University Press, 1970.

Paton, Rob. 1978. *Some Problems in Co-operative Organization.* Co-operatives Research Monograph 3. Milton Keynes, England: The Open University.

Perrow, Charles. 1970. *Organizational Analysis: A Sociological View.* Belmont, Calif.: Wadsworth.

———. 1976. "Control in Organizations: The Centralized-Decentralized Bureaucracy." Paper presented at the annual meeting of the American Sociological Association.

Perry, Stewart E. 1978. *San Francisco Scavengers.* Berkeley: University of California Press.

Quinn, Robert, and Graham Staines. 1977. *The 1977 Quality of Employment Survey.* Ann Arbor, Mich.: University of Michigan, Survey Research Center.

Reinharz, Shulamit. 1984. "Alternative Settings and Social Change." In *Psychology and Community Change,* ed. K. Heller, R. Price, S. Reinharz, S. Riger, and A. Wandersman. Homewood, Ill.: Dorsey Press.

Rhoades, Rosemary. 1981. "Milkwood Cooperative, Ltd." Cooperative Research Unit, Case Study no. 4. Milton Keynes, England: The Open University.

Rifkin, Jeremy. 1977. *Own Your Own Job.* New York: Bantam Books.

Rosen, Corey, Katherine Klein, and Karen Young. 1986. *Employee Ownership in America: The Equity Solution.* Lexington, Mass.: Lexington Books.

Rosner, Menachem. 1983. "Participatory Political and Organizational Democracy and the Experience of the Israeli Kibbutz." *International Yearbook of Organizational Democracy,* 1:455–484. Chichester: John Wiley & Sons.

Rosner, Menachem, and Joseph Blasi. 1981. "Theories of Participatory Democracy and the Kibbutz." In *Festschrift Volume for Professor S.N. Eisenstadt,* ed. Erik Cohen et al. Boulder, Colo.: Westview Press.

Rothschild-Whitt, Joyce. 1976a. "Problems of Democracy." *Working Papers for a New Society,* 4:41–45.

1976b. "Conditions Facilitating Participatory-Democratic Organizations." *Sociological Inquiry,* 46:75–86.

1979a. "The Collectivist Organization: An Alternative to Rational-Bureaucratic Models." *American Sociological Review,* 44:509–527.

1979b. "Worker Ownership and Control in the U.S." School of Industrial and Labor Relations, Cornell University. Photocopy.

1981. "There's More Than One Way to Run a Democratic Enterprise: Self-Management from the Netherlands." *Sociology of Work and Occupations,* 8:201–223.

1983. "Worker Ownership in Relation to Control: A Typology of Work Reform." *International Yearbook of Organizational Democracy,* 1:389–406.

1984. "Worker Ownership: Collective Response to an Elite-Generated Crisis." *Research on Social Movements, Conflict and Change,* 6:99–118.

Rueschmeyer, Detrich. 1977. "Structural Differentiation, Efficiency, and Power." *American Journal of Sociology,* 83 (July):1–25.

Rus, Veljko. 1970. "Influence Structure in Yugoslav Enterprise." *Industrial Relations,* 9:148–160.

Russell, Raymond. 1982a. "Rewards of Participation in the Worker-Owned Firm." In *Workplace Democracy and Social Change,* ed. F. Lindenfeld and J. Rothschild-Whitt, pp. 109–124. Boston: Porter Sargent.

1984. "The Role of Culture and Ethnicity in the Degeneration of Democratic Firms." *Economic and Industrial Democracy,* 5:73–96.

1985. *Sharing Ownership in the Workplace.* Albany: State University of New York Press.

Russell, Raymond, Art Hochner, and Stewart Perry. 1977. "San Francisco's Scavengers Run Their Own Firm." *Working Papers for a New Society,* 5:30–36.

1979. "Participation, Influence and Worker-Ownership." *Industrial Relations,* 18:330–341.

Sale, Kirkpatrick. 1980. *Human Scale.* New York: Coward, McCann & Geoghegan.

Sandkull, Bengt. 1982. "Managing the Democratization Process in Work Cooperatives." University of Linkoping, Sweden. Photocopy.

Sarason, Seymour. 1972. *The Creation of Settings and the Future Societies.* San Francisco: Jossey-Bass.

Satow, Roberta. 1975. "Value-Rational Authority and Professional Organizations: Weber's Missing Type." *Administrative Science Quarterly,* 20:526–531.

Scaff, Lawrence. 1981. "Max Weber and Robert Michels." *American Journal of Sociology,* 86:1269–1286.

Schlesinger, Melinda, and Pauline Bart. 1982. "Collective Work and Self-Identity: Working in a Feminist Illegal Abortion Collective." In *Workplace Democracy and Social Change,* ed. F. Lindenfeld and J. Rothschild-Whitt, pp. 139–153. Boston: Porter Sargent.

Schumacher, E. F. 1973. *Small Is Beautiful: Economics As If People Mattered.* New York: Harper & Row.

Schumpeter, Joseph. 1942. *Capitalism, Socialism, and Democracy.* New York: Harper and Brothers.

Scott, W. Richard. 1981. *Organizations: Rational, Natural, and Open Systems.* Englewood Cliffs, N.J.: Prentice-Hall.

1949. *TVA and the Grass Roots.* Berkeley: University of California Press.

Shirom, Arie. 1972. "The Industrial Relations System of Industrial Cooperatives in the U.S., 1880–1935." *Labor History,* 13:533–551.

Sills, David. 1957. *The Volunteers.* New York: The Free Press.

Simmons, John, and William Mares. 1983. *Working Together: Participation from the Shop Floor to the Boardroom.* New York: Knopf.

Simpson, Richard. 1972. "Beyond Rational Bureaucracy: Changing Values and Social Integration in Post-Industrial Society." *Social Forces,* 51:1–6.

Sonquist, John, and Thomas Koenig. 1975. "Interlocking Directorates in the Top U.S. Corporations: A Graph Theory Approach." *Insurgent Sociologist,* 5:196–229.

Stein, Barry. 1974. *Industrial Efficiency and Community Enterprise.* Cambridge: Center for Community Economic Development.

Stephens, Evelyne, and John Stephens. 1982. "The Labor Movement, Political Power, and Workers' Participation in Western Europe." In *Political Power and Social Theory,* Vol. 3, pp. 215–249. Greenwich, Conn.: JAI Press.

Stern, Robert, and Tove Hammer. 1978. "Buying Your Job: Factors Affecting the Success or Failure of Employee Acquisition Attempts." *Human Relations,* 31:1101–1117.

Sullivan, Teresa. 1978. *Marginal Workers, Marginal Jobs: The Underutilization of American Workers.* Austin: University of Texas Press.

Swidler, Ann. 1976. "Teaching in a Free School." *Working Papers for a New Society,* 4:30–34.

1979. *Organization without Authority: Dilemmas of Social Control in Free Schools.* Cambridge, Mass.: Harvard University Press.

Taylor, Rosemary. 1976. "Free Medicine." *Working Papers for a New Society,* 4:21–23, 83–94.

Terkel, Studs. 1975. *Working.* New York: Avon.

Thompson, James D., and Arthur Tuden. 1959. "Strategies, Structures, and Processes of Organizational Decision." Chap. 12 of *Comparative Studies in Administration,* ed. James Thompson. Pittsburgh: University of Pittsburgh Press, 1959.

Thornley, Jenny. n.d. "The Product Dilemma for Workers' Co-operatives in Britain, France and Italy." Co-operatives Research Occasional Paper no. 1. Milton Keynes, England: Co-operatives Research Unit, The Open University.

Thurow, Lester. 1980. *The Zero-Sum Society: Distribution and Possibilities for Economic Change.* New York: Penguin Books.

Toffler, Alvin. 1970. *Future Shock.* New York: Random House.

Torbert, William. 1973. "An Experimental Selection Process for a Collaborative Organization." *Journal of Applied Behavioral Science,* 9:331–350.

Useem, Michael. 1984. *The Inner Circle.* New York: Oxford University Press.

U.S. Department of Health, Education, and Welfare (HEW). 1973. *Work in America.* Cambridge, Mass.: MIT Press.

Vail, David. 1975. "The Case for Rural Industry: Economic Factors Favoring Small Scale,

Decentralized, Labor-Intensive Manufacturing." Institute on Science, Technology, and Development, Cornell University. Photocopy.
Vanek, Juroslav. 1977. "Some Fundamental Considerations in Financing and Form of Ownership under Labour Management." In *The Labour Managed Economy*. Ithaca, N.Y.: Cornell University Press.
Vocations for Social Change. 1976. *No Bosses Here: A Manual on Working Collectively*. Cambridge, Mass.: Vocations for Social Change.
Vroom, V. 1964. *Work and Motivation*. New York: John Wiley.
Walton, Richard. 1979. "Work Innovations in the United States." *Harvard Business Review* (July-August):88–98.
Ward, Colin. 1966. "The Organization of Anarchy." In *Patterns of Anarchy*, ed. Leonard Krimerman and Lewis Perry, pp. 386–396. Garden City, N.Y.: Anchor Books.
1972. "The Anarchist Contribution." In *The Case for Participatory Democracy*, ed. C. George Benello and Dimitrios Roussopoulos, pp. 283–294. New York: Viking Press.
Webb, Beatrice. 1921. "The Co-operative Movement of Great Britain and Its Recent Developments." *International Labour Review*, November, pp. 227–256.
Webb, Eugene. 1970. "Unconventionality, Triangulation, and Inference." In *Sociological Methods*, ed. Norman Denzin. Chicago: Aldine.
Webb, Sidney, and Beatrice Webb. 1920. *A Constitution for the Socialist Commonwealth of Great Britain*. London: Longmans.
Weber, Max. 1946. *From Max Weber: Essays in Sociology*. Trans. and ed. Hans Gerth and C. Wright Mills. New York: Oxford University Press.
1947. *The Theory of Social and Economic Organization*. Trans. A. M. Henderson and Talcott Parsons. Glencoe, Ill.: The Free Press.
1954. *Max Weber on Law in Economy and Society*. Cambridge, Mass.: Harvard University Press.
1968. *Economy and Society*. Ed. Guenther Roth and Claus Wittich. New York: Bedminster Press.
Wenig, M. 1982/83. "Still Serving that Dream: An Interview with Virginia Blaisdell." *Communities*, 56:15–22.
Whitehead, Alfred North. 1925. *Science and the Modern World*. New York: Macmillan.
Whitt, J. Allen. 1982. *Urban Elites and Mass Transportation: The Dialectics of Power*, Princeton: Princeton University Press.
Whyte, Martin King. 1973. "Bureaucracy and Modernization in China: The Maoist Critique." *American Sociological Review*, 38:149–163.
Whyte, William Foote. 1978. *Congressional Record*. House of Representatives. Washington, D.C.: Government Printing Office, June 19.
1979. "On Making the Most of Participant Observation." *The American Sociologist*, 14:56–66.
Woodcock, George. 1962. *Anarchism: A History of Libertarian Ideas and Movements*. New York: World.
Wright, E. 1976. "Class Boundaries in Advanced Capitalist Societies." *New Left Review*, 98:3–41.
Zablocki, Benjamin. 1971. *The Joyful Community: An Account of the Bruderhof*. Baltimore: Penguin Books.
Zald, Mayer, and Roberta Ash. 1966. "Social Movement Organizations: Growth, Decay, and Change." *Social Forces*, 44:327–341.
Zald, Mayer, and John McCarthy. 1975. "Organizational Intellectuals and the Criticism of Society." *Social Service Review*, 49:344–362.
Zeitlin, Maurice. 1970. *Revolutionary Politics and the Cuban Working Class*. New York: Harper & Row.

Zipp, John, Paul Luebke, and Richard Landerman. 1984. "The Social Bases of Support for Workplace Democracy." Washington University, St. Louis, Mo. Photocopy.

Zwerdling, Daniel. 1975. "Shopping Around: Nonprofit Food." *Working Papers for a New Society,* 3:21–31.

1976. "The Day the Workers Took Over." *New Times,* December 10, pp. 38–45.

1979. "The Uncertain Revival of Food Cooperatives." In *Co-ops, Communes and Collectives,* ed. John Case and Rosemary Taylor, pp. 89–111. New York: Pantheon Books.

INDEX

Other books in the Arnold and Caroline Rose Monograph Series of the American Sociological Association

J. Milton Yinger, Kiyoshi Ikeda, Frank Laycock, and Stephen J. Cutler. *Middle Start: An Experiment in the Educational Enrichment of Young Adolescents*

James A. Geschwender: *Class, Race, and Worker Insurgency: The League of Revolutionary Black Workers*

Paul Ritterband: *Education, Employment, and Migration: Israel in Comparative Perspective*

John Low-Beer: *Protest and Participation: The New Working Class in Italy*

Orrin E. Klapp: *Opening and Closing: Strategies of Information Adaptation in Society*

Rita James Simon: *Continuity and Change: A Study of Two Ethnic Communities in Israel*

Marshall B. Clinard: *Cities with Little Crime: The Case of Switzerland*

Steven T. Bossert: *Tasks and Social Relationships in Classrooms: A Study of Instructional Organization and Its Consequences*

Richard E. Johnson: *Juvenile Delinquency and Its Origins: An Integrated Theoretical Approach*

David R. Heise: *Understanding Events: Affect and the Construction of Social Action*

Ida Harper Simpson: *From Student to Nurse: A Longitudinal Study of Socialization*

Stephen P. Turner: *Sociological Explanation as Translation*

Janet W. Salaff: *Working Daughters of Hong Kong: Filial Piety or Power in the Family?*

Joseph Chamie: *Religion and Fertility: Arab Christian–Muslim Differentials*

William Friedland, Amy Barton, and Robert Thomas: *Manufacturing Green Gold: Capital, Labor, and Technology in the Lettuce Industry*

Richard N. Adams: *Paradoxical Harvest: Energy and Explanation in British History, 1870–1914*

Mary F. Rogers: *Sociology, Ethnomethodology, and Experience: A Phenomenological Critique*

James R. Beniger: *Trafficking in Drug Users: Professional Exchange Networks in the Control of Deviance*

Jon Miller: *Pathways in the Workplace: The Effects of Race and Gender on Access to Organizational Resources*

Andrew J. Weigert, J. Smith Teitge, and Dennis W. Teitge: *Society and Identity: Toward a Sociological Psychology*

Michael A. Faia: *The Strategy and Tactics of Dynamic Functionalism*